Lecture Notes in Computer Science 15179

The series Lecture Notes in Computer Science (LNCS), including its subseries Lecture Notes in Artificial Intelligence (LNAI) and Lecture Notes in Bioinformatics (LNBI), has established itself as a medium for the publication of new developments in computer science and information technology research, teaching, and education.

LNCS enjoys close cooperation with the computer science R & D community, the series counts many renowned academics among its volume editors and paper authors, and collaborates with prestigious societies. Its mission is to serve this international community by providing an invaluable service, mainly focused on the publication of conference and workshop proceedings and postproceedings. LNCS commenced publication in 1973.

Smita Ghosh · Zhao Zhang
Editors

Algorithmic Aspects in Information and Management

18th International Conference, AAIM 2024
Virtual Event, September 21–23, 2024
Proceedings, Part I

Springer

Editors
Smita Ghosh
Santa Clara University
Santa Clara, CA, USA

Zhao Zhang
Zhejiang Normal University
Jinhua, China

ISSN 0302-9743 ISSN 1611-3349 (electronic)
Lecture Notes in Computer Science
ISBN 978-981-97-7797-6 ISBN 978-981-97-7798-3 (eBook)
https://doi.org/10.1007/978-981-97-7798-3

This Springer imprint is published by the registered company Springer Nature Singapore Pte Ltd.
The registered company address is: 152 Beach Road, #21-01/04 Gateway East, Singapore 189721, Singapore

Preface

This book serves as the proceedings of the 18th International Conference on Algorithmic Aspects in Information and Management (AAIM 2024), September 21–23, 2024. The conference was held online and was based at the University of Texas at Dallas, USA. The book consists of 45 presented papers, selected from 76 submissions. Among them, there were no negative reports for 42 papers. 3 of them were accepted after serious discussions between negative and positive arguments. Actually, there were 6 papers with quite different reports. After serious discussions, 3 of them were accepted and 3 of them were rejected. The topics cover most aspects of algorithmic aspects in information and management, such as combinatorial optimization and applications pertaining to computing. All accepted papers were selected by an international program committee consisting of a large number of scholars from various countries and regions, distributed over the world, including Asia, North America, Europe, and Australia. Each paper was required to be submitted in double-blind style and was evaluated by at least three reviewers. The decision was made based on these evaluations through a process that included a discussion period. Authors of selected papers come from the following countries and regions: Japan, India, Romania, China (including Hong Kong and Macau), USA, France, and Brazil. Many of these papers represent reports of continuing research, and it is expected that most of them will appear in a more polished and complete form in scientific journals.

We wish to thank all who have made this meeting possible and successful: the authors for submitting papers, the program committee members for their excellent work in reviewing papers, the sponsors, the local organizers (software supporters), and Springer for their support and assistance. We are especially grateful to Ke Su, who made tremendous efforts in setting up the online system.

September 2024

Smita Ghosh
Zhao Zhang

Preface

Organizing Committee

Steering Co-chairs

Yinfeng Xu	Xi'an Jiaotong University, China
Binhai Zhu	Montana State University, USA

General Co-chairs

Ding-Zhu Du	University of Texas at Dallas, USA
Zhao Zhang	Zhejiang Normal University, China

PC Co-chair

Smita Ghosh	Santa Clara University, USA

Web Co-chair

Ke Su	University of Texas at Dallas, USA

Program Committee Members

Sergey Bereg	University of Texas at Dallas, USA
Sergiy Butenko	Texas A&M University, USA
Zhipeng Cai	Georgia State University, USA
Mihaela Cardei	Florida Atlantic University, USA
Tiantian Chen	Santa Clara University, USA
Xujin Chen	Chinese Academy of Sciences, China
Yong Chen	Hangzhou Dianzi University, China
Zhizhong Chen	Tokyo Denki University, Japan
Yongxi Cheng	Xi'an Jiaotong University, China
Dingzhu Du	University of Texas at Dallas, USA
Donglei Du	University of New Brunswick, Canada
Hongwei Du	Harbin Institute of Technology Shenzhen, China

Ling Gai	University of Shanghai for Science and Technology, China
Xiaofeng Gao	Shanghai Jiao Tong University, China
Priyanshi Garg	University of Texas at Dallas, USA
Suning Gong	Qingdao University, China
Shuyang Gu	Texas A&M University - Central Texas, USA
Jianxiong Guo	Beijing Normal University, China
Hejiao Huang	Harbin Institute of Technology Shenzhen, China
Liying Kang	Shanghai University, China
Michael Khachay	Krasovsky Institute of Mathematics and Mechanics, Russia
Ralf Klasing	CNRS and University of Bordeaux, France
Chia-Wei Lee	University of Taipei, Taiwan
Joong-Lyul Lee	University of North Carolina at Pembroke, USA
Bo Li	Hong Kong Polytechnic University, China
Deying Li	Renmin University of China, China
Jianping Li	Yunnan University, China
Min Li	Shandong Normal University, China
Minming Li	City University of Hong Kong, China
Xianyue Li	Lanzhou University, China
Xiao Li	University of Texas at Dallas, USA
Bin Liu	Ocean University of China, China
Wen Liu	Hebei Normal University, China
Chuanwen Luo	Beijing Forestry University, China
Manki Min	Louisiana Tech University, USA
Viet Hung Nguyen	University of Clermont Auvergne, France
Qiufen Ni	Guangdong University of Technology, China
Qingqing Nong	Ocean University of China, China
Arash Rafiey	Indiana State University, USA
Ke Su	University of Texas at Dallas, USA
Guangmo Tong	University of Delaware, USA
Weitian Tong	Eastern Michigan University, USA
Guanghua Wang	University of Texas at Dallas, USA
Lusheng Wang	City University of Hong Kong, China
Chenchen Wu	Tianjin University of Technology, China
Lidong Wu	University of Texas at Dallas, USA
Han Xiao	Ocean University of China, China
Mingyu Xiao	University of Electronic Science and Technology of China, China
Dachuan Xu	Beijing University of Technology, China
Yicheng Xu	Chinese Academy of Sciences, China
Boting Yang	University of Regina, Canada

Wenguo Yang	University of Chinese Academy of Sciences, China
Jing Yuan	University of North Texas, USA
Guochuan Zhang	Zhejiang University, China
Nan Zhang	Xidian University, China
Yapu Zhang	Beijing University of Technology, China
Yong Zhang	Chinese Academy of Sciences, China
Zhao Zhang	Zhejiang Normal University, China
Fay Zhong	California State University, East Bay, USA
Yisheng Zhou	University of Texas at Dallas, USA
Binhai Zhu	Montana State University, USA
Jianming Zhu	University of Chinese Academy of Sciences, China

Contents – Part I

Contents – Part II

Quantum and Others

Optimization and Applications

Maximum Core Spanning Tree Insertion Maintenance for Large Dynamic Graphs

Xiaowei Lv[1], Yongcai Wang[1(✉)], Deying Li[1], and Haodi Ping[2]

[1] Renmin University of China, Beijing, China
{lvxiaowei,ycw,deyingli}@ruc.edu.cn
[2] School of Computer Sciences, Beijing University of Technology, Beijing, China
haodi.ping@bjut.edu.cn

Abstract. The Maximum Core Spanning Tree (MCST) is a representative cohesive structure generated based on k-core, which is the maximum edge weight spanning tree indicating the "staired coreness hierarchy" in each connected component. The edge weight here is defined as $w_{uv} = \min\{core(u), core(v)\}$, and $core(x)$ is the corness of vertex x. Unlike the insertion maintenance problem of Maximum Spanning Tree (MST) which has known efficient algorithms, MCST insertion maintenance raises special challenges, which is mainly due to the cascaded vertex coreness changes after single-edge insertion. In this paper, we show a series of properties of MCST and MCST insertion maintenance problem and propose a LoopFree algorithm to maintain the MCST efficiently. In particular, the time complexity for MCST maintenance for edge insertion is bounded by $O(|E^*| + |V|)$, where E^* is the edge set whose edge weight changes after insertion. Through extensive evaluations, we show the proposed MCST insertion maintenance algorithm has good efficiency on real-world datasets.

Keywords: Dense subgraph mining · k-core · Maximum Core Spanning Tree · Dynamic insertion maintenance

1 Introduction

Dense subgraph mining of complex networks is widely used in various graph structures, including social networks [6], biology [18], and advertisement in e-commerce, etc. Among the dense subgraph models, k-core [20] extracts connected subgraphs in which vertices have degree at least k, and serves as a foundation for other cohesive graph models.

The Maximum Core Spanning Tree (MCST) [4] based on k-core further reveals the inner structure of the graph. In particular, a MCST is the maximum edge weight spanning tree, where the edge weight is defined as $w_{uv} = min\{core(u), core(v)\}$. Note that $core(u)$ represents the coreness of vertex u. We show a MCST has a stair structure, where the tree edges connecting the vertices with the same coreness can be considered on the same stair. In MCST

S. Ghosh and Z. Zhang (Eds.): AAIM 2024, LNCS 15179, pp. 3–15, 2024.
https://doi.org/10.1007/978-981-97-7798-3_1

generation process, MCST will greedily connect all the vertices in a higher stair before connecting to the low stair vertices. So a MCST indicates a "staired coreness hierarchy" in each connected component, which can be considered as a different hierarchical representation of the k-cores.

In large scale networks, when the vertices and connections are dynamic, how to maintain the extracted cohesive structures efficiently is a crucial problem [13,15,16,22,23]. For tracking the inner structure of k-cores, the insertion maintenance of MCST is important as well, which is however not well investigated in existing studies. In comparison, the insertion maintenance of traditional MST (Maximum Spanning Tree) has attracted a lot of attentions and a series of efficient algorithms are proposed [2,3,7,8]. But existing MST insertion maintenance methods are usually designed for weight change of a single edge. Whereas the insertion maintenance of MCST raises special challenges due to the special property of k-core. A single-edge insertion in the graph may trigger a significant number of cascaded vertex coreness changes [13], so the MCST insertion maintenance must consider simultaneous weight changes of a set of edges after single edge insertion.

We show the main challenges to maintain MCST are:

(1) How to determine which set of edges will change their weights?
(2) How to update the MCST according to the edge-weight changes?

In response to above challenges, we investigate a series of properties of MCST and for MCST insertion maintenance, including (1) the stair-like structure of MCST; (2) Unique determination of MCST by a given Kruskal-order; and (2) Various principles to narrow down the maintenance scope. Using these properties, we design complexity-bounded maintenance algorithm. In particular, a LoopFree (LF) algorithm is designed for edge insertion, which maintains MCST by iteratively deleting the smallest weight edge in the loop. Extensive evaluations show the proposed MCST insertion maintenance algorithm have good efficiency on real-world datasets.

2 Related Work

Dynamic Graph Structure Analysis. With the characteristics of real-time processing of different graph structures, the research on dynamic graph maintenance is becoming more and more extensive. Including but not limited to k-truss [1,16], k-clique [9,17], densest subgraph [5], etc. And the maintenance of the k-core [13,22,23] has also been conducted by many researches. Based on the dynamic maintenance of k-core, Lin et al. [15] also proposes algorithms for the dynamic maintenance of the hierarchical structure of k-core.

Spanning Tree Problem. The maximum spanning tree is a variant of the minimum spanning tree (MST) problem, and the two methods have similarities, so we only mention the MST problem below. The method of the last century [2,7,8] is aimed at the MST maintenance under single-edge insertion. Cattaneo et al.

Table 1. Summary of Notations.

Notation	Definition
$G = (V, E)$	undirected simple graph
$n = \|V\|$; $m = \|E\|$	number of vertices; number of edges
$N_G(v)$	the set of neighbors of v
$deg_G(v)$	the degree of v
C^k	a k-core
$core(v, G)$	the coreness of v in graph G
k_{max}	the maximum coreness value of all vertices
$w_{uv}(G)$ or $w_e(G)$	$min\{core(v), core(u)\}$
$G_w = (V, E, W)$	the weighted graph of G according to w_{uv}
$G_c = (V, E + E', W_c)$	the weighted graph with inserted edges
$T(G)$	MCST of graph G
$mcst(G)$	edge set in MCST of graph G
$cw(G)$	the total weight of the MCST
E^*	$\{e \in E(G) \| w_e(G) \neq w_e(G^*)\}$

[3] presented an experimental study on several algorithms for the MST problem, which also proposed an efficient implementation of the algorithm of [10]. Compared with MST dynamic insertion maintenance, MCST's dynamic insertion maintenance can take advantage of the properties of k-core for optimization.

3 Preliminaries

We define an undirected simple graph as $G = (V, E)$, where $n = |V|$ represents the number of vertices, and $m = |E|$ represents the number of edges. The neighbors of v in G is defined as $N_G(v)$, and the degree of vertex v is denoted as $deg_G(v) = |N_G(v)|$. When the context is clear, we simply use $N(v)$ and $deg(v)$ for clarity. All notations are summarized in Table 1. The specific definitions of k-core and coreness are as follows.

Definition 1 (k-Core). *Given an integer $k (k \geq 0)$, the k-core of G, denoted by C^k, is the maximal subgraph of G, such that $\forall v \in C^k, deg_{C^k}(v) \geq k$.*

Definition 2 (Coreness). *Given a graph G, the coreness value of a vertex $v \in V(G)$, denoted by $core(v)$, is the largest k such that v is in the k-core, i.e., $core(v) = max_{v \in C^k}\{k\}$, where C^k is set as k-core.*

3.1 Maximum Core Spanning Tree

Let $G_w = (V, E, W)$ be a weighted graph, which has the same vertices and edges as G, and the edge weight is defined based on the coreness values of the vertices. For $(u, v) \in E$, $w_{uv} \in W$ is defined as $w_{uv} = min\{core(u), core(v)\}$. Then we can get the MCST on G_w.

Definition 3 (Maximum Core Spanning Tree). *Given the weighted graph G_w generated from G, the maximum core spanning tree (MCST) T of G, denoted by $T(G)$ is a maximum spanning tree in G_w, and the maximum means the spanning tree of the largest sum weight.*

To quickly get the MCST, we can use the degeneracy order [14] of the vertices to assign a partial order relationship to the vertices, which can help to quickly extract MCST.

Definition 4 (DEGENERACY ORDERING [14]). *Given a graph G, a permutation $(v_1, v_2, ...v_n)$ of all vertices of G is a degeneracy ordering of G if every vertex v_i has the minimum degree in the subgraph of G induced by $\{v_i, ...v_n\}$.*

As shown in Algorithm 1, we can use the method in [4] to generate the MCST in $O(m)$ time complexity, where *seq* represents the degeneracy order corresponding to the vertex. After generating the MCST $T(G)$, the set of edges selected in MCST is denoted as $mcst(G_w)$. To handle graph dynamics, we consider the maintenance of MCST.

Definition 5 (Maximum Core Spanning Tree Insertion Maintenance). *Given a graph G, let E' be the edge set we want to insert to G. Let G_w denote the original weighted graph obtained by G. Let G_c denote the changed weighted graph, i.e., $G_c = (V, E+E', W_c)$, where W_c is the updated edge weights. Maximum Core Spanning Tree Insertion Maintenance is to update $T(G_w)$ to $T(G_c)$, and updates the MCST edge set $mcst(G_w)$ to $mcst(G_c)$ and the total weight of MCST from $cw(G_w)$ to $cw(G_c)$.*

Since the insertion of vertices can be regarded as the insertion of multiple edges, and the response to multiple edge insertion can be accomplished by calling single edge maintenance multiple times, so we only consider the insertion of single edge here. Besides, $T(G)$ and $cw(G)$ can be easily obtained from $mcst(G)$, so we mainly maintain $mcst(G)$.

Algorithm 1: Maximum Core Spanning Tree

Input : A graph $G = (V, E)$, a degeneracy order *seq* of vertices, and core numbers core(.) of vertices

Output: A maximum core spanning tree T of G

$T \leftarrow (V, \emptyset)$;
Initialize a disjoint-set data structure $\mathscr{F}$ for V;
foreach *vertex u in seq in reverse order* **do**
    foreach *neighbor v of u in G that appear after u in seq* **do**
        if *v and u are not in the same set in $\mathscr{F}$* **then**
            Union u and v in $\mathscr{F}$;
            Add (u, v) with weight $core(u)$ into T;
return T;

4 Properties

We firstly investigate the properties of MCST and the properties for MCST insertion maintenance. Let G_w be the weighted graph of G. Let G_c represent the weighted graph updated from G_w after edge insertion into G, where $G_c = (V, E + (x_1, x_2), W_c)$.

4.1 Properties of MCST

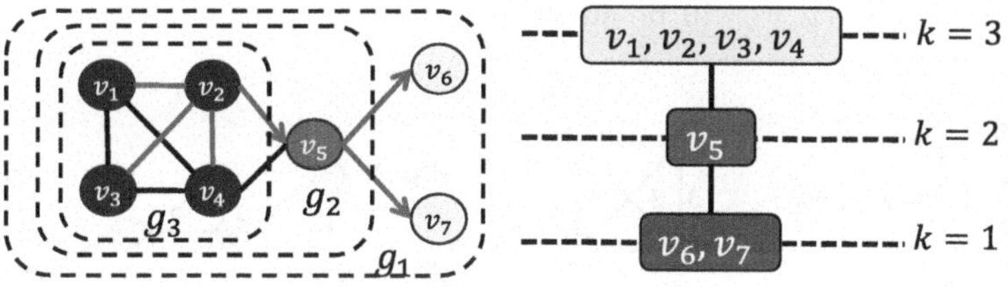

(a) Maximum Core Spanning Tree (b) Stair-like Structure

Fig. 1. Example to show the stair-like structure

1. MCST has a stair-like structure. If we consider the vertices with the same coreness are on the same stair, then the MCST forms a natural stair structure during its construction. By the greedy construction algorithm of MST, the higher stair (weight) tree edges will be added into MCST before the lower stair (weight) tree edges. We give an example in Fig. 1 to show stair-like structure.

In Fig. 1a, g_3, g_2, and g_1 are 3-core, 2-core, and 1-core, respectively, where the red edges represent the edges in MCST. For a tree edge (u, v), if $core(u) > core(v)$, we convert (u, v) to be a directed edge from u to v, then the vertices connected by the undirected tree edges have the same coreness value. Each directed tree edge is going down from a higher stair to a lower stair. We can further construct the stair structure as shown in Fig. 1b. It can be considered as a different representation of the peel hierarchies [4] of k-core.

2. A MCST is uniquely determined by a given Kruskal-order. Given a Kruskal-order defined below, the result MSCT is unique. The same graph may produce different MCST and Kruskal-order due to the traversal order of vertices, but the Kruskal-order determines the unique traversal order of edges, so it can uniquely determine the MCST.

Definition 6 (Kruskal Order). *Given a graph G, the order in which edges are traversed for the first time in Algorithm 1 is the Kruskal-order. That is to say, $e_i \succeq e_j$ if (i) $w_{e_i} > w_{e_j}$ or (ii) $w_{e_i} = w_{e_j}$ when e_i is earlier visited than e_j in Algorithm 1.*

2. Edge weight changes at most one after single edge insertion. The edge whose edge weight changes can be determined by the vertexes whose coreness changes. Single-edge insertion will only change the coreness of each vertex by at most one, so the edge weight changes at most one. Based on this property we can use the state-of-the-art method [23] to update the vertex coreness efficiently. We define V^* as the set of vertices whose coreness values have changed after inserting the edge (x_1, x_2), where V^* may be empty. By the property, we know that for $v \in V^*$ and $core(v, G_w) = K$, there is $core(v, G_c) = K + 1$.

Then, we use V^* to find the edge set E^* whose edge weight changes by visiting the neighbor $N_{G_c}(v)$ of each vertex v in V^*. If $min(core(u), core(v))$ changes, then edge (u, v) will be added to E^*.

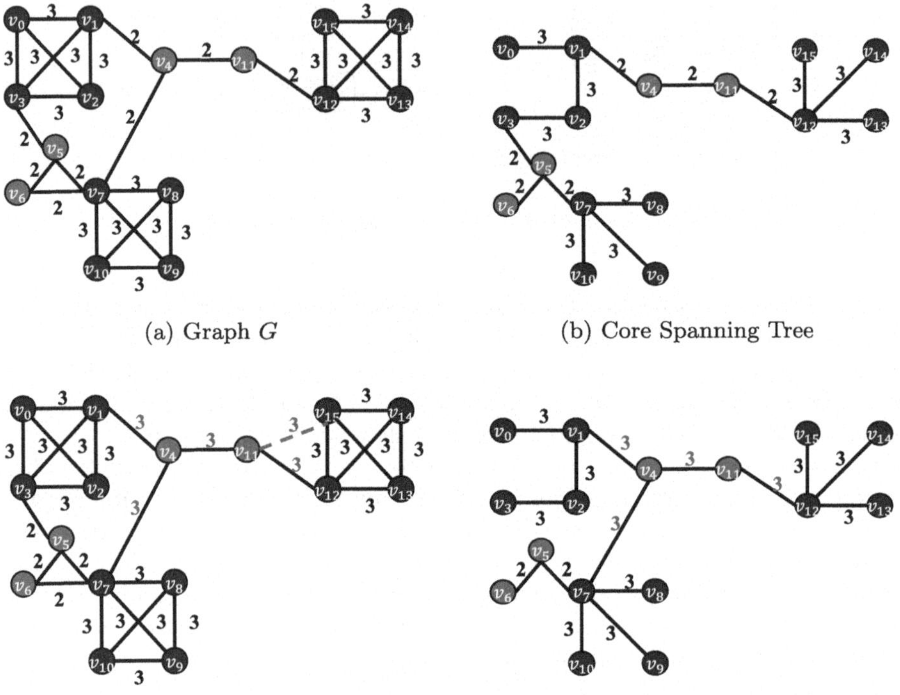

(a) Graph G (b) Core Spanning Tree

(c) Insert edge (v_{11}, v_{15}) into G (d) Updated Core Spanning Tree

Fig. 2. An Example for Core Spanning Tree Insertion Maintenance

4.2 Insertion Properties of MCST

Inheriting the properties of MST, not all changes will affect MCST when a single-edge (x_1, x_2) in G_w is inserted. In this case, $G_c = (V, E + (x_1, x_2), W_c)$.

Theorem 1 (Pure weight increase of tree edges will not change the edge set of MCST). *If* $E^* \subseteq mcst(G_w)$, *then* $mcst(G_c) = mcst(G_w)$, $T(G_c) =$

$T(G_w)$ and $cw(G_c) = cw(G_w) + |E^*|$, where $|E^*|$ denotes the number of edges in E^*.

Proof. We denote the updated edge as (u, v). Its Kruskal order before insertion is O_1, and after insertion is O_2, where $O_2 > O_1$. Ignoring the changes of other edges, according to the enumeration order of the Kruskal algorithm, the order of edge enumeration before O_1 and after O_2 remains unchanged. Since $O_2 > O_1$, edge (u, v) must be enumerated first in the new graph. Then it will still become a tree edge and connect the original parts of the MCST where u and v are located, which is the same as its contribution in O_1 order. Thus, the edge set in the MCST remains unchanged. However, due to the change in edge weight from K to $K + 1$, $cw(G_c)$ increases by one accordingly. As a result, a total of $|E^*|$ edges will make $cw(G_c) = cw(G_w) + |E^*|$, so the above problem is proved.

Theorem 1 makes sure that when inserting a single edge, if only some tree edges have increased their weights, they will not change the edge set in the MCST and only increase the total weight of MCST. Therefore, we mainly need to deal with the inserted edge and the weight-increased non-tree edges.

MCST can be maintained by removing the loops under single edge insertion. We will introduce this property in detail in Sect. 5.2.

5 Maintenance of Edge Insertion

5.1 Insertion Overview

Our main idea to rapidly maintain MCST under single-edge insertion is to delete the smallest weight edge forming loop with the tree edges. We first look for edge sets whose edge weight changes according to the coreness change mentioned in Sect. 4.1. Secondly, an efficient insertion maintenance algorithm is proposed for whether the changed edge is a tree or non-tree edge. We add the inserted edge and non-tree edges into the new MCST, and then we update the MCST by removing the corresponding edges forming loop with tree edges. It is also necessary to maintain the Kruskal order after the insertion operation.

5.2 The Impact of the Inserted Edge and Edge Weight Changes

We discuss the following two cases for the weight-increased edges.

Case-1: According to Theorem 1, if the weight-increased edges all belong to tree edge, the edge set in the MCST will not change. So in **Case-1**, we only need to increase the weight of the MCST. But we still need to process the inserted edge (x_1, x_2), which is a non-tree edge.

Case-2: The weight-increased edges contain tree edges and non-tree edges. We first can easily get that the weight-increased tree edges only increase the total weight of the MCST according to the Theorem 1. Then, we focus on dealing with the non-tree weight increased edges.

In **Case-2**, the edges in $E_{nonTree}$ will be added to the new MCST candidate edge set. The candidate edge set in the MCST consists of three parts. The first part is the inserted edge (x_1, x_2) (**Case-1** is the same) because its insertion may become a candidate edge of the $mcst(G_c)$. The second part is the non-tree edges with increased weight. Due to the increase in edge weight of non-tree edge (u, v), $O_c(u, v)$ correspondingly moves forward to make it become a candidate edge of the $mcst(G_c)$. The third part is the edges of the $mcst(G_w)$. Because they were selected as tree edges of MCST before the update, we must analyze whether they will be removed after inserting (x_1, x_2). We record all the above three parts' edges as E_{cand}. E_{cand} may form extra loops, and our next step is to remove the loops.

Theorem 2 MCST can be maintained by removing loops. *If* $E' \subseteq E_{cand}$ *constitute a loop, then* $(u, v) \notin mcst(G_c)$ *if* w_{uv} *is the smallest among all the edges in* E'*.*

Proof. According to the definition of the MCST, it cannot contain loops. Therefore, edges must be removed until they become free of loops. According to the definition of the MCST, the edge with the smallest weight in the loop should be deleted. By considering all loops and examining all candidate sets, the selected edge set is deemed optimal.

According to Theorem 2, we maintain the MCST by removing the loops with the smallest edge weight in the E_{cand}. In actual operation, we can use the idea of the union-find set to traverse all candidate edges whose edge weight is from large to small and judge whether their two vertices are in the same union-find set. If not, they will be added to the result set. Otherwise, the edge is skipped.

Example 1. In Fig. 2, after inserting (v_{11}, v_{15}), the corresponding $E^* = \{(v_1, v_4), (v_4, v_7), (v_4, v_{11}), (v_{11}, v_{12}), (v_{11}, v_{15})\}$. So all the edges in E^* plus the edges of the original CST should be in E_{cand}. Then we found that two loops were generated in the graph: $\{v_{11}, v_{12}, v_{15}\}$, and the other is $\{v_1, v_2, v_3, v_4, v_5, v_7\}$. We need to remove the edges with the smallest weight in the loops. Here, we remove $\{(v_3, v_5), (v_{11}, v_{15})\}$, and the remained edges are the edge set of the CST.

5.3 LoopFree Algorithm

Based on the above analysis, we propose a LoopFree algorithm to implement the dynamic update of the MCST after inserting (x_1, x_2), as shown in Algorithm 2. In Algorithm 2, lines 1–2 are the initialization part, where *cnt_mcst* is used to count the number of edges in the MCST. Lines 3–8 find the corresponding E^* set according to V^*. Among them, Line 8 processes the tree edges in **Case-1**. Lines 9–17 deal with **Case-2**, where line 14 determines whether u, v belongs to different sets in $\mathscr{F}$ to determine whether it forms a loop with $mcst(G_c)$. Since the edges are traversed in decreasing order, if (u, v) forms a loop with $mcst(G_c)$, (u, v) must be the edge with the smallest weight. Early stop can be performed by counting the number of edges during MCST insertion maintenance because the number of edges of MCST increases by at most one after insertion. The time complexity of Algorithm 2 is $\mathcal{O}(|E^*| + |V|)$, where $\mathcal{O}(|E^*|) = \mathcal{O}(deg(v) \times |V^*|)$.

Theorem 3. *The time complexity of Algorithm 2 is* $\mathcal{O}(|E^*| + |V|)$.

Proof. The time complexity has three parts. The first part comes from the update of coreness [23] and can be omitted due to it is much smaller than the latter two parts. The second part is to calcualte E^*, where $\mathcal{O}(|E^*|) = \mathcal{O}(|V^*| \times deg(v))$, and the time complexity of mergesort is $\mathcal{O}(|E^*| + |V| - 1)$. The time complexity of the third part for selecting the edge set in $MCST$ is also $\mathcal{O}(|E^*| + |V| - 1)$. Adding up the three parts, the final time complexity is $\mathcal{O}(|E^*| + |V|)$.

Algorithm 2: LoopFree Algorithm

Input : A graph G_w, edge set of MCST $mcst(G_w)$, inserted edge (x_1, x_2)
Output: Updated MCST edge set $mcst(G_c)$

$E^* \leftarrow \emptyset$, $cnt\_mcst \leftarrow 0$, $old\_tree\_num \leftarrow |mcst(G_w)|$;
$V^* \leftarrow$ vertices with coreness changed by inserting (x_1, x_2) to G;
/* calculate E^* */
foreach *vertex* $v \in V^*$ **do**
  foreach *vertex* $u \in N_{G_w}(v)$ **do**
    if $(u, v) \in E^*$ **then** continue;
    if $min(core(u), core(v))$ *changes* **then**
      $E^* \leftarrow E^* \cup (u, v)$;
      /* Case-1 */
      if $(u, v) \in mcst(G_w)$ **then** $E^* \leftarrow E^* - (u, v)$;
/* Case-2 */
$E_{cand} \leftarrow$ an empty queue;
Mergesort all edges in E^* and $mcst(G_w)$ according to the edge weight and output the result to E_{cand};
Initialize a disjoint-set data structure $\mathcal{F}$ for V;
while $E_{cand} \neq \emptyset \wedge cnt\_mcst \leq old\_tree\_num + 1$ **do**
  $(u, v) \leftarrow E_{cand}$.dequeue();
  if (u, v) *will not form a loop with* $mcst(G_c)$ **then**
    Union u and v in $\mathcal{F}$;
    Add edge (u, v) with w_{uv} into $mcst(G_c)$;
    $cnt\_mcst \leftarrow cnt\_mcst + 1$;
return $mcst(G_c)$;

6 Experiments

In this section, experiments are carried out on 13 large real-world graphs. We conducted comparative experiments in Sect. 6.1 to prove the efficiency of our algorithm.

Settings. All algorithms are implemented in C++ and compiled by g++ at -O3 optimization level, and they are conducted on a machine with Inter(R) Xeon(R) CPU E5-2667 v4@3.20 GHz processor and 128 GB memory, with Ubuntu OS installed. We call the Snap [12] library to implement the graph structure.

Table 2. Datasets

Dataset	$n = \|V\|$	$m = \|E\|$	avg.deg	max k
scc_reality	6,809	4,714,485	1,385	1,235
twitter_copen	8,580	473,614	110	582
infect_dublin	10,972	175,573	32	83
human_gene2	14,340	9,027,024	1,259	1,902
Facebook	63,731	817,035	25.64	52
BlogCatalog	88,784	2,093,195	47	222
Gowalla	196,591	950,327	9.67	51
DBLP	317,080	1,049,866	6.62	112
BerkStan	685,230	6,649,470	19.41	201
Google	875,713	4,322,051	9.87	44
pokec	1,632,803	30,622,564	37.51	47
orkut	3,072,441	117,185,083	76.28	253
LiveJournal	4,847,571	68,993,773	28.46	372

Datasets. We use 13 publicly available datasets listed in Table 2. These datasets contain different types of actual application scene graphs, among which Facebook [21] is an online social network that can be downloaded on Konect. The following datasets infect_dublin, twitter_copen, scc_reality, human_gene2, BlogCatalog are all from [19]. The first four are temporal graphs, and the last two are biological network and social network, respectively. The rest of the datasets can be downloaded from the SNAP [11], including social networks (DBLP, pokec, LiveJournal and orkut), network graphs (BerkStan and Google), and location-based social networks (Gowalla). The statistics are shown in Table 2, including the number of edges and vertices, average degree and maximum coreness value.

6.1 Comparative Experiment

We compare the proposed LoopFree (LF) Algorithm (Algorithm 2) with the static recalculation (SR) (Algorithm 1) and coreness update in [23] with static recalculation MCST (CSR). The SR algorithm recalculates the degeneracy order after each insertion and calls Algorithm 1 to obtain the MCST. Compared with SR, CSR saves the time of calculating the degeneracy order by dynamically maintain coreness.

Table 3. Performance on All Datasets.

Datasets	Insert (s)		
	LF	CSR	SR
scc_reality	**0.441**	14.272	24.911
twitter_copen	**0.132**	1.976	3.308
infect_dublin	**0.258**	1.278	2.247
human_gene2	**0.865**	42.212	83.532
Facebook	**1.355**	7.224	15.158
BlogCatalog	**1.905**	12.690	27.010
Gowalla	**4.015**	18.319	36.53
DBLP	**6.743**	30.116	54.971
BerkStan	**14.818**	77.153	161.669
Google	**20.531**	105.378	242.581
pokec	**45.479**	400.832	1126.494
orkut	**101.631**	1569.060	6501.904
LiveJournal	**178.170**	1073.798	3207.107

We repeatedly insert 500 edges to test the algorithm's performance. For the temporal graphs (infect_dublin, twitter_dublin, and scc_reality), we take the 500 edges with the maximum timestamps and randomly select 500 edges from all edges for other datasets. For each graph, we measure the sum time cost for the MCST maintenance under 500 edges insertion. The results are shown in Table 3. We compare our proposed algorithm LF with CSR and SR. The MCST maintenance time of LF significantly outperforms the CSR and SR algorithms in all tested datasets. For human_gene2, the speed up even achieves up to 48 times and 96 times compared to CSR and SR.

7 Conclusions

In this work, we study the problem of MCST maintenance regarding dynamic edge insertion. We investigate a series of properties, such as the stair-like structure; the unique determination of MCST by a given Kruskal-order; and various principles to narrow down the maintenance cost. Using these properties, we propose the LF algorithm for MCST maintenance under single-edge insertion. The key idea is to iteratively remove corresponding edges from MSCT. We conduct experiments to prove the effectiveness of our algorithm on 13 real-world datasets.

Acknowledgments. Dr. Li is supported in part by the National Natural Science Foundation of China Grant No.12071478. Dr. Wang is supported in part by the National Natural Science Foundation of China Grant No. 61972404, Public Computing Cloud, Renmin University of China, and the Blockchain Lab. School of Information, Renmin University of China.

References

1. Akbas, E., Zhao, P.: Truss-based community search: a truss-equivalence based indexing approach. Proc. VLDB Endow. **10**(11), 1298–1309 (2017)
2. Amato, G., Cattaneo, G., Italiano, G.F.: Experimental analysis of dynamic minimum spanning tree algorithms. In: SODA, vol. 97, pp. 314–323. Citeseer (1997)
3. Cattaneo, G., Faruolo, P., Petrillo, U.F., Italiano, G.F.: Maintaining dynamic minimum spanning trees: an experimental study. In: Mount, D.M., Stein, C. (eds.) ALENEX 2002. LNCS, vol. 2409, pp. 111–125. Springer, Heidelberg (2002). https://doi.org/10.1007/3-540-45643-0_9
4. Chang, L., Qin, L.: Cohesive Subgraph Computation Over Large Sparse Graphs: Algorithms, Data Structures, and Programming Techniques. Springer, Cham (2018). https://doi.org/10.1007/978-3-030-03599-0
5. Epasto, A., Lattanzi, S., Sozio, M.: Efficient densest subgraph computation in evolving graphs. In: Proceedings of the 24th International Conference on World Wide Web, pp. 300–310 (2015)
6. Fang, Y., Cheng, R., Li, X., Luo, S., Hu, J.: Effective community search over large spatial graphs. Proc. VLDB Endow. **10**(6), 709–720 (2017)
7. Frederickson, G.N.: Data structures for on-line updating of minimum spanning trees. In: Proceedings of the Fifteenth Annual ACM Symposium on Theory of Computing, pp. 252–257 (1983)
8. Henzinger, M.R., King, V.: Maintaining minimum spanning trees in dynamic graphs. In: Degano, P., Gorrieri, R., Marchetti-Spaccamela, A. (eds.) ICALP 1997. LNCS, vol. 1256, pp. 594–604. Springer, Heidelberg (1997). https://doi.org/10.1007/3-540-63165-8_214
9. Himmel, A.S., Molter, H., Niedermeier, R., Sorge, M.: Adapting the Bron-Kerbosch algorithm for enumerating maximal cliques in temporal graphs. Soc. Netw. Anal. Min. **7**, 1–16 (2017)
10. Holm, J., De Lichtenberg, K., Thorup, M.: Poly-logarithmic deterministic fully-dynamic algorithms for connectivity, minimum spanning tree, 2-edge, and biconnectivity. J. ACM (JACM) **48**(4), 723–760 (2001)
11. Leskovec, J., Krevl, A.: SNAP Datasets: Stanford large network dataset collection, June 2014. http://snap.stanford.edu/data
12. Leskovec, J., Sosič, R.: SNAP: a general-purpose network analysis and graph-mining library. ACM Tran. Intell. Syst. Technol. (TIST) **8**(1), 1 (2016)
13. Li, R.H., Yu, J.X., Mao, R.: Efficient core maintenance in large dynamic graphs. IEEE Trans. Knowl. Data Eng. **26**(10), 2453–2465 (2013)
14. Lick, D.R., White, A.T.: k-degenerate graphs. Can. J. Math. **22**(5), 1082–1096 (1970)
15. Lin, Z., Zhang, F., Lin, X., Zhang, W., Tian, Z.: Hierarchical core maintenance on large dynamic graphs. Proc. VLDB Endow. **14**(5), 757–770 (2021)
16. Luo, Q., Yu, D., Cheng, X., Sheng, H., Lyu, W.: Exploring truss maintenance in fully dynamic graphs: a mixed structure-based approach. IEEE Trans. Comput. **72**(3), 707–718 (2022)
17. Ottosen, T.J., Vomlel, J.: Honour thy neighbour: clique maintenance in dynamic graphs. On Probabilistic Graphical Models, p. 201 (2010)
18. Palla, G., Derényi, I., Farkas, I., Vicsek, T.: Uncovering the overlapping community structure of complex networks in nature and society. Nature **435**(7043), 814–818 (2005)

19. Rossi, R.A., Ahmed, N.K.: The network data repository with interactive graph analytics and visualization. In: AAAI (2015). https://networkrepository.com
20. Seidman, S.B.: Network structure and minimum degree. Soc. Netw. **5**(3), 269–287 (1983)
21. Viswanath, B., Mislove, A., Cha, M., Gummadi, K.P.: On the evolution of user interaction in Facebook. In: Proceedings of the 2nd ACM SIGCOMM Workshop on Social Networks (WOSN 2009), August 2009
22. Wen, D., Qin, L., Zhang, Y., Lin, X., Yu, J.X.: I/O efficient core graph decomposition at web scale. In: 2016 IEEE 32nd International Conference on Data Engineering (ICDE), pp. 133–144. IEEE (2016)
23. Zhang, Y., Yu, J.X., Zhang, Y., Qin, L.: A fast order-based approach for core maintenance. In: 2017 IEEE 33rd International Conference on Data Engineering (ICDE), pp. 337–348. IEEE (2017)

NFTO: DAG-Based Task Offloading and Energy Optimization Algorithm

Luchuan Zeng[1], Kunming Jin[1], Zichen Wang[1], Chen Zhang[2], and Hongwei Du[1(✉)]

[1] School of Computer Science and Technology, Harbin Institute of Technology (Shenzhen), Shenzhen, China
{22S151164,23S051005}@stu.hit.edu.cn, hongwei.du@ieee.org
[2] The Hang Seng University of Hong Kong, Hong Kong, China
czhang@hsu.edu.hk

Abstract. Mobile Edge Computing emerges as a key technology to address the challenges of real-time data processing by extending computing capabilities to the network edge, reducing latency, and enhancing service reliability. Existing research often inadequately addresses offloading tasks with Directed Acyclic Graph structures and fails to consider the dynamic MEC environment, limiting practical applicability. This paper proposes a DAG Task Computation Offloading Algorithm based on Network Flow Theory. The core innovation lies in applying network flow theory to optimize task scheduling in dynamic environments, considering service caching and device energy constraints. We validate the algorithm's advantages through comprehensive simulations.

Keywords: Mobile Edge Computing · Computation Offloading · Directed Acyclic Graph · Task Scheduling

1 Introduction

In the wave of the digital era, the rapid evolution of mobile internet and IoT technologies is profoundly reshaping our social structure and daily lives. As these technologies become widespread, massive amounts of data are continuously generated and require real-time processing, presenting unprecedented challenges to computing resources. Against this backdrop, Mobile Edge Computing (MEC) emerges as a key technology to address these challenges [1]. By extending computing power from centralized cloud to the network edge, MEC effectively reduces data processing latency, enhances service immediacy and reliability, and provides users with a richer and more flexible computing experience [2].

Despite MEC offering new solutions for handling massive data, effectively managing and scheduling edge computing resources remains an unresolved issue. In the MEC architecture, computation offloading is a critical technique that

L. Zeng and K. Jin—These authors contributed equally to this work.

S. Ghosh and Z. Zhang (Eds.): AAIM 2024, LNCS 15179, pp. 16–27, 2024.
https://doi.org/10.1007/978-981-97-7798-3_2

allows mobile devices to offload computation-intensive tasks to edge servers, thereby reducing the computational burden on local devices, saving battery energy, and optimizing overall computational performance. Existing research has explored optimal strategies for computation offloading through various algorithms, such as linear programming, queuing theory, or heuristic methods. However, these methods often neglect the computation offloading problem for tasks with Directed Acyclic GraphDAG structures and fail to fully consider the complex dynamics of the MEC environment, especially practical factors like service cache capacity, network state fluctuations, and device energy limitations, limiting their applicability in real-world scenarios.

To address this challenge, we proposes a computation offloading strategy based on a network flow theory-based algorithm for DAG tasks. The core of the algorithm lies in applying network flow theory to the decision-making process of computation offloading, thereby more accurately simulating and optimizing task scheduling in dynamic environments. Specifically, our algorithm not only takes into account the physical constraints of service cache and device energy but also considers real-time changes in network state, ensuring the shortest task completion time while optimizing energy efficiency.

Our main contributions are as follows:

- Utilizing network flow theory to address the computation offloading problem for DAG tasks in MEC, optimizing both task completion time and energy efficiency.
- Focusing the algorithm on the dual physical constraints of service cache and user device energy, enhancing the algorithm's applicability in practical MEC environments.
- Through comprehensive simulations, this study validates the algorithm's advantages in optimizing task completion time and energy consumption, comparing it with existing technologies and demonstrating its superior performance in dynamic and uncertain environments.

The remainder of this paper is organized as follows: Sect. 2 presents the system model and problem formalization. Section 3 presents the problem analysis. Section 4 details the algorithm. Section 5 discusses experimental setups and results. Section 6 reviews related works; Sect. 7 concludes the paper.

2 System Model and Problem Formalization

2.1 System Overview

In a MEC system, there are multiple edge servers $S = \{s_1, s_2, \ldots, s_M\}$ distributed across different geographical locations, connected to multiple mobile user devices $U = \{u_1, u_2, \ldots, u_N\}$. Each edge server is equipped with a base station, and the mobile devices connect to the edge servers via a wireless network, while the edge servers are interconnected through a wired network. The MEC system primarily consists of two components: user devices and edge servers.

2.2 Computation Model

A DAG task T is modeled as a DAG, represented as $T = (V, E)$, where V is the set of subtasks and E is the set of edges representing task dependencies. The DAG task consists of a set of subtasks $V = \{v_1, v_2, \ldots, v_n\}$, with $E = \{(v_i, v_j)\}$ representing the dependencies between tasks, meaning subtask v_i must be completed before subtask v_j can start. Each subtask $v_i \in V$ has a data size $D_{v(i)}$ and a computational demand $C_{v(i)}$.

Computation Processing Time. Each subtask v_i must be executed on either a local device or an edge server s_k that has cached the corresponding service. This can be represented by a cache indicator function $\mathbb{C}(v_i, s_k)$, which is 1 if subtask v_i can be executed on server s_k, and 0 if the server does not have the corresponding service cached.

During the computation offloading process, each subtask v_i can be offloaded to one of the following locations: 1. User equipment u_j, if the device has enough energy to execute the task. 2. An edge server s_k that has the corresponding service cached. 3. If the user equipment lacks the energy to execute the task, and no edge server has the corresponding service cached, the most suitable edge server s_{best} is selected, and the required service is downloaded from the cloud. The computational capability of the mobile user device is $P_u(u_j)$, and the computational capability of the edge server is $P_s(s_k)$. The workload of server s_k affects its available computational capability $P_{avail}(s_k) = P_s(s_k) \times (1 - W(s_k))$. The server's workload is dynamic and can be adjusted based on the real-time system state. Execution on the mobile device u_j is given by: $T^u_{proc}(v_i, u_j) = \frac{C_i}{P_u(u_j)}$. Execution on an edge server s_k with the corresponding service cached is given by: $T^s_{proc}(v_i, s_k) = \frac{C_i}{P_{avail}(s_k)}$. If the service needs to be downloaded from the cloud, the edge server s_{best} is chosen to minimize $T_{dl}(s_{best}) + T^s_{proc}(v_i, s_{best})$. For simplification, we assume the service download time is a proportion of the task execution time. Thus, for an edge server s_{best} that requires service download, the subtask's service download time can be represented as: $T_{dl}(s_{best}) = \beta \cdot T^s_{proc}(v_i, s_{best})$. Therefore, for tasks executed on an edge server without the corresponding service cached: $T^s_{proc}(v_i, s_k) = \frac{C_i}{P_{avail}(s_k)} + T_{dl}(s_k)$.

This way, the computation offloading strategy considers the energy constraints of the user equipment and the service caching status of edge servers, while also accounting for the option of downloading services from the cloud when necessary. This flexible offloading strategy helps optimize overall computation efficiency and response time, ensuring that tasks are effectively handled under the constraints of energy and service availability.

Completion Time of the Entire DAG Task. The longest dependency chain in the DAG is calculated using topological sorting. The completion time T_{DAG} is the sum of the total computation times on the longest dependency chain, i.e., $T_{DAG} = \max_{v_i \in V}(EFT(v_i))$, where $EFT(v_i)$ is the earliest finish time of subtask v_i.

For each subtask v_i in the DAG, its computation time $T_{comp}(v_i)$ includes computation processing time $T_{proc}(v_i)$ and result transmission time $T_{trans}(v_i)$. Thus:

$$T_{comp}(v_i) = T_{proc}(v_i) + T_{trans}(v_i). \tag{1}$$

Considering the execution location of subtask v_i, the computation processing time $T_{proc}(v_i)$ can be expressed with the following conditional formula:

$$T_{proc}(v_i) = \mathbb{I}_{\approxeq}(v_i) \cdot T^u_{proc}(v_i, u_j) + \mathbb{I}_{\sim}(v_i, s_k) \cdot T^s_{proc}(v_i, s_k) \tag{2}$$

where the indicator function $\mathbb{I}_{\approxeq}(v_i)$ represents whether subtask v_i is executed on the user equipment. When $\mathbb{I}_{\approxeq}(v_i) = 0$, the subtask is not executed on the user equipment; when $\mathbb{I}_{\approxeq}(v_i) = 1$, the subtask is executed on the user equipment. The indicator function $\mathbb{I}_{\sim}(v_i, s_k)$ indicates whether subtask $v(i)$ is executed on the edge server; if $\mathbb{I}_{\sim}(v_i, s_k) = 0$, the subtask is not executed on the edge server s_k; if $\mathbb{I}_{\sim}(v_i, s_k) = 1$, the subtask is executed on s_k.

The result transmission time $T_{trans}(v_i)$ is calculated based on the result transmission time of all predecessor subtasks. It can be expressed by the following formula:

$$T_{trans}(v_i) = \max_{v_j \in \text{pred}(v_i)} \left(T_{comp}(v_j) + \frac{D_{trans}(v_j, v_i)}{R_{trans}(v_j, v_i)} \right). \tag{3}$$

where, $\text{pred}(v_i)$ is the set of all predecessor subtasks of subtask v_i. $T_{comp}(v_j)$ is the completion time of the predecessor subtask v_j. $D_{trans}(v_j, v_i)$ is the amount of data to be transmitted from subtask v_j to v_i. $R_{trans}(v_j, v_i)$ is the network transmission rate from subtask v_j to v_i, which may vary due to network conditions.

Initialize the earliest start time $EST(v_i) = 0$ and the earliest finish time $EFT(v_i) = T_{comp}(v_i)$ for all subtasks. For each subtask v_i in the DAG, iterate according to the topological order. Update the earliest start time and earliest finish time of each subtask:

$$EST(v_i) = \max_{(v_j, v_i) \in E} (EFT(v_j)), \tag{4}$$

$$EFT(v_i) = EST(v_i) + T_{comp}(v_i). \tag{5}$$

The completion time T_{DAG} of the DAG task is equal to the maximum of the earliest finish times of all subtasks:

$$T_{DAG} = \max_{v_i \in V} (EFT(v_i)). \tag{6}$$

2.3 Energy Consumption Model

The total energy consumption E_{total} includes the energy consumed by the user equipment E_u and the edge servers E_s:

$$E_{total} = E_u + E_s. \tag{7}$$

Energy Consumption of User Equipment. The total energy consumption of the user equipment E_u includes the computation energy consumption E^u_{comp} and the data transmission energy consumption E^u_{trans}. It can be expressed as:

$$E_u = \sum_{v_i \in V} \left(\mathbb{I}_{\approxeq}(v_i) \cdot (E^u_{comp}(v_i) + E^u_{trans}(v_i)) \right). \tag{8}$$

The energy consumption of subtask v_i when executed locally can be expressed as:

$$E^u_{comp}(v_i) = \zeta_u \cdot f_u^2 \cdot \frac{C_{v(i)}}{f_u} = \zeta_u \cdot f_u \cdot C_{v(i)}. \tag{9}$$

where the average energy consumption per CPU cycle is $\zeta_u f_u^2$, ζ_u is the parameter describing the effective capacitance, f_u is the CPU frequency, and $C_{v(i)}$ is the computational demand of the subtask.

Considering the power consumption during data transmission, the unit data transmission energy consumption E_{unit} can be approximately expressed as $\frac{P_{trans}}{B_u}$, where P_{trans} is the average power consumption during data transmission, usually equal to the transmission signal power of the user equipment p_u. The data transmission rate B_u is calculated according to Shannon's theorem:

$$B_u = r_u \cdot \log_2 \left(1 + \frac{p_u \cdot \theta_u}{\sigma^2} \right). \tag{10}$$

The data transmission energy consumption of subtask v_i, $E^u_{trans}(v_i)$, can be expressed as:

$$\begin{aligned} E^u_{trans}(v_i) = E_{unit} \cdot \Bigg(& \sum_{v_j \in \text{pred}(v_i)} D_{trans}(v_j, v_i) \\ & + \sum_{v_k \in \text{succ}(v_i)} D_{trans}(v_i, v_k) \Bigg). \end{aligned} \tag{11}$$

where pred(v_i) and succ(v_i) represent the sets of direct predecessor and successor subtasks of v_i, respectively. $D_{trans}(v_j, v_i)$ and $D_{trans}(v_i, v_k)$ are the data amounts that need to be transmitted from v_j to v_i and from v_i to v_k, respectively. This formula allows us to estimate the energy consumption due to data transmission when executing subtask v_i on the user equipment.

Energy Consumption of Edge Servers. Since edge servers are interconnected through wired connections, the energy consumption of data transmission between them can be ignored. Therefore, the total energy consumption of edge servers E_s only needs to consider the computation energy consumption E^s_{comp}, which can be expressed as:

$$E_s = \sum_{v_i \in V} \left(\mathbb{I}_{\sim}(v_i, s_k) \cdot (E^s_{comp}(v_i)) \right). \tag{12}$$

The computation energy consumption of a subtask v_i executed on an edge server is calculated similarly to the computation energy consumption of the user equipment, but it takes into account the specific parameters of the edge server, such as the server's CPU frequency and computational demand. The following is the formula for the computation energy consumption of an edge server:

$$E_{comp}^{s}(v_i) = \zeta_s \cdot f_s^2 \cdot \frac{C_{v(i)}}{f_s}. \tag{13}$$

where ζ_s is the parameter describing the effective capacitance of the edge server's CPU, related to the server's microarchitecture and manufacturing process, f_s is the CPU frequency of the edge server, and $C_{v(i)}$ is the computational demand of subtask v_i (e.g., the number of required computation cycles). This formula assumes that the energy consumption per CPU cycle is proportional to the square of the CPU frequency, while also considering the computational demand of the subtask. Through this formula, the energy consumption of executing a specific subtask on the edge server can be estimated.

2.4 Problem Formalization

Based on the above system model, we present the formal definition of the problem. Our goal is to minimize the completion time of the DAG tasks and the energy consumption of the edge computing system through reasonable computation offloading decisions. Given a weight factor λ (where $\lambda \in \{0, 1\}$), the objective function to be solved can be expressed as:

$$\min\left(\lambda \cdot T_{DAG} + (1 - \lambda) \cdot E_{total}\right) \tag{14}$$

s.t.

$$\mathbb{I}_{\approxeq}(v_i) = \{0, 1\}, \quad \forall v_i \in V, \tag{15}$$

$$\mathbb{I}_{\sim}(v_i, s_k) = \{0, 1\}, \quad \forall v_i \in V, \tag{16}$$

$$\mathbb{I}_{\approxeq}(v_i) + \mathbb{I}_{\sim}(v_i, s_k) = 1, \quad \forall v_i \in V, \tag{17}$$

$$\mathbb{C}(v_i, s_k) = \{0, 1\}, \quad \forall v_i \in V, \tag{18}$$

$$\mathbb{C}(v_i, s_k) = 1 \vee T_{dl}(s_k) > 0 \quad \Rightarrow \mathbb{I}_s(v_i, s_k) = 1, \tag{19}$$

$$\sum_{v_i \in V} \mathbb{I}_u(v_i) \cdot E_{comp}^{u}(v_i) \leq E_{rmax}^{u}, \quad \forall v_i \in V, \tag{20}$$

$$\mathbb{I}_u(v_i) = \mathbb{I}_u(v_j), EFT(v_i) \leq EST(v_j), \forall v_i, v_j \in V, i \neq j, \tag{21}$$

$$\mathbb{I}_u(v_i) = \mathbb{I}_u(v_j), EFT(v_j) \leq EST(v_i), \forall v_i, v_j \in V, i \neq j, \tag{22}$$

$$\mathbb{I}_s(v_i, s_k) = \mathbb{I}_s(v_j, s_k), EFT(v_i) \leq EST(v_j), \forall v_i, v_j \in V, i \neq j, \tag{23}$$

$$\mathbb{I}_s(v_i, s_k) = \mathbb{I}_s(v_j, s_k), EFT(v_j) \leq EST(v_i), \forall v_i, v_j \in V, i \neq j, \tag{24}$$

$$\sum_{v_i \in V} \mathbb{I}_s(v_i, s_k) \cdot C_{v(i)} \leq W_{\max}(s_k), \quad \forall s_k \in S, \tag{25}$$

$$\sum_{v_i \in V} \mathbb{I}_u(v_i) \cdot C_{v(i)} \leq W_{\max}(u_j), \quad \forall u_j \in U. \tag{26}$$

3 Problem Analysis

To transform the computation offloading problem in mobile edge computing into a Minimum Cost Maximum Flow (MCMF) problem, we construct a network flow model that, according to the topology of the DAG task, includes several network flows where subtasks of the DAG task that can be executed in parallel constitute a network flow graph. Taking one of these network flow graphs as an example, it contains four levels of nodes: source nodes, DAG subtask nodes that can be executed in parallel, computation nodes (including user equipment and edge servers), and sink nodes. The source node represents the start of the process, while the sink node represents the end. Subtask nodes represent the subtasks to be executed, and computation nodes represent the devices and servers that can execute these subtasks. $V = \{v_1, v_2, \ldots, v_n\}$: The set of subtasks in the DAG. $V' = \{L_1, L_2, \ldots, L_k\}$: The list of subtasks ordered by dependency, divided into multiple levels $L_1, L_2, \ldots, L_k$, where each level contains subtasks that can be processed in parallel. U: User equipment. $S = \{s_1, s_2, \ldots, s_m\}$: The set of edge servers. $T^u_{comp}(v_i)$: The computation time of subtask v_i on user equipment. $T^s_{comp}(v_i, s_k)$: The computation time of subtask v_i on edge server s_k. $E^u_{comp}(v_i)$: The energy consumption of subtask v_i on user equipment. $E^s_{comp}(v_i, s_k)$: The energy consumption of subtask v_i on edge server s_k. λ: The weight coefficient between time and energy consumption, with a value of 0 or 1. For convenience, let d_j represent user equipment or edge servers, $d_j \in S \cup U$, and $C(d_j)$ is the remaining computational capacity of d_j, equal to the maximum workload allowed by the device minus the current workload. $f(v_i, d_j)$ represents the flow through the edge (v_i, d_j). Each edge from subtask node v_i to computation node (user equipment or server s_k) has a capacity limit set to 1 (indicating the subtask can be executed on one device). The cost of each edge $Cost(v_i)$ is the weighted sum of the time and energy required to complete the subtask, specifically $\lambda \cdot T^u_{comp}(v_i) + (1-\lambda) \cdot E^u_{comp}(v_i)$ or $\lambda \cdot T^s_{comp}(v_i, s_k) + (1-\lambda) \cdot E^s_{comp}(v_i, s_k)$.

Our goal is to minimize the total weight of the entire DAG task:

$$\min \sum_{L_k \in V'} \sum_{v_i \in L_k} Cost(v_i) \cdot f(v_i, d_j) \tag{27}$$

s.t.

$$0 \le f(v_i, d_j) \le c(v_i, d_j) \quad \forall v_i \in L_k, d_j \in S \cup U, \tag{28}$$

$$f(v_i, d_j) \in \{0, 1\} \quad \forall v_i \in L_k, d_j \in S \cup U, \tag{29}$$

$$0 \le f(v_i, d_j) \le c(d_j) \quad \forall v_i \in L_k, d_j \in S \cup U, \tag{30}$$

$$\mathbb{I}_u(v_i) + \sum_{s_k \in S} \mathbb{I}_s(v_i, s_k) = 1, \quad \forall v_i \in V, \tag{31}$$

$$\sum_{v_i \in V} \mathbb{I}_s(v_i, s_k) \cdot C_{v(i)} \le W_{\max}(s_k), \quad \forall s_k \in S, \tag{32}$$

$$\sum_{v_i \in V} \mathbb{I}_u(v_i) \cdot C_{v(i)} \le W_{\max}(u_j), \quad \forall u_j \in U. \tag{33}$$

In this way, the computation offloading problem is transformed into a network flow problem, finding a DAG subtask allocation scheme that minimizes the weighted sum of the completion time and energy consumption of the entire DAG task. This model can be solved using existing Minimum Cost Maximum Flow algorithms. By using the MCMF algorithm, our goal is to find a configuration of the flow network that minimizes the overall cost (i.e., the weighted sum of the completion time and energy consumption of all subtasks) while meeting the execution requirements and resource constraints of all subtasks. Thus, the computation offloading problem is transformed into a standard network flow problem that can be efficiently solved using existing algorithms.

4 Algorithm Description

Algorithm 1: Network Flow Based Task Offloading Algorithm (NFTO)

Input: DAG $V = \{v_1, v_2, \ldots, v_n\}$,$D = S \cup U$,$C(v_i)$ for each task v_i ,$C(d_j)$ for each device d_j ,$Cost(v_i)$ for each pair (v_i, d_j)

Output: Optimal assignment of tasks to devices

Minimum total cost Z across all layers

Initialize $Z = 0$ and $F =$ empty set for storing flows
Topologically sort the DAG to get a list V' of tasks
Divide V' into layers $L_1, L_2, \ldots, L_k$ for parallel processing
Initialize flow network graph G with source node s and sink node t
for *each layer* L_i **do**
    Add tasks in L_i to G
    for *each task* v_i *in* L_i **do**
        Add edge from s to v_i in G with cupacity 1 and weight 0
    for *each pair* (v_i, d_j) **do**
        if $C(v_i) \leq C(d_j)$ **then**
            Add edge from v_i to d_j in G with capacity 1 and weight $Cost(v_i)$
    for *each device* d_j **do**
        Add edge from d_j to t in G with capacity $C(d_j)$ and weight 0
    Apply Minimum-Cost Maximum-Flow algorithm on G
    Update Z and F with the flow and cost from the algorithm
    Update G for the next layer
Extract task assignment information from F
return Z and the optimal assignment of tasks to servers

The Network Flow Based Task Offloading Algorithm (NFTO) is described in Algorithm 1. This algorithm is designed to solve the computation offloading problem for DAG tasks in mobile edge computing. It decomposes the DAG task into hierarchical levels $L_1, L_2, \ldots, L_k$ that can be processed in parallel through topological sorting, and constructs a flow network graph G that includes a source node s and a sink node t. For each level of tasks L_i, the algorithm adds edges between the source node and task nodes, as well as between task nodes and computation nodes (devices d_j). The capacity of these edges is usually set to 1, indicating that a single subtask can be executed on one device. The weight of the edges is determined by the computation cost $Cost(v_i)$ of the subtask on a specific device. The algorithm then applies the minimum cost maximum flow algorithm to determine the optimal allocation scheme for tasks in each level, and updates the total cost Z and flow F. Finally, the algorithm outputs the optimal allocation scheme for each subtask to devices and the minimum total cost across all levels. If completion time is considered, the primary focus is on the maximum completion time in each level, and the completion time of the DAG task is the sum of the longest completion times of subtasks in each level. If energy consumption is considered, the energy consumption of the DAG task is the sum of the energy consumption of all subtasks in each level.

5 Experiment and Results Analysis

5.1 Experiment Settings and Comparison Algorithms

In this section, we use simulation experiments to verify the effectiveness of the proposed algorithm. In the simulation experiments, we compare the NFTO algorithm with the following existing algorithms:

- HEFT: This algorithm first sorts all subtasks in the DAG using the B-level method and then uses a greedy approach to select the server that completes the current node's computation the fastest, according to the sorted order [3].
- DCDS: This algorithm schedules using servers that just meet the requirements from the perspective of energy saving and deadline, and considers the impact of the predecessor deployment scheme on the successor [4].
- Random: Randomly selects an edge server to offload subtasks in the DAG.
- Local: Offloads the subtasks in the DAG to local execution.

Table 1. Experimental Simulation Parameters

Definition	Parameter	Range of Parameter
Subtask Size	M	1000–100000 bit (uniform distribution)
Workload per Bit	W	1000 cycle/bit
Bandwidth Size	r_u	50–100 MHz (uniform distribution)
Cache Rate	ρ	0.2–0.5 (uniform distribution)
CPU Frequency	f	1–2 GHz
Default Number of Edge Servers	N	10
Default Number of Subtasks	V	10

To ensure the generality of the experimental results, we take the average completion time of scheduling 1000 randomly generated DAG tasks in the experiment. The remaining parameters of the experiment are set as shown in Table 1.

5.2 Experimental Results

From Fig. 1(a) and 1(b), it can be seen that the NFTO algorithm always has a shorter task completion time compared to other algorithms as the number of DAG tasks increases. From Fig. 1(c), it can be seen that the NFTO algorithm performs better compared to other algorithms when the DAG task density is high, but its performance is slightly worse than the DCDS and HEFT algorithms when the DAG task density is low. This indicates that our proposed algorithm is more suitable for computational offloading scenarios with high-density DAG tasks. Figure 1(d) shows that when the service cache rate is between 0.2 and 0.86, the task completion time of the NFTO algorithm is better than other algorithms. When the service cache rate is higher than 0.86, the performance of the NFTO algorithm is slightly worse than the DCDS and HEFT algorithms. However, in practical edge computing scenarios, the storage capacity of each edge server is limited and cannot achieve such a high cache rate. This indicates that the NFTO algorithm can make better computational offloading decisions under scenarios with limited edge server caching capabilities, effectively reducing task completion time.

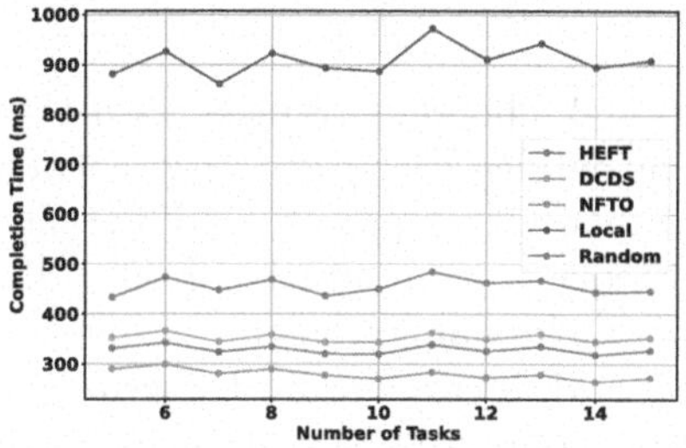

(a) Comparison of task completion time with the number of tasks.

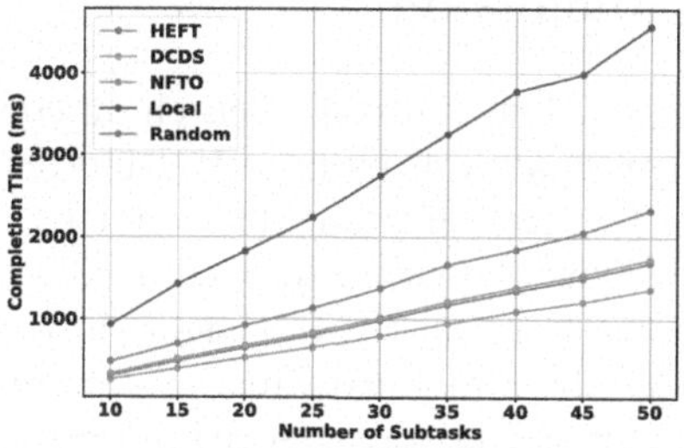

(b) Comparison of task completion time with the number of DAG subtasks.

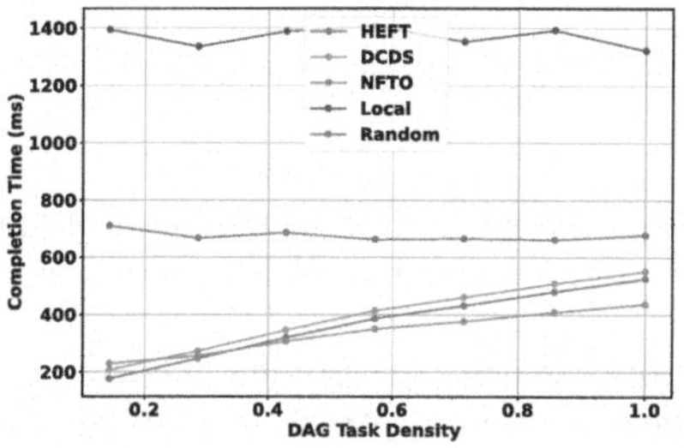

(c) Comparison of task completion time with DAG task density.

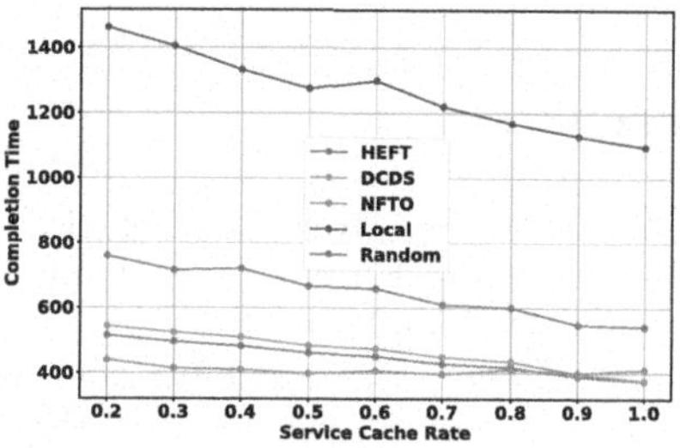

(d) Comparison of task completion time with service cache rate.

Fig. 1. The experimental comparison results of different algorithms

6 Related Works

Under the premise of not considering task correlation, tasks can mainly be divided into two parts: the part of the task executed locally and the part that needs to be offloaded. The study [5] models the task offloading problem as an M/G/1 queuing model, analyzes the problem theoretically from a game theory perspective, and provides the optimal offloading strategy for each MD. The study [6] considers network congestion issues in multi-hop networks and uses heuristic algorithms to determine the partial task offloading ratio, achieving good results. [7] proves that there is a threshold structure in partial offloading problems, determining whether to fully or minimally offload tasks based on users' channel gains and energy consumption.

Due to the complexity of various applications today, many application tasks are divided using a DAG structure, where subtasks are interrelated. The aforementioned studies do not perform well in offloading related tasks in this problem. Some current works also study the offloading of related tasks [8]. In [9], the problem equation is relaxed, converting the original non-convex equation into a solvable one, and the offloading probability calculated is used for task offloading. In [4], the authors use a heuristic approach to solve the related tasks, improving the method proposed in [3] and proposing a two-stage scheduling scheme for energy savings.

7 Conclusion

Considering the lack of research on computational offloading of tasks with DAG structures, as well as practical factors such as service cache capacity, network status fluctuations, and device energy limitations, we propose the NFTO algorithm. This algorithm applies network flow theory to computational offloading decisions. Specifically, our algorithm optimizes task completion time while considering physical constraints of service caching and device energy. Experimental results show that our algorithm performs better compared to other algorithms.

Acknowledgments. This work is supported by National Natural Science Foundation of China (No. 62172124)

References

1. Li, L., Quek, T.Q., Ren, J., Yang, H.H., Chen, Z., Zhang, Y.: An incentive-aware job offloading control framework for multi-access edge computing. IEEE Trans. Mob. Comput. **20**(1), 63–75 (2019)
2. Liu, J., Zhang, Q.: Reliability and latency aware code-partitioning offloading in mobile edge computing. In: 2019 IEEE Wireless Communications and Networking Conference (WCNC), pp. 1–7. IEEE (2019)
3. Topcuoglu, H., Hariri, S., Wu, M.-Y.: Performance-effective and low-complexity task scheduling for heterogeneous computing. IEEE Trans. Parallel Distrib. Syst. **13**(3), 260–274 (2002)

4. Lou, J., Tang, Z., Zhang, S., Jia, W., Zhao, W., Li, J.: Cost-effective scheduling for dependent tasks with tight deadline constraints in mobile edge computing. IEEE Trans. Mob. Comput. **22**, 5829–5845 (2022)
5. Zhang, T., Chen, W.: Computation offloading in heterogeneous mobile edge computing with energy harvesting. IEEE Trans. Green Commun. Netw. **5**(1), 552–565 (2021)
6. Sahni, Y., Cao, J., Yang, L., Ji, Y.: Multi-hop multi-task partial computation offloading in collaborative edge computing. IEEE Trans. Parallel Distrib. Syst. **32**(5), 1133–1145 (2020)
7. You, C., Huang, K., Chae, H., Kim, B.-H.: Energy-efficient resource allocation for mobile-edge computation offloading. IEEE Trans. Wireless Commun. **16**(3), 1397–1411 (2016)
8. Kao, Y.-H., Krishnamachari, B., Ra, M.-R., Bai, F.: Hermes: latency optimal task assignment for resource-constrained mobile computing. IEEE Trans. Mob. Comput. **16**(11), 3056–3069 (2017)
9. Sundar, S., Liang, B.: Offloading dependent tasks with communication delay and deadline constraint. In: IEEE INFOCOM 2018-IEEE Conference on Computer Communications, pp. 37–45. IEEE (2018)

Approximation Algorithm for the Maximum Interval Multi-cover Problem

Yuxuan Xu, Yingli Ran, and Zhao Zhang(✉)

School of Mathematical Sciences, Zhejiang Normal University, Jinhua 321004, Zhejiang, China
{wlxyx1223,hxhzz}sina.com, ranyingli@zjnu.edu.cn

Abstract. Given a set of points $\mathcal{P}$ on a line, a collection of intervals $\mathcal{I}$ along the line, and a positive integer $K \leq |\mathcal{I}|$, each point $p \in \mathcal{P}$ is associated with a covering requirement q_p, the goal of the *maximum interval multi-cover* (MaxIMC) problem is to find a sub-collection of intervals $\mathcal{I}' \subseteq \mathcal{I}$ with $|\mathcal{I}'| \leq K$ to maximize the number of fully-covered points, where a point p is fully-covered by $\mathcal{I}'$ if it belongs to at least q_p intervals of $\mathcal{I}'$. In this paper, we present a $\frac{1}{2}$-approximation algorithm for the MaxIMC problem.

Keywords: maximum cover · multi-cover · dynamic programming · approximation ratio

1 Introduction

The *minimum set cover* (MinSC) problem has wide applications in the real world, such as resource allocation [12], facility location [4], crew scheduling [18], job scheduling [2] and so on. Given a set of elements E, a collection of sets $\mathcal{S} \subseteq 2^E$, the goal of MinSC is to find a smallest sub-collection of sets $\mathcal{S}' \subseteq \mathcal{S}$ to cover all elements, where element e is covered by $\mathcal{S}'$ if e belongs to a set in $\mathcal{S}'$. MinSC is also a classic combinatorial problem that plays a fundamental role in the field of computational complexity and approximation algorithm [5,8].

There are many variants of MinSC, among which the following two are very basic: the *minimum set multi-cover* (MinSMC) problem [20] and the *maximum coverage* (MaxC) problem [9]. They are wide used in real applications [1,3,21], and there are extensive studies on them, which have achieved fairly tight results [6,9,19]. A natural question is: what if we consider fault-tolerant cover within a budget? This leads to the *maximum multi-cover* (MaxMC) problem.

Given $E, \mathcal{S}$, a set of integers $\{q_e\}_{e \in E}$ and a budget K, find a sub-collection of sets $\mathcal{S}' \subseteq \mathcal{S}$ such that $|\mathcal{S}'| \leq K$ and the number of elements fully-covered by $\mathcal{S}'$

NSFC (U20A2068).

S. Ghosh and Z. Zhang (Eds.): AAIM 2024, LNCS 15179, pp. 28–34, 2024.
https://doi.org/10.1007/978-981-97-7798-3_3

is maximized, where an element e is *fully-covered* by $\mathcal{S}'$ if it belongs to at least q_e sets of $\mathcal{S}'$.

A special case of this problem was studied in [13], in which all q_e equal to a constant k. It was proposed as a model for designing peptide vaccines for COVID-19. Experiments showed that it is superior to 29 other published designs. Heuristic algorithms were used in [13]. As far as we know, there is no theoretical studies on this natural and important variant of the set cover problem.

Our studies show that MaxMC includes the famous *Densest l subgraph* (DlS) problem [11] as a special case. So, studying MaxMC is highly challenging in a general setting. Then, a natural question arises: what if we restrict our attention to some typical setting, such as a line? This is the motivation of the study of this paper.

1.1 Related Work

The MaxC problem admits approximation within factor $(1-\frac{1}{e})$ [15]. The *budgeted set cover* (BSC) problem is a cost generalization of the MaxC problem. Khuller et al. [10] presented a $(1-\frac{1}{e})$-approximation algorithm for BSC. They also asserted that there is no approximation algorithm with performance guarantee $(1-\frac{1}{e}+\epsilon)$ for any $\epsilon > 0$ unless $NP \subseteq DTIME(n^{O(\log\log n)})$.

For the MinSMC problem, Dobson [6] presented an $H(\Delta)$-approximation algorithm by a greedy strategy, where Δ is the maximum size of a multi-set and $H(\cdot)$ is the Harmonic number. Rajagopalan and Vazirani [17] gave a parallel algorithm attaining the same approximation ratio, using dual fitting analysis. Approximation ratio $H(\Delta)$ is already tight because even for the MinSC problem, Feige [7] has proved a lower bound of $(1-\epsilon)\ln n$ under the assumption that $NP \not\subseteq DTIME(n^{O(\log\log n)})$, where n is the number of elements.

The MaxMC problem is a natural combination of the maximum coverage problem and the multi-cover requirement. Surprisingly, we only find one paper [13] on this problem in which heuristic algorithms were used without theoretical guarantee. The reason might be: the MaxMC problem is very hard. In fact, it includes the DlS problem. Manurangsi [14] has proved that DlS cannot be approximated within factor $n^{\frac{1}{(\log\log n)^c}}$ under the unique games conjecture. As a consequence, the MaxMC problem also has $n^{\frac{1}{(\log\log n)^c}}$ as a lower bound for its approximation ratio.

1.2 Contribution

In this paper, we propose a 1/2-approximation algorithm for the MaxIMC problem. This is the first approximation algorithm for MaxIMC with a theoretical guarantee.

The rest of the paper is organized as follows. Section 2 gives the algorithm and its theoretical guarantee. Section 3 concludes the paper and points out some problems that deserve further exploration.

2 The Algorithm and Its Theoretical Analysis

In this section, we give a dynamic programming algorithm for the MaxIMC problem. The formal definition of the problem is given as follows.

Definition 1 (maximum interval multi-cover (MaxIMC) problem). Given a set of points $\mathcal{P}$ on a line, a set of intervals $\mathcal{I}$ along the line, and a positive integer $K \leq |\mathcal{I}|$, each point p is associated with a covering requirement q_p, the goal of MaxIMC is to find a sub-collection of intervals $\mathcal{I}' \subseteq \mathcal{I}$ such that $|\mathcal{I}'| \leq K$ and the number of points fully-covered by $\mathcal{I}'$ is maximized, where a point p is fully-covered by $\mathcal{I}'$ if and only if it belongs to at least q_p intervals of $\mathcal{I}'$.

Some Notations. In this paper, it is assumed that all points and intervals are in generic positions, i.e., no intervals have common endpoints, and no point lies at the endpoint of an interval. For an interval I, we denote its left and right endpoints by $\ell(I)$ and $r(I)$, respectively. When $r(I_1) < \ell(I_2)$, intervals I_1, I_2 are disjoint. For two disjoint intervals I_1 and I_2, let

$$B(I_1, I_2) = \{I \in \mathcal{I} : \ell(I_1) < \ell(I) < r(I_1), r(I_1) < r(I) < r(I_2)\}. \tag{1}$$

Note that the definition of $B(I_1, I_2)$ uses strict inequalities, thus $B(I_1, I_2)$ does not include I_1, I_2 themselves. Also note that $B(I_1, I_2)$ does not contain those intervals included in I_1 or I_2. Consider Fig. 1 for an instance, we have $B(I_1, I_2) = \{I_3, I_4, I_6\}$.

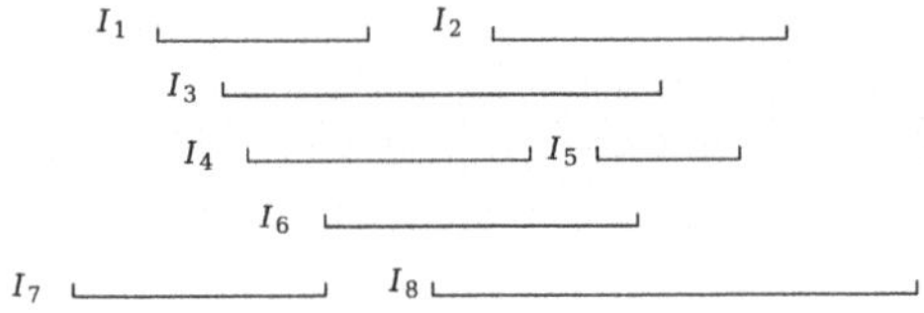

Fig. 1. An illustration for $B(I_1, I_2)$, $L(I_1, I_2, p)$ and $R(I_1, I_2, p)$.

Suppose there are t intervals in $B(I_1, I_2)$. Order them as $I_{j_1}, I_{j_2}, \ldots, I_{j_t}$ satisfying $\ell(I_{j_1}) < \ell(I_{j_2}) < \ldots < \ell(I_{j_t})$. They can also be ordered as $I_{i_1}, I_{i_2}, \ldots, I_{i_t}$ satisfying $r(I_{i_1}) > r(I_{i_2}) > \ldots > r(I_{i_t})$. For an integer $p \leq t$, let

$$L(I_1, I_2, p) = \{I_{j_r} : 1 \leq r \leq p\}, \tag{2}$$

$$R(I_1, I_2, p) = \{I_{i_r} : 1 \leq r \leq p\}. \tag{3}$$

In words, $L(I_1, I_2, p)$ contains the leftmost p intervals of $B(I_1, I_2)$ ordered according to left endpoint and $R(I_1, I_2, p)$ contains the rightmost p intervals of $B(I_1, I_2)$ ordered according to right endpoints. For the example in Fig. 1, $L(I_1, I_2, 2) = \{I_3, I_4\}$ and $R(I_1, I_2, 2) = \{I_3, I_6\}$.

Constructing a DAG. Note that a dynamic programming algorithm for a maximization problem is usually equivalent to finding a longest path in a *directed acyclic graph* (DAG). In the following, we will construct an auxiliary DAG.

Order the intervals in $\mathcal{I}$ as $I_1, \ldots, I_m$ such that $r(I_1) < r(I_2) < \cdots < r(I_m)$. Let I_0, I_{m+1} be two *virtual intervals* satisfying $r(I_0) < \ell(I_1)$ and $\ell(I_{m+1}) > r(I_m)$. We may assume that I_0, I_{m+1} do not cover any point. The vertices of the DAG correspond to the configurations defined as follows.

Definition 2 (configuration). A *configuration* is formed by a 5-tuple $\left(I_{prev}, I_{curr}, \tilde{\ell}, \tilde{r}, \tilde{k}\right)$, where the parameters satisfy the following conditions,

(i) $I_{prev}, I_{curr} \in \mathcal{I} \cup \{I_0, I_{m+1}\}$,
(ii) $r(I_{prev}) < \ell(I_{curr})$ or $I_{prev} = I_{curr} = I_0$ or I_{m+1},
(iii) $\tilde{\ell}$, $\tilde{r}$ are integers satisfying $0 \leq \tilde{\ell} + \tilde{r} \leq |B(I_{prev}, I_{curr})|$,
(iv) $\tilde{k}$ is an integer satisfying $\tilde{\ell} + \tilde{r} + |C(I_{prev}) \cup C(I_{curr}))| \leq \tilde{k} \leq |\mathcal{I}_{\leq I_{curr}}|$.

For a configuration $u = \left(I_{prev}, I_{curr}, \tilde{\ell}, \tilde{r}, \tilde{k}\right)$, denote S^u as

$$S^u = C(I_{prev}) \cup C(I_{curr}) \cup L\left(I_{prev}, I_{curr}, \tilde{\ell}\right) \cup R(I_{prev}, I_{curr}, \tilde{r}),$$

and call S^u as a *configuration set.*

The following is the construction of the DAG.

Definition 3 (DAG G). The vertex set V of G consists of all configurations, a source vertex $u_{src} = (I_0, I_0, 0, 0, 0)$, and some sink vertices $u_{sink}^{K'} = (I_{m+1}, I_{m+1}, 0, 0, K')$ with $K' \leq K$. The arcs are constructed as follows:

(*Arc between two configurations*) Configuration $u = \left(I_{prev}^u, I_{curr}^u, \tilde{\ell}^u, \tilde{r}^u, \tilde{k}^u\right)$ is linked to configuration $v = \left(I_{prev}^v, I_{curr}^v, \tilde{\ell}^v, \tilde{r}^v, \tilde{k}^v\right)$ if and only if the following conditions hold:

(*i*) $I_{curr}^u = I_{prev}^v$,
(*ii*) $\tilde{k}^v = \tilde{k}^u + |S^v \setminus S^u|$.

The weight on such an arc (u, v) is set to be $w(u, v) = |P(u, v)|$, where $P(u, v)$ is the set of points between $\ell\left(I_{prev}^v\right)$ and $\ell\left(I_{curr}^v\right)$ that are fully-covered by $S^u \cup S^v$.

(*Arcs from source vertex to configurations*) Arcs (u_{src}, v) and their weights $w(u_{src}, v)$ are constructed using the same rules as the above.

(*Arcs from configurations to sink vertices*) Link each configuration u of the form $u = (I_{prev}^u, I_{m+1}, 0, 0, \tilde{k}^u)$ to sink vertex $u_{sink}^{K'} = (I_{m+1}, I_{m+1}, 0, 0, K')$ with $K' = \tilde{k}^u$, and set the weight $w\left(u, u_{sink}^{K'}\right) = 0$.

To get a feeling of the construction, we use the example in Fig. 2 to show that a feasible solution $\mathcal{I}'$ can be traced by a directed path in the DAG constructed as the above. Suppose $K = 6$ and $q_p = 2$ for any point p. The blackened intervals form a feasible solution $\mathcal{I}'$ that fully-covers four points. Table 1 lists a sequence of nodes on a source to sink dipath in the DAG.

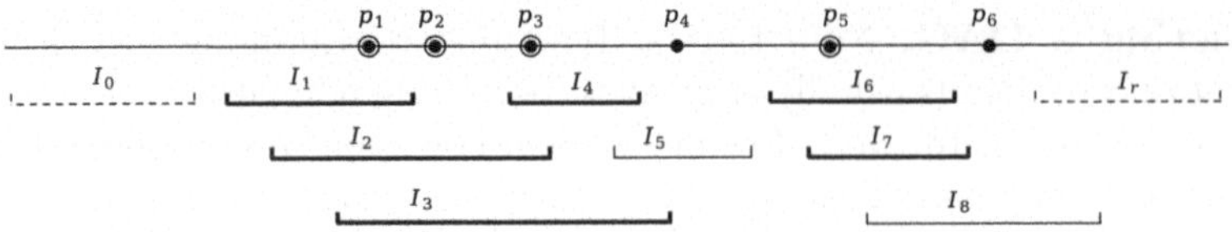

Fig. 2. An illustration for the construction of the DAG

Table 1. The sequence of nodes on a source to sink dipath of the DAG corresponding to the set of blackened intervals $\mathcal{I}'$ in Fig. 2. In this table, L^{u_i} and R^{u_i} denote the set of leftmost $\tilde{\ell}$ intervals and the set of rightmost $\tilde{r}$ intervals added into the solution.

i	configuration u_i	S^{u_i}	$C(I_{i-1}) \cup C(I_i)$	L^{u_i}	R^{u_i}	$P(u_{i-1}, u_i)$	$w(u_{i-1}, u_i)$
0	$(I_0, I_0, 0, 0, 0)$	I_0	$\emptyset$	$\emptyset$	$\emptyset$	$\emptyset$	0
1	$(I_0, I_1, 0, 0, 1)$	I_0, I_1	$\emptyset$	$\emptyset$	$\emptyset$	$\emptyset$	0
2	$(I_1, I_4, 1, 0, 4)$	I_1, I_4, I_2, I_3	I_3	I_2	$\emptyset$	p_1, p_2	2
3	$(I_4, I_6, 0, 0, 5)$	I_4, I_6, I_3	I_3	$\emptyset$	$\emptyset$	p_3	1
4	$(I_6, I_r, 1, 0, 6)$	I_6, I_r, I_7	$\emptyset$	I_7	$\emptyset$	p_5	1

It should be noted that the number of vertices in G is $O(K^3m^2)$, where m is the number of intervals and K is the budget. So the size of G is polynomial in the input size of the MaxIMC instance.

Relation Between an Optimal Solution of MaxIMC and a Longest Source-Sink Path in the DAG. The following lemma shows a structural property of an optimal solution.

Lemma 1. *There exists an optimal solution OPT for MaxIMC such that if $I \in OPT$, then all the intervals in $C(I)$ also belong to OPT.*

In view of this lemma, it suffices to find an optimal solution satisfying the condition of Lemma 1. The next lemma shows that a source-sink dipath in the auxiliary graph corresponds to a feasible solution to the MaxIMC instance.

Lemma 2. *Let $Q = u_{src}u_1 \cdots u_t u_{sink}$ be a source-sink path in G. Then the number of intervals in the set $\mathcal{I}_Q = \bigcup_{i=1}^{t} S^{u_i}$ is at most K and $w(Q) = f(\mathcal{I}_Q)$.*

It should be noted that an optimal solution to a MaxIMC instance might not correspond to a source-sink dipath in the auxiliary graph. However, we have successfully constructed a source-sink dipath whose weight is at least half that of the optimal value, as shown in the following lemma.

Lemma 3. *There is a source-sink dipath Q^* in the DAG G with $w(Q^*) \geqslant \frac{1}{2} opt$, where opt is the number of points that can be fully-covered by an optimal solution.*

Sketch of the Construction. Suppose $\mathcal{I}^*$ is an optimal solution satisfying Lemma 1. Construct representative intervals of $\mathcal{I}^*$ formed of disjoint intervals as follow. Order intervals of $\mathcal{I}^*$ from left to right *according to their right ends.* Let I_1 be the first interval in $\mathcal{I}^*$. For $i \geq 2$, let I_i be the first interval of $\mathcal{I}^*$ that is disjoint from $I_1 \cup \cdots \cup I_{i-1}$. Adding the two virtual intervals I_0 and I_{t+1}, we obtain a sequence of representative intervals $\{I_0, I_1, \ldots, I_t, I_{t+1}\}$. Then we construct two sequences of configurations $\{u_i\}_{i=1}^t$ and $\{v_i\}_{i=1}^t$. Let $n_i = |B(I_{i-1}, I_i) \cap \mathcal{I}^*|$ be the number of intervals of $\mathcal{I}^*$ falling into $B(I_{i-1}, I_i)$. In the construction of sequence $\{u_i\}_{i=1}^t$, for consecutive representative intervals (I_{i-1}, I_i), select intervals from $B(I_{i-1}, I_i)$ obeying the following rules: if i is even, then take the rightmost n_i intervals, and if i is odd, then take the leftmost n_i intervals. While in the construction of sequence $\{v_i\}_{i=1}^t$, "leftmost" and "rightmost" are swapped. Let $u_0 = v_0 = (I_0, I_0, 0, 0, 0)$ be the source vertex, and let $u_{t+1} = (I_{t+1}, I_{t+1}, 0, 0, \tilde{k}^{u_t})$, $v_{t+1} = (I_{t+1}, I_{t+1}, 0, 0, \tilde{k}^{v_t})$. There are two source-sink dipaths $Q_1^* = u_0 u_1 \ldots u_t u_{t+1}$ and $Q_2^* = v_0 v_1 \ldots v_t v_{t+1}$ in G. One may find that any point $p \in (r(I_{2k-1}), r(I_{2k})) \cap P_{opt}$ is fully-covered by $S^{u_{2k}} \cup S^{u_{2k+1}}$, and any point $p \in (r(I_{2k}), r(I_{2k+1})) \cap P_{opt}$ is fully-covered by $S^{v_{2k+1}} \cup S^{v_{2k+2}}$. Such an observation leads to $\max(w(Q_1^*), w(Q_2^*)) \geq \frac{1}{2} opt$.

Combining the above three lemmas, finding a shortest source-sink dipath in the auxiliary DAG corresponds to a 1/2-approximate solution to the MaxIMC problem. Since computing a longest source-sink dipath in a DAG on $|V|$ vertices and $|E|$ edges takes time $O(|V| + |E|)$ [5], our algorithm has time complexity $O(K^6 m^4)$ because of $|E| = O(|V|^2)$.

Theorem 1. *A longest source-sink path in the auxiliary digraph G corresponds to a feasible solution to the MaxIMC instance with approximation ratio at least $\frac{1}{2}$, and the solution can be computed in time $O(K^6 m^4)$.*

3 Conclusion

In this paper, we present a $\frac{1}{2}$-approximation algorithm for the MaxIMC problem. The algorithm employs a dynamic programming method. Because intervals in an optimal solution may appear in a rather irregular manner, our method can only obtain an approximate solution. Similar to the DlS problem [16], whether MaxIMC is NP-hard is still an open question. Note that our method does not work for the weighted version of the MaxIMC problem. Also note that studying the MaxMC problem on the plane might be very challenging. New techniques have to be further explored to study these problems.

References

1. Ali, J., Dyo, V.: Coverage and mobile sensor placement for vehicles on predetermined routes: a greedy heuristic approach. In: Proceedings of the 14th International Joint Conference on e-Business and Telecommunications (ICETE 2017) - WINSYS, pp. 83–88. INSTICC, SciTePress (2017)
2. Bansal, N., Batra, J.: Non-uniform geometric set cover and scheduling on multiple machines. In: Proceedings of the 2021 ACM-SIAM Symposium on Discrete Algorithms (SODA) (2021)

3. Bhattacharya, B.B., Nandy, S.C.: New variations of the maximum coverage facility location problem. Eur. J. Oper. Res. **224**(3), 477–485 (2013)
4. Boonmee, C., Arimura, M., Asada, T.: Facility location optimization model for emergency humanitarian logistics. Int. J. Disaster Risk Reduct. **24**, 485–498 (2017)
5. Cormen, T.H., Leiserson, C.E., Rivest, R.L., Stein, C.: Introduction to Algorithms. MIT Press, Cambridge (2022)
6. Dobson, G.: Worst-case analysis of greedy heuristics for integer programming with nonnegative data. Math. Oper. Res. **7**(4), 515–531 (1982)
7. Feige, U.: A threshold of lnn for approximating set cover. J. ACM (JACM) **45**(4), 634–652 (1998)
8. Hochbaum, D.S.: Approximation algorithms for the set covering and vertex cover problems. SIAM J. Comput. **11**(3), 555–556 (1982)
9. Hochbaum, D.S.: Approximating covering and packing problems: set cover, vertex cover, independent set, and related problems. In: Approximation Algorithms for NP-Hard Problems, pp. 94–143. (1996)
10. Khuller, S., Moss, A., Naor, J.(Seffi): The budgeted maximum coverage problem. Inf. Process. Lett. **70**(1), 39–45 (1999)
11. Kortsarz, G., Peleg, D.: On choosing a dense subgraph (extended abstract). In: DBLP (1993)
12. Li, X., Pan, T., Alim, M.A., Thai, M.T.: Resource allocation in highly dynamic device-to-device communication: an adaptive set multi-cover approach. IEEE Trans. Wirel. Commun. **18**(10), 4607–4619 (2019)
13. Liu, G., Dimitrakakis, A., Carter, B., Gifford, D.: Maximum n-times coverage for vaccine design. In: The Tenth International Conference on Learning Representations, ICLR 2022, Virtual Event, 25–29 April (2022)
14. Manurangsi, P.: Almost-polynomial ratio ETH-hardness of approximating densest k-subgraph. In: Proceedings of the 49th Annual ACM SIGACT Symposium on Theory of Computing (2016)
15. Nemhauser, G.L., Wolsey, L.A., Fisher, M.L.: An analysis of approximations for maximizing submodular set functions–I. Math. Program. **14**, 265–294 (1978)
16. Nonner, T.: PTAS for densest k-subgraph in interval graphs. Algorithmica **74**(1), 528–539 (2016)
17. Rajagopalan, S., Vazirani, V.V.: Primal-dual RNC approximation algorithms for set cover and covering integer programs. SIAM J. Comput. **28**(2), 525–540 (1998)
18. Rubin, J.: A technique for the solution of massive set covering problems, with application to airline crew scheduling. Transp. Sci. **7**(1), 34–48 (1973)
19. Vazirani, V.V.: Approximation Algorithms. Springer, Heidelberg (2001). https://doi.org/10.1007/978-3-662-04565-7
20. Jia, X., Zhou, K., Sixu, W., Dai, H., Lijie, X., Liu, L.: Robust fault-tolerant placement of wireless chargers for directional charging. IEEE Trans. Mob. Comput. **23**, 5295–5309 (2023)
21. Yang, X., Zhang, Z., Zhang, L., Fan, X., Ye, Q., Liyong, F.: Global superpixel-merging via set maximum coverage. Eng. Appl. Artif. Intell. **127**, 107212 (2024)

Minsum Problem for Discrete and Weighted Set Flow on Dynamic Path Network

Bubai Manna[1], Bodhayan Roy[1], and Vorapong Suppakitpaisarn[2(✉)]

[1] IIT Kharagpur, Kharagupur, India
[2] The University of Tokyo, Tokyo, Japan
vorapong@is.s.u-tokyo.ac.jp

Abstract. In this research, we examine the minsum flow problem in dynamic path networks where flows are represented as discrete and weighted sets. The minsum flow problem has been widely studied for its relevance in finding evacuation routes during emergencies such as earthquakes. However, previous approaches often assume that individuals are separable and identical, which does not adequately account for the fact that some groups of people, such as families, need to move together and that some groups may be more important than others. To address these limitations, we modify the minsum flow problem to support flows represented as discrete and weighted sets. We also propose a 2-approximation pseudo-polynomial time algorithm to solve this modified problem for path networks with uniform capacity.

Keywords: Minsum Bin Packing · Dynamic Flow · Approximation Algorithm

1 Introduction

Flow problems on dynamic graphs [7] are considered by many researchers (e.g. [13,15]) because of many reasons. One of the reasons is their relevance in finding evacuation routes during emergencies such as earthquakes or fires [12]. In those applications, we aim to move persons in the ways that they arrive at aiding facilities as soon as possible.

A common objective function for those problems is minmax, which aims to minimize the time until all persons arrive at facilities. In this work, however, we consider another common objective function called minsum, which aims to minimize the summation of time that each individual needs for their trips.

This research was partly conducted during Bubai Manna's and Bodhayan Roy's visit to The University of Tokyo. The visit was hosted by Prof. Reiji Suda and was supported by the JST Sakura Science Program. Vorapong Suppakitpaisarn was partially supported by KAKENHI Grant 23H04377. The authors would like to thank the reviewers for their comments, which significantly improved this paper. The full version of this paper is available at https://arxiv.org/abs/2407.02177.

S. Ghosh and Z. Zhang (Eds.): AAIM 2024, LNCS 15179, pp. 35–47, 2024.
https://doi.org/10.1007/978-981-97-7798-3_4

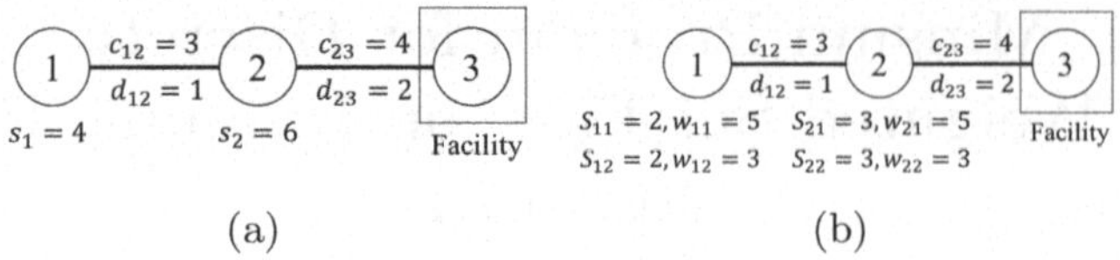

Fig. 1. (a) An instance of the dynamic flow problem (b) An instance of the problem for discrete and weighted set flow on dynamic network

Example 1. In Fig. 1a, there are 4 people at node 1 and 6 people at node 2. These 10 people need to be transported to the aid facility at node 3. Both edges have capacity constraints: a maximum of 3 people can be moved on the edge between nodes 1 and 2 in one unit of time, and a maximum of 4 people can be moved on the edge between nodes 2 and 3 in one unit of time. It takes 1 unit of time to travel from node 1 to node 2 and 2 units of time to travel from node 2 to node 3. At time 1, we can move 3 people from node 1 to node 2 and 4 people from node 2 to node 3. This leaves 1 person at node 1, 5 people at node 2, and 4 people in the middle of the edge between nodes 2 and 3. At time 2, the remaining person at node 1 is moved to node 2, and 4 people at node 2 are moved to node 3. This results in 4 people from node 2 arriving at the facility within 2 units of time, 4 people arriving within 3 units of time, and 2 people arriving within 4 units of time. The maximum time was 4 units of time. The summation of times was $4 \times 2 + 4 \times 3 + 2 \times 4 = 28$. The move which we discussed here minimized both the maximum time and the summation.

It has been shown that both objective functions of flow problems can be solved using time-expanded networks [7,8]. However, these temporal graphs can be exponentially large in relation to the input size, making the algorithm pseudo-polynomial. For minmax problems, polynomial-time algorithms have been developed for paths [3] and trees [4]. There are also FPTAS for general graphs when the number of facilities is constant [1]. In contrast, minsum problems have only been shown to have polynomial-time algorithms for path graphs [2].

All known algorithms assume that individuals are distinct and identical, meaning that we can move any number of people over a particular edge as long as the total number does not exceed the edge's capacity. However, this may not always be possible in practice. For example, some groups of people, such as families, must be moved together, and some groups may require emergency aid and should therefore be given higher priority. These considerations must be taken into account when determining how to move people from one location to another.

1.1 Our Contributions

In short, we modify the minsum flow problem to support flows represented as discrete and weighted sets. We also propose a 2-approximation pseudo-polynomial time algorithm to solve this modified problem for path networks with uniform capacity.

We illustrate the ideas of the modified problem in the following example.

Example 2. In Fig. 1b, there are two groups of people at node 1, each with 2 people. The weight of the first group is 5, while the weight of the second group is 3. There are also two groups of 3 people at node 2, with weights of 5 and 3, respectively. We refer to the group with size S_{ij} and weight w_{ij} as G_{ij}. At time 1, we move group G_{11} from node 1 to node 2 and group G_{21} from node 2 to node 3. Before time 2, group G_{12} is at node 1, groups G_{11} and G_{22} are at node 2, and group G_{21} is in the middle of the edge between nodes 2 and 3. At time 2, we move group G_{12} to node 2 and group G_{11} to node 3. At time 3, we move group G_{12} from node 2 to node 3, and at time 4, we move group G_{22}. As a result, group G_{21} arrives at time 2, group G_{11} arrives at time 3, group G_{12} arrives at time 4, and group G_{22} arrives at time 5. The weighted summation of arrival time is then $2w_{21} + 3w_{11} + 4w_{12} + 5w_{22} = 52$.

It is clear that the modified problem is harder than the original version. Indeed, we can show that it is NP-hard by a reduction to the partition problem. The formal definition of this problem with its NP-hardness proof can be found in Sect. 2.

We discuss in Sect. 3 that when we have two nodes, our problem is equivalent to the weighted minsum bin packing problem [5,6]. To support the case that we have more than two nodes, we need to derive a bin-packing algorithm that can support items with different arrival times. Suppose that t_i is the arrival time of item i. The item cannot be packed in the first $(t_i - 1)$-th bag. We show that the algorithm is a 2-approximation.

As there is PTAS proposed in [5] for the minsum bin packing problem, one may think that we can extend that PTAS to support items with arrival times. Unfortunately, by the requirement that we cannot insert particular items in some bags, we strongly believe that the extension is not straightforward. We are aiming to give that extension as our future work.

In Sect. 4, we extend the bin packing algorithm presented in Sect. 3 to address our main problem. We demonstrate that the extended algorithm is a 2-approximation when all capacities are uniform, and there is only one facility. It is worth noting that several works in dynamic network flows also make this assumption of uniform capacities [11,14] and a single facility [9,10].

2 Problem Definitions

In this section, we define our problem called minsum problem for discrete and weighted set flow on a dynamic path network (MS-DWSF).

Consider a path graph with n nodes, denoted by $P_n = (V = \{1, \ldots, n\}, E = \{\{i, i+1\} : 1 \leq i \leq n-1\})$. Each node i has m_i sets of persons to evacuate. Those sets of persons are denoted by $G_{i,1}, \ldots, G_{i,m_i}$. For group G, the size of G is denoted by $S(G) \in \mathbb{Z}_+$ and the weight of G is denoted by $w(G) \in \mathbb{Z}_+$. The capacity of all edges is $C \in \mathbb{Z}_+$. Each edge e has distance $d(e) \in \mathbb{Z}_+$, which is the time that persons need to move between two terminals of the edge.

Suppose that the single aiding facility is located at $a \in V$. People originally at node $i < a$ must move in a direction that increases the node number they are at, while people originally at node $i > a$ must move in a direction that decreases the node number they are at in any optimal solution.

Let us denote the collection of groups that are at node i at time t by $\mathcal{S}_i^{(t)}$. We select from $\mathcal{S}_i^{(t)}$ which groups to be sent along the edge $\{i, i+1\}$ for $i < a$ and along the edge $\{i-1, i\}$ for $i > a$. We denote the collection of groups that we choose to send by $D_i^{(t)}$. The summation of group sizes in $D_i^{(t)}$ must not be larger than C, i.e. $\sum_{G \in D_i^{(t)}} S(G) \leq C$.

For $t = 0$, we have $\mathcal{S}_i^{(0)} = \{G_{i,1}, \ldots, G_{i,m_i}\}$ for all i. Let denote $A_i^{(t)}$ be a collection of groups arriving at i from node $i-1$ at time t and denote $B_i^{(t)}$ be a collection of groups arriving at i from node $i+1$ at time t. We have $A_i^{(t)} = D_{i-1}^{(t-d(\{i-1,i\}))}$ when $1 < i \leq a$ and $t \geq d(\{i-1, i\})$ and $A_i^{(t)} = \emptyset$ otherwise. Similarly, $B_i^{(t)} = D_{i+1}^{(t-d(\{i,i+1\}))}$ when $a \leq i < n$ and $t \geq d(\{i, i+1\})$ and $B_i^{(t)} = \emptyset$ otherwise. Then, $\mathcal{S}_i^{(t)} = \mathcal{S}_i^{(t-1)} \backslash D_i^{(t)} \cup A_i^{(t)} \cup B_i^{(t)}$.

The arrival time of G, denoted by $\alpha(G)$ is the earliest time that the group is at a, i.e. $\min\{t : G \in \mathcal{S}_a^{(t)}\}$. In the MS-DWSF, we aim to minimize $\sum_G w(G)\alpha(G)$.

We show that the problem is NP-hard in the full version of this paper available at https://arxiv.org/abs/2407.02177.

3 Minsum Bin Packing Problem for Weighted Items with Different Ready Times

To address the MS-DWSF problem, we first introduce a related problem called the minsum bin packing problem for weight items with different ready times (MS-BPWRT). In this section, we present a 2-approximation pseudo-polynomial time algorithm for the MS-BPWRT problem. We will then use the solution obtained from this algorithm to develop a 2-approximation pseudo-polynomial time algorithm for the MS-DWSF in the following section.

3.1 Definition of MS-BPWRT

The MS-BPWRT problem can be defined in the following definition:

Definition 1 (MS-BPWRT(τ)). *Given a collection of groups $\mathcal{G} = \{G_1, \ldots, G_m\}$. Each group G_i has size $S(G_i)$, weight $w(G_i)$, and ready time $\tau(G_i)$. We find a way to pack those groups into a set of bins $B_1, \ldots, B_T \subseteq \mathcal{G}$ with capacity C with the following constraints: 1) $\bigcup_{1 \leq j \leq m} B_j = \{G_1, \ldots, G_m\}$, 2) $B_j \cap B_{j'} = \emptyset$ for $j \neq j'$, 3) For all $1 \leq j \leq T$, $\sum_{G \in B_j} S(G) \leq C$, and 4) Denote $t(G) = j$ when $G \in B_j$, we must have $t(G) \geq \tau(G)$. We aim to minimize $\sum_G w(G)t(G)$.*

When $w(G) = 1$ for all G and we do not have the fourth constraint, the MS-BPWRT is equivalent to the minsum bin packing problem [5]. We use some ideas from the minsum bin packing problem to provide an algorithm and prove the approximation ratio for the MS-BPWRT.

We have included the weight of group G, denoted as $w(G)$, in the problem formulation because we recognize that different groups may have varying levels of importance. The ready time, $\tau(G)$, signifies that group G cannot be placed in any bin with an index less than $\tau(G)$. In other words, group G is not ready to be inserted until time $\tau(G)$.

3.2 Approximation Algorithm for MS-BPWRT

The approximation algorithm for MS-BPWRT is described in Algorithm 1. The collection $\mathcal{G}$ contains groups that have not been placed in any bin, while the collection $\mathcal{G}'_j$ is a candidate set for bin B_j. If any remaining group can be considered a candidate for B_j by replacing the code in Line 3 with $\mathcal{G}'_j \leftarrow \mathcal{G}'$, then Algorithm 1 becomes the next fit decreasing algorithm [16], based on the ratio $w(G)/S(G)$. It is worth noting that the minsum bin packing algorithm in [5] uses the next fit increasing algorithm based on $s(G)$ (or the next fit decreasing algorithm based on $1/s(G)$). The criteria for the next fit algorithm is how we apply the weights $w(G)$ to the minsum bin packing problem.

Algorithm 1: 2-Approximation Algorithm for MS-BPWRT

Input: Collection of all groups $\mathcal{G}$, size of each group $S : \mathcal{G} \to \mathbb{Z}_+$, weight of each group $w : \mathcal{G} \to \mathbb{Z}_+$, ready time of each group $\tau : \mathcal{G} \to \mathbb{Z}_+$, capacity of each bin $C \in \mathbb{Z}_+$

Output: Groups in bins $B_1, \dots, B_T \subseteq \mathcal{G}$

1 $\mathcal{G}' \leftarrow \mathcal{G}$; $j \leftarrow 1$
2 **while** $\mathcal{G}' \neq \emptyset$ **do**
3     $\mathcal{G}'_j \leftarrow \{G \in \mathcal{G}' : \lceil (\tau(G) - 1)/2 \rceil \times 2 + 1 \leq j\}$
4     $G' \leftarrow \arg\max_{G \in \mathcal{G}'_j} w(G)/S(G)$
5   **if** $\sum_{G \in B_j} S(G) + S(G') > C$ **then**
6         $j \leftarrow j + 1$
7   **else**
8         $B_j \leftarrow B_j \cup \{G'\}, \mathcal{G}' \leftarrow \mathcal{G}' \setminus \{G'\}$
9   **end**
10 **end**

We consider the ready times at Line 3 of the algorithm. The collection $\mathcal{G}'_j$ is the collection of groups that have not been added to any bin of which the ready time $\tau(G)$ satisfies $\lceil (\tau(G) - 1)/2 \rceil \times 2 + 1 \leq j$. We know that $\lceil (\tau(G) - 1)/2 \rceil \times 2 + 1 = \tau(G)$ when $\tau(G)$ is odd and $\lceil (\tau(G) - 1)/2 \rceil \times 2 + 1 = \tau(G) + 1$ when $\tau(G)$ is even. Recall that $\mathcal{G}'_j$ is the candidate to be added to B_j. For G such that

$\tau(G)$ is odd, we add G to the candidate set of B_j for any $j \geq \tau(G)$ that matches with the ready time constraint. On the other hand, for G such that $\tau(G)$ is even, we do not add G to the candidate set of B_j for $j = \tau(G)$, but add only when $j \geq \tau(G) + 1$. Informally, we delay the addition of G by one bin here.

3.3 Proof for Approximation Ratio

We prove that the algorithm in the previous subsection is a two-approximation algorithm for the MS-BPWRT problem. First, we define the relaxed version of MS-BPWRT in the following definition:

Definition 2 (MS-BPWRT-REAL(τ)). *Suppose we have a collection of groups* $\mathcal{G} = G_1, \ldots, G_m$, *where each group* G_i *has a size of* $S(G_i)$, *a weight of* $w(G_i)$, *and a ready time of* $\tau(G_i)$. *We are given a capacity* C. *For each* $1 \leq i \leq m$ *and* $1 \leq j \leq T$, *we find* $x_{ij} \in [0, 1]$ *such that 1)* $\sum_j x_{ij} = 1$ *for all* i, *2)* $\sum_i S(G_i)x_{ij} \leq C$ *for all* j, *and, 3) for each* $x_{ij} > 0$, *we must have* $j \geq \tau(G_i)$. *Our goal is to minimize* $\sum_{i,j} j \cdot w(G_i) \cdot x_{ij}$.

Informally speaking, in the MS-BPWRT-REAL problem, each group can be partially assigned to each bin. The variable x_{ij} represents the proportion of group G_i assigned to bin B_j. Let $OPT(\tau), OPT_R(\tau)$ be the optimal value of the MS-BPWRT(τ) and MS-BPWRT-REAL(τ) problems. We have the following properties:

Proposition 1. *$OPT_R(\tau) \leq OPT(\tau)$.*

Proof. Let $B_1^*, \ldots, B_T^*$ be an optimal solution of MS-BPWRT(τ), $T^*(G_i) = j$ if $G_i \in B_j^*$, and let $x'_{ij} = 1$ if $G_i \in B_j^*$ and $x'_{ij} = 0$ otherwise. We know that $\langle x'_{ij} \rangle_{i,j}$ is a solution of MS-BPWRT-REAL(τ) because 1) $\sum_j x'_{ij} = 1$ for all i because each G_i is a member of exactly one bin by constraints 1) and 2) of MS-BPWRT(τ), 2) For all j, $\sum_i S(G_i)x'_{ij} = \sum_{G_i \in B_j^*} S(G_i) \leq C$ by the third constraint of MS-BPWRT(τ), 3) When $x'_{ij} > 0$, $G_i \in B_j^*$ and, by the fourth constraint of MS-BPWRT(τ), $j \geq \tau(G)$. The objective value of $\langle x'_{ij} \rangle_{i,j}$ is $\sum_{i,j} w(G_i)(j \cdot x'_{ij}) = \sum_i w(G_i) \sum_j (j \cdot x'_{ij}) = \sum_i w(G_i)T^*(G_i) = OPT(\tau)$. We then know that there is a solution of MS-BPWRT-REAL(τ) with objective value $OPT(\tau)$. The optimal value of MS-BPWRT-REAL(τ) must not be larger than $OPT(\tau)$, i.e. $OPT_R(\tau) \leq OPT(\tau)$. □

Proposition 2. *Let* τ, τ' *be a function such that* $\tau(G) \leq \tau'(G)$ *for all* $G \in \mathcal{G}$. *Then,* $OPT_R(\tau) \leq OPT_R(\tau')$.

Proof. Let $\langle x^*_{ij} \rangle_{i,j}$ be an optimal solution of the MS-BPWRT-REAL(τ') problem. It is straightforward to show that $\langle x^*_{ij} \rangle_{i,j}$ satisfies the first and the second constraints of MS-BPWRT-REAL(τ). Also, because for each $x_{ij} > 0$, we have $j \geq \tau'(G_i) \geq \tau(G_i)$, we know that $\langle x^*_{ij} \rangle_{i,j}$ also satisfies the third constraint of MS-BPWRT-REAL(τ), and is a feasible solution of MS-BPWRT-REAL(τ).

The objective value of $\langle x^*_{ij}\rangle$ is $OPT_R(\tau')$. We then know that there is a solution of MS-BPWRT-REAL(τ) with objective value $OPT_R(\tau')$. The optimal value of MS-BPWRT-REAL(τ) must not be larger than $OPT_R(\tau')$, i.e. $OPT_R(\tau) \leq OPT_R(\tau')$. □

Denote a solution from Algorithm 1 by $B'_1, \ldots, B'_T$. Let $x'_{ij} = 1$ when $G_i \in B'_{2j-1} \cup B'_{2j}$. It is clear that $\langle x'_{ij}\rangle_{i,j}$ is not a feasible solution of MS-BPWRT-REAL(τ'). We prove a property of $\langle x'_{ij}\rangle_{i,j}$ in the following proposition:

Proposition 3. *For all j such that $B'_{2j} \neq \emptyset$, $\sum_i S(G_i)x'_{ij} > C$.*

Proof. At Line 3 of Algorithm 1, we define the set $\mathcal{G}'_j$. We can observe that both $\mathcal{G}_{2j-1} = \{G \in \mathcal{G}' : \tau(G) \leq 2j-1\}$ and $\mathcal{G}_{2j} = \{G \in \mathcal{G}' : \tau(G) \leq 2j-1\}$ are defined in the same manner. This implies that, for any integer k, even if we increase k from $2k-1$ to $2j$ at Line 6, the set of groups considered remains unchanged.

Let G' be the first element added to the bin B'_{2j}. It is a group in $G'_{2j}\backslash(B'_1 \cup \cdots \cup B'_{2j-1})$ which maximizes $w(G)/S(G)$.

Since the set of groups considered for bins B'_{2j-1} and B'_{2j} are the same, G' must have already been considered for inclusion in B'_{2j-1}. However, it was not added to B'_{2j-1} because doing so would result in $\sum\limits_{G \in B'_{2j-1}} S(G) + S(G') > C$.

Therefore, by the definition of x'_{ij}, we have $\sum_i S(G_i) \cdot x'_{ij} = \sum_{G \in B'_{2j-1}} S(G) + \sum_{G \in B'_{2j}} S(G) \geq \sum_{G \in B'_{2j-1}} S(G) + S(G') > C$. This completes the proof. □

Next, we define a problem called weight maximization problem (WM) as follows:

Definition 3 (WM($\tau', C_1, \ldots, C_{T'}$)). *Suppose we have a collection of groups $\mathcal{G} = G_1, \ldots, G_m$, where each group G_i has a size of $S(G_i)$, a weight of $w(G_i)$, and a ready time of $\tau(G_i)$. We are given a capacity C. For each $1 \leq i \leq m$ and $1 \leq j \leq T'$, we find $x_{ij} \in [0,1]$ such that: 1) $\sum\limits_j x_{ij} \leq 1$ for all i,*

2) $\sum\limits_i S(G_i)x_{ij} \leq C_j$ for all j, and, 3) for each $x_{ij} > 0$, we must have $j \geq \tau(G_i)$. We aim to maximize $\sum\limits_{i,j} w(G_i) \cdot x_{ij}$.

Let $C'_j = \sum_i S(G_i)x'_{ij}$, and let τ' be a function such that, for all $G_i \in \mathcal{G}$, $\tau'(G_i) = \lceil \tau(G_i)/2 \rceil$. We then can show the following property:

Proposition 4. *For all $1 \leq T' \leq T$, $\langle x'_{ij}\rangle_{j \leq T',i}$ is an optimal solution of WM($\tau', C'_1, \ldots, C'_{T'}$).*

Proof. We prove this proposition by induction on T'.

Let us examine the scenario where $T' = 1$. Remember from Algorithm 1 that the bins B_1 and B_2 contain groups from the set $\{G \in \mathcal{G} : \tau(G) = 1\}$ that maximize the ratio $w(G)/S(G)$. Using the greedy algorithm, any collection of groups $\mathcal{D} \subseteq \{G \in \mathcal{G} : \tau'(G) = 1\}$ with $\sum\limits_{G \in \mathcal{D}} S(G) \leq C'_1$ must satisfy $\sum\limits_{G \in \mathcal{D}} w(G) \leq$

$\sum_{G \in B_1 \cup B_2} w(G) = \sum_i w(G_i) \cdot x'_{i1}$. Therefore, we can deduce that the sequence $\langle x'_{ij} \rangle_{j=1,i}$ represents an optimal solution for $\mathrm{WM}(\tau', C'_1)$.

Next, let us assume the proposition holds true for all $T' \leq \mathsf{T}$. We will assume, aiming for a contradiction, that the sequence $\langle x'_{ij} \rangle_{j \leq \mathsf{T},i}$ does not represent an optimal solution for $\mathrm{WM}(\tau', C'_1, \ldots, C'_{\mathsf{T}})$. Let us say an optimal solution for $\mathrm{WM}(\tau', C'_1, \ldots, C'_{\mathsf{T}})$ is represented by the sequence $\langle x^*_{ij} \rangle_{j \leq \mathsf{T},i}$.

Based on our assumption that the sequence $\langle x'_{ij} \rangle_{j \leq \mathsf{T}-1,i}$ is an optimal solution for $\mathrm{WM}(\tau', C'_1, \ldots, C'_{\mathsf{T}-1})$, it follows that $\sum_{j \leq \mathsf{T}-1,i} w(G_i) \cdot x'_{ij} \geq \sum_{j \leq \mathsf{T}-1,i} w(G_i) \cdot x^*_{ij}$. To satisfy the condition $\sum_{j \leq \mathsf{T},i} w(G_i) \cdot x'_{ij} \geq \sum_{j \leq \mathsf{T},i} w(G_i) \cdot x^*_{ij}$, it is necessary to have $\sum_i w(G_i) x'_{i\mathsf{T}} < \sum_i w(G_i) x^*_{i\mathsf{T}}$.

Recall from the construction that $x'_{ij} \in \{0,1\}$ for all i. To have $\sum_i w(G_i) x'_{i\mathsf{T}} < \sum_i w(G_i) x^*_{i\mathsf{T}}$, there must be i^* such that $x'_{i^*\mathsf{T}} = 0$, $x^*_{i^*\mathsf{T}} > 0$, and $w(G_{i^*})/S(G_{i^*}) > w(G_i)/S(G_i)$ for all i such that $x'_{i\mathsf{T}} = 1$.

Consider the case that $x'_{i^*\mathsf{T}'} = 1$ for some $\mathsf{T}' < \mathsf{T}$. Then, in the solution $\langle x^*_{ij} \rangle_{j \leq \mathsf{T},i}$, we move G_{i^*} to the bin $B'_{2\mathsf{T}'-1} \cup B'_{2\mathsf{T}'}$. Let $s = \min\{S(G_{i^*}) x^*_{i^*\mathsf{T}}, C'_{\mathsf{T}'} - \sum_i S(G_i) x^*_{i\mathsf{T}'}\}$. If $s > 0$, we have more spaces to put the group G_{i^*}. We then decrease the value of $x^*_{i^*,\mathsf{T}}$ by $s/S(G_{i^*})$ and increase the value of $x^*_{i^*,\mathsf{T}'}$ by the same value.

If $C'_{\mathsf{T}'} - \sum_i S(G_i) x^*_{i\mathsf{T}'} = 0$, there is no space left in the bin $B_{\mathsf{T}'}$. We then have to swap G_{i^*} with some other groups. There is an item i' such that $x^*_{i'\mathsf{T}'} > 0$ while $x'_{i'\mathsf{T}'} = 0$. Let $s = \min\{S(G_{i'}) x^*_{i'\mathsf{T}'}, S(G_{i^*}) x^*_{i^*\mathsf{T}}\}$. We can update the value of $\langle x^*_{ij} \rangle_{i,j}$ in the following ways: 1) decrease the value of $x^*_{i^*\mathsf{T}}$ by $s/S(G_{i^*})$, 2) decrease the value of $x^*_{i'\mathsf{T}'}$ by $s/S(G_{i'})$, 3) increase the value of $x^*_{i^*\mathsf{T}'}$ by $s/S(G_{i^*})$, and 4) increase the value of $x^*_{i'\mathsf{T}}$ by $s/S(G_{i'})$. Informally, we exchange s units of G_{i^*} in bin T with an equal mass of $G_{i'}$ in bin T'. This updated result continues to be a feasible solution for $\mathrm{WM}(\tau', C'_1, \ldots, C'_T)$, and the objective value remains unchanged.

We can iterate the update in the previous paragraph until there is no i^* such that $x'_{i^*\mathsf{T}'} = 1$ for some T'. It is sufficient to only consider the case when, for all such i^*, we have not included the group G_{i^*} to bins $B_1, \ldots, B_{2\mathsf{T}-2}$. However, by the assumption that $w(G_{i^*})/S(G_{i^*}) > w(G_i)/S(G_i)$ for all i such that $x'_{i\mathsf{T}} = 1$, the greedy algorithm must have already included the group i^* to the bin $B_{2\mathsf{T}-1}$. This gives $x'_{i^*\mathsf{T}} = 1$, which contradicts our assumption that $x'_{i^*\mathsf{T}} = 0$. □

From the next proposition, let us consider the problem $\mathrm{WM}(\tau', C_1, \ldots, C_{T'})$ where $C_1 = C_2 = \cdots = C_{T'} = C$. We denote the optimal solution of the problem by $OPT_{\mathrm{WM}}(T')$.

Proposition 5. *For all* $1 \leq T' \leq T$, $OPT_{WM}(T') \leq \sum_{j \leq T',i} w(G_i) \cdot x'_{ij}$.

Proof. Let T' be such that $B'_{2T'} \neq \emptyset$. Using the definition of C_j and Proposition 3, we have $C_j = \sum_i S(G_i) x'_{ij} > C$. An optimal solution for $\mathrm{WM}(\tau', C, \ldots, C)$ is a feasible solution for $\mathrm{WM}(\tau', C'_1, \ldots, C'_{T'})$. Therefore, the objective value of an optimal solution for $\mathrm{WM}(\tau', C'_1, \ldots, C'_{T'})$, which is $\sum_{j \le T', i} w(G_i) \cdot x'_{ij}$ according to Proposition 4, cannot be less than $OPT_{\mathrm{WM}}(T')$.

When $B'_{2T'} = \emptyset$, it implies that all items have been allocated to the bins $B'_1, \ldots, B'_{2T'-1}$. In this case, the value of $\sum_{j \le T', i} w(G_i) x'_{ij}$ is equal to $\sum_i w(G_i)$, which is greater than or equal to the sum of the weights of any feasible solution. Hence, we have $\sum_{j \le T', i} w(G_i) x'_{ij} = OPT_{\mathrm{WM}}(T')$. □

The next lemma gives a relationship between the sequence $\langle x'_{ij} \rangle_{i,j}$ and the MS-BPWRT-REAL problem.

Lemma 1. $OPT_R(\tau') \ge \sum_{i,j} j \cdot w(G_i) \cdot x'_{ij}$

Proof. Let $\mathcal{X}$ be a collection of all feasible solutions of the MS-BPWRT-REAL(τ'), and let $W = \sum_i w(G_i)$. By Proposition 5, we have that $OPT_R(\tau') = \min_{\langle x_{ij} \rangle_{i,j} \in \mathcal{X}} \sum_j \sum_i j \cdot w(G_i) \cdot x_{ij} = \min_{\langle x_{ij} \rangle_{i,j} \in \mathcal{X}} \sum_{T'} \sum_{j \ge j', i} w(G_i) \cdot x_{ij} \ge \sum_{T'} \min_{\langle x_{ij} \rangle_{i,j} \in \mathcal{X}} \left[W - \sum_{j < T', i} w(G_i) \cdot x_{ij} \right] = \sum_{T'} \left[W - \max_{\langle x_{ij} \rangle_{i,j} \in \mathcal{X}} \sum_{j < T', i} w(G_i) \cdot x_{ij} \right] \ge \sum_{T'} \left[W - \sum_{j < T', i} w(G_i) \cdot x'_{ij} \right] = \sum_{T'} \sum_{j \ge j', i} w(G_i) \cdot x'_{ij} = \sum_{i,j} j \cdot w(G_i) \cdot x'_{ij}$. □

We are now ready to prove the main theorem of this section.

Theorem 1. *The bin $B'_1, \ldots, B'_T$ obtained from Algorithm 1 is a 2-approximation solution for MS-BPWRT.*

Proof. Let SOL be an objective value of $B'_1, \ldots, B'_T$. We have that $SOL = \sum_j \sum_{G_i \in B'_j} j \cdot w(G_i) \le \sum_j \sum_{G_i \in B'_{2j-1} \cup B'_{2j}} 2j \cdot w(G_i) \le 2 \cdot \sum_{i,j} j \cdot w(G_i) \cdot x'_{ij} \le 2 \cdot OPT_R(\tau') \le 2 \cdot OPT_R(\tau) \le 2 \cdot OPT(\tau)$. The inequality at Line 3 of the chain is obtained from the definition of $\langle x'_{ij} \rangle_{i,j}$. The inequality at Line 4 is obtained from Lemma 1, the inequality at Line 5 is obtained from Proposition 2, and the inequality at Line 6 is obtained from Proposition 1. □

4 Approximation Algorithm for MS-DWSF

In this section, we will develop an approximation algorithm for our main problem, MS-DWSF, utilizing the findings presented in the previous section.

4.1 Algorithm

Our two-approximation algorithm for the MS-DWSF is shown in Algorithm 2. The algorithm addresses congestion on the busiest edge. Specifically, when the destination node is denoted as a, the most congested edges are $\{a-1, a\}$ and

Algorithm 2: 2-Approximation Algorithm for MS-DWSF

Input: 1) A path graph $V = \{1, \ldots, n\}$ and $E = \{\{i, i+1\} : 1 \leq i \leq n-1\}$,
2) for each $1 \leq i \leq n$, a set of groups $S_i^{(0)} = \{G_{i,1}, \ldots, G_{i,m_i}\}$,
3) for each group $G_{i,j}$, the size of $G_{i,j}$ denoted by $S(G_{i,j})$ and the weight of $G_{i,j}$ denoted by $w(G_{i,j})$.
4) for each edge $e \in E$, a distance of e denoted by $d(e)$.
5) the capacity of all edges denoted by C.
6) the destination node denoted by $a \in V$

Output: Groups to move from node i at time t denoted by $D_i^{(t)}$

1 For $i < a$, let $d(i, a-1) = \sum_{v=i}^{a-2} d(\{v, v+1\})$.
2 Execute Algorithm 1 under the following conditions: $\mathcal{G}$ is a set of groups $\{G_{i,j} : i < a\}$, where $\tau(G_{i,j}) = d(i, a-1)$, and the size and weight of each group correspond to the input parameters of MS-DWSF. Suppose that the output of the algorithm is $B_1', \ldots, B_T'$
3 $D_i^{(t)} \leftarrow B'_{t+d(i,a-1)} \cap \{G_{i',j} : i' \leq i\}$ for all $i < a$ and t.
4 For $i > a$, let $d(i, a+1) = \sum_{v=i}^{a} d(\{v, v+1\})$.
5 Execute Algorithm 1 under the following conditions: $\mathcal{G}$ is a set of groups $\{G_{i,j} : i > a\}$, where $\tau(G_{i,j}) = d(i, a+1)$, and the size and weight of each group correspond to the input parameters of MS-DWSF. Suppose that the output of the algorithm is $B_1', \ldots, B_T'$
6 $D_i^{(t)} \leftarrow B'_{t+d(i,a+1)} \cap \{G_{i',j} : i' \geq i\}$ for all $i > a$ and t.

$\{a, a+1\}$. To tackle this issue, we can examine separate strategies for each of these edges. It is worth noting that the concepts behind both strategies are the same, so we will only elaborate on the approach for edge $\{a-1, a\}$ here.

To transmit all groups in the set $\{G_{i,j} : i < a\}$ through the edge $\{a-1, a\}$, we rely on the results of Algorithm 1 (denoted by $B_1', \ldots, B_T'$) to determine the appropriate timing for each item. At time T', items within bin $B_{T'}$ are dispatched along the $\{a-1, a\}$ edge.

The MS-DWSF constraint requires that all groups $G_{i,j} \in B_{T'}$ be present at node $a-1$ during transmission. To satisfy this condition, if $i \leq a-2$, the group is sent from node $a-2$ to $a-1$ at the time $T' - d(\{a-1, a-2\})$. Similarly, if $i \leq v$, the group is sent from node v to $a-1$ at time $T' - d(v, a-1)$, following the same idea.

The collection of groups transmitted from node i at time t is obtained by taking the intersection of $B'_{t+d(i,a-1)}$ with $\{G_{i',j} : i' \leq i\}$, as assigned in Line 3 of the algorithm.

Since group $G_{i,j}$ is initially located at node i, it cannot reach node $a-1$ before time $d(i, a-1)$. Thus, it is not possible to assign group $G_{i,j}$ to bin B_j for $j < d(i, a-1)$. This is why we set $\tau(G_{i,j}) = d(i, a-1)$ in Line 2 of the algorithm.

4.2 Feasibility and Approximation Ratio

In this subsection, we show that Algorithm 2 always gives a feasible solution. Then, we show that it is a two-approximation ratio for MS-DWSF.

Theorem 2. *$\langle D_i^{(t)} \rangle_{i,t}$ in Algorithm 2 is a feasible solution to MS-DWSF.*

Proof. To show that the solution of Algorithm 2 is feasible, we need to show that, for all i and t, $\sum_{G \in D_i^{(t)}} S(G) \leq C$ and $D_i^{(t)} \subseteq S_i^{(t)}$. The first inequality can be shown by the fact that $D_i^{(t)} \subseteq B'_{t+d(i,a-1)}$ and $\sum_{G \in B'_{t+d(i,a-1)}} S(G) \leq C$ by the constraint of MS-BPWRT.

We will now demonstrate that $D_i^{(t)} \subseteq S_i^{(t)}$. Suppose we have a group $G_{i',j} \in D_i^{(t)}$. Since $G_{i',j} \in D_i^{(t)}$, we have $G_{i',j} \in B'_{t+d(i,a-1)}$. By the MS-BPWRT constraint, we know that $G_{i',j}$ cannot be assigned to any bin $B_{T'}$ for $T' < t + d(i, a-1)$. As a result, $G_{i',j}$ is not in $D_i^{(t')}$ for any $t' < t$. Hence, for $i' = i$, we conclude that $G_{i',j} \in S_i^{(t)}$.

For $i' < i$, we have that $G_{i',j} \in B'_{t+d(i,a-1)} = B'_{t-d(\{i-1,i\})+d(i-1,a-1)}$. For $t' = t - d(\{i-1,i\})$, we have $G_{i',j} \in B'_{t'+d(\{i-1,a-1\})}$ and $G_{i',j} \in D_{i-1}^{(t')}$. By the problem definition of MS-DWSF, we know that $G_{i',j} \in S_i^{(t)}$ when $G_{i',j} \in D_{i-1}^{(t-d(\{i-1,i\})}$. □

The next theorem will show that Algorithm 2 is a two-approximation algorithm for MS-DWSF.

Theorem 3. *Let $\alpha'(G)$ be the time that G arrives at the node a in Algorithm 2, and let OPT_D be an optimal solution of the MS-DWSF problem. We have that $\sum_G w(G)\alpha'(G) \leq 2 \cdot OPT_D$.*

Proof. We can construct a feasible solution of MS-BPWRT from an optimal solution of MS-DWSF by setting $B_{T'}$ to $D_{a-1}^{(T')}$ for all T'. Let $W_D = d(\{a-1,a\}) \sum_G w(G)$. As a group $G_{i,j} \in D_{a-1}^{(T')}$ arrives at the destination node a at time $T' + d(\{a-1,a\})$, we have that $\sum_{T'} \sum_{G \in B_{T'}} T' \cdot w(G) = \sum_{T'} \sum_{G \in B_{T'}} (\alpha(G) - d(\{a-1,a\})) \cdot w(G) = \sum_G \alpha(G)w(G) - W_D = OPT_D - W_D$. If OPT_B is an optimal value of MS-BPWRT, we have that $OPT_B \leq OPT_D - W_D$.

Algorithm 1 gives a solution of MS-BPWRT of which the objective function, denoted by SOL_B, is not larger than $2 \cdot OPT_B$. From that solution, we can construct a solution of MS-DWSF using Line 3 of Algorithm 2. The objective value of the MS-DWSF solution, denoted by SOL_D, is $SOL_B + W_D$. We then obtain that $SOL_D = SOL_B + W_D \leq 2SOL_B + W_D \leq 2OPT_D$. □

5 Conclusion

This paper presents an extension of the minsum bin packing problem, which considers items with varying ready times and weights. We propose a 2-approximation algorithm for this new problem and apply it to develop an evacuation method for non-separable groups of individuals. At present, our algorithm is limited to path graphs with a single destination. However, we are actively working on expanding its capabilities to handle multiple destinations and non-path network structures.

References

1. Belmonte, R., Higashikawa, Y., Katoh, N., Okamoto, Y.: Polynomial-time approximability of the k-sink location problem. arXiv preprint arXiv:1503.02835 (2015)
2. Benkoczi, R., Bhattacharya, B., Higashikawa, Y., Kameda, T., Katoh, N.: Minsum k-sink problem on path networks. Theoret. Comput. Sci. **806**, 388–401 (2020)
3. Bhattacharya, B., Golin, M.J., Higashikawa, Y., Kameda, T., Katoh, N.: improved algorithms for computing k-sink on dynamic flow path networks. In: WADS 2017. LNCS, vol. 10389, pp. 133–144. Springer, Cham (2017). https://doi.org/10.1007/978-3-319-62127-2_12
4. Chen, D., Golin, M.: Minmax centered k-partitioning of trees and applications to sink evacuation with dynamic confluent flows. Algorithmica (2022, in press)
5. Epstein, L., Johnson, D.S., Levin, A.: Min-sum bin packing. J. Comb. Optim. **36**(2), 508–531 (2018)
6. Epstein, L., Levin, A.: Minimum weighted sum bin packing. In: Kaklamanis, C., Skutella, M. (eds.) WAOA 2007. LNCS, vol. 4927, pp. 218–231. Springer, Heidelberg (2008). https://doi.org/10.1007/978-3-540-77918-6_18
7. Ford, L.R., Jr., Fulkerson, D.R.: Constructing maximal dynamic flows from static flows. Oper. Res. **6**(3), 419–433 (1958)
8. Fu, N., Suppakitpaisarn, V.: Clustering 1-dimensional periodic network using betweenness centrality. Comput. Soc. Netw. **3**(1), 1–20 (2016)
9. Fujie, T., Higashikawa, Y., Katoh, N., Teruyama, J., Tokuni, Y.: Minmax regret 1-sink location problems on dynamic flow path networks with parametric weights. In: Uehara, R., Hong, S.-H., Nandy, S.C. (eds.) WALCOM 2021. LNCS, vol. 12635, pp. 52–64. Springer, Cham (2021). https://doi.org/10.1007/978-3-030-68211-8_5
10. Higashikawa, Y., et al.: Minimax regret 1-sink location problem in dynamic path networks. Theoret. Comput. Sci. **588**, 24–36 (2015)
11. Higashikawa, Y., Golin, M.J., Katoh, N.: Minimax regret sink location problem in dynamic tree networks with uniform capacity. In: Pal, S.P., Sadakane, K. (eds.) WALCOM 2014. LNCS, vol. 8344, pp. 125–137. Springer, Cham (2014). https://doi.org/10.1007/978-3-319-04657-0_14
12. Higashikawa, Y., Katoh, N.: A survey on facility location problems in dynamic flow networks. Rev. Socionetw. Strateg. **13**(2), 163–208 (2019)
13. Higashikawa, Y., Katoh, N., Teruyama, J.: Almost linear time algorithms for some problems on dynamic flow networks. In: Katoh, N., et al. (eds.) Sublinear Computation Paradigm, pp. 65–85. Springer, Singapore (2022). https://doi.org/10.1007/978-981-16-4095-7_5
14. Kamiyama, N., Katoh, N., Takizawa, A.: An efficient algorithm for evacuation problem in dynamic network flows with uniform arc capacity. IEICE Trans. Inf. Syst. **89**(8), 2372–2379 (2006)

15. Klinz, B., Woeginger, G.J.: Minimum-cost dynamic flows: the series-parallel case. Netw. Int. J. **43**(3), 153–162 (2004)
16. Rhee, W.T.: Probabilistic analysis of the next fit decreasing algorithm for bin-packing. Oper. Res. Lett. **6**(4), 189–191 (1987)

Efficient Approximation Algorithms for Parallel Batch Machine Scheduling of Malleable Jobs

Fenghe Xia[1], Longkun Guo[1](✉), and Xiaoyan Zhang[2]

[1] School of Mathematics and Statistics, Fuzhou University, Fuzhou 350116, China
lkguo@fzu.edu.cn

[2] School of Mathematical Science and Institute of Mathematics, Nanjing Normal University, Jiangsu 210023, China

Abstract. Cloud computing environments impose challenges in incorporating the maximum degree of parallelism constraints into Parallel Batch Machine Scheduling (PBMS). Unlike traditional PBMS that considers only fixed job widths, this paper studies generalized PBMS with malleable jobs that allow job width to be changed during the job execution, provided it does not exceed its maximum degree of parallelism. We propose a fast $O(n \log n)$ approximation algorithm by extending the state-of-the-art PBMS algorithm by setting each job's width to its maximum degree of parallelism, where n is the number of jobs. Due to the unique nature of malleable jobs, previous ratio proofs are inapplicable. We develop new proof techniques and establish that our algorithm achieves a ratio of $(4 - \frac{2}{Bm})$, where B and m denote machine capacities and numbers, respectively. Furthermore, by exploiting the relationship between maximum job demand and average processing capacity per processor, we refine the algorithm and achieve an improved ratio of $(4 - \frac{4}{Bm})$ while preserving $O(n \log n)$ runtime complexity. In addition, we fine-tune our algorithm to achieve a ratio of $(3 - \frac{2}{Bm})$ when jobs are with identical release times.

Keywords: Parallel batch scheduling · Cloud computing · Approximation algorithm

1 Introduction

Parallel Batch Machine Scheduling (PBMS), as a batch scheduling method, has been widely applied in various fields [10], including manufacturing sectors such as ceramics, semiconductor chip production, and steel manufacturing [8], as well as textile industries like fabric dyeing [16], and logistics domains such as vehicle transportation [1]. Compared to traditional scheduling problems, PBMS presents greater complexity. Batch scheduling involves grouping jobs into batches, ensuring that the total job width within each batch does not exceed the machine capacity, and then assigning each batch to machines to minimize the maximum completion time.

S. Ghosh and Z. Zhang (Eds.): AAIM 2024, LNCS 15179, pp. 48–59, 2024.
https://doi.org/10.1007/978-981-97-7798-3_5

With the widespread application of parallel computing in cloud computing, modern high-performance computers contain numerous processors. Jobs can be processed simultaneously by multiple processors. Each job has a maximum degree of parallelism, which denotes the maximum number of processors that can simultaneously process the job to achieve linear speedup [6]. To optimize processor utilization, it is crucial to develop scheduling algorithms that consider the specific characteristics of parallel jobs. If the number of processors allocated to the job can be changed during the job execution, it is called malleable; otherwise, it is termed moldable [14]. Therefore, the extension of PBMS problems incorporating malleable job properties (i.e., allowing job width to be changed during the job execution) presents further challenges compared to traditional PBMS problems.

In this paper, our work tackles a key challenge in cloud computing by using the PBMS model with malleable jobs. It is crucial for several reasons: it adapts to dynamic resource allocation, respects practical parallelism limits, and is suitable for batch processing tasks such as data processing and simulations. By integrating various scheduling aspects, this method effectively optimizes cloud resource management.

1.1 Problem Statement

In our study, we address a scheduling problem involving a set of jobs $J = \{J_1, J_2, \ldots, J_n\}$ and a set of parallel batch machines $M = \{M_1, M_2, \ldots, M_m\}$ (e.g., supercomputing centers, cloud systems). Each job J_j is characterized by a 3-tuple $< r_j, D_j, k_j >$, where r_j represents the release time, D_j denotes the demand (i.e., workload), $k_j \in \mathbb{Z}^+$ indicates the maximum degree of parallelism (i.e., the maximum number of processors that can simultaneously process the job to achieve linear speedup [6]).

Malleable jobs allow the job width (i.e., the number of processors assigned to the job, also known as the degree of parallelism) to be changed during the job execution. We introduce the decision variable s_j to represent the width of job J_j, constrained by $s_j \leq k_j$. We consider the linear speedup model a special case of the malleable job scheduling problem. Given that the demand of each job is constant, the processing time of each job decreases as its width increases, expressed as $p_j = \frac{D_j}{s_j}$. Each machine has an identical, finite capacity, denoted as B (i.e., the total number of processors per machine), and can process multiple jobs simultaneously as a batch, provided that the total width of jobs in the batch does not exceed the machine's capacity. Without loss of generality, we assume that $B \geq 2$, $B \in \mathbb{Z}^+$ and $s_j \in \mathbb{Z}^+$.

Each batch has an identical capacity B. Let B_h denote the h-th batch. The processing time of batch B_h is equal to the maximum processing time among all jobs in the batch, expressed as $P_h = \max_{j \in B_h} p_j$. The start time of the batch must be no earlier than the latest release time of any job in the batch, represented as $S_h \geq \max_{j \in B_h} r_j$. We consider the non-preemptive PBMS problem, where job processing is non-interruptible, jobs cannot be split across batches, and all jobs within the same batch have the same start and completion times. Let C_j represent the completion time of job J_j. The makespan of a job set J is defined as the maximum completion time among all jobs within the set, expressed as $C_{\max}(J) = \max_{j \in J} C_j(J)$.

Our objective is to determine the job widths subject to parallelism constraints and to find a feasible schedule for the jobs in J on m parallel batch machines, in order to minimize the makespan of the job set J. Inspired by Graham's scheduling notation [5], we define our problem as $P_m \mid p\text{-}batch, r_j, D_j, k_j, B, s_j \leq k_j \mid C_{\max}$. For brevity, we use $j \in X$ to denote $J_j \in X$ and $i \in M$ to denote $M_i \in M$.

We define the following problems:

Definition 1. *Parallel Batch Machine Scheduling with Malleable Jobs (PBMSM) represents the problem* $P_m \mid p\text{-}batch, r_j, D_j, k_j, B, s_j \leq k_j \mid C_{\max}$. *Specifically,* $PBMSM_1$ *represents the problem* $P_m \mid p\text{-}batch, r_j = r, D_j, k_j, B, s_j \leq k_j \mid C_{\max}$.

1.2 Related Work

In recent decades, the PBMS problem has garnered significant attention and in-depth research within the academic community. There are some insightful surveys of the research outcomes in parallel batch scheduling [4,11]. Our review focuses on the literature concerning PBMS problems, which encompass scenarios involving non-identical job widths and malleable jobs, along with related issues in parallel batch scheduling.

In the field of approximate algorithms for PBMS problems with non-identical job widths, the $1 \mid s_j, p\text{-}batch, B \mid C_{\max}$ problem has been extensively studied. Initially, four heuristic algorithms were introduced by Uzsoy [12] without theoretical analysis. Subsequently, Zhang et al. [15] conducted a theoretical study on these heuristics. Dósa et al. [3] further improved the approximation ratio to $\frac{17}{10}$ theoretically. For the parallel machine case $P_m \mid s_j, p\text{-}batch, B \mid C_{\max}$, a $\frac{8}{3}$-approximation algorithm with a time complexity of $O(n \log n)$ was presented by Cheng et al. [2]. Dósa et al. [3] later demonstrated that it is impossible to achieve an approximation ratio better than 2 in polynomial time unless P = NP, even under the condition where each job has a unit processing time ($p_j = 1$), and designed an algorithm with absolute performance ratio $(2+\varepsilon)$, where ε is an arbitrarily small positive number. For the more general $P_m \mid r_j, s_j, p\text{-}batch, B \mid C_{\max}$ problem, a $(2 + \varepsilon)$-approximation algorithm was provided by Li [9], where ε is an arbitrarily small positive number.

Previous research has extensively addressed the scheduling challenges of non-identical job widths, where job widths remain fixed during the job execution. However, recent attention has shifted towards scheduling malleable jobs in response to the growing demand for batch data processing in cloud computing [13]. In the domain of Bounded Flexible Scheduling (BFS), the objective is to select a subset of jobs, ensuring completion before their respective deadlines while maximizing total job value. Several approximation algorithms for BFS have been proposed in [6,7]. Although the BFS and PBMSM problems differ, both address malleable jobs that allow for linear speedup within their maximum parallelism degree range. To the best of our knowledge, the PBMS problem with malleable jobs has not been extensively theoretically analyzed, and no approximation ratios have been established. We introduce a PBMSM model, which extends existing models.

Table 1. Our main results

Problem	Past Result			Our Result		
	Ratio	Complexity	Result	Ratio	Complexity	Result
$P_m \mid p\text{-}batch, r_j, D_j, k_j, B, s_j \leq k_j \mid C_{\max}$	-			$(4-\frac{2}{Bm})$	$O(n\log n)$	Theorem 1
$P_m \mid p\text{-}batch, r_j, D_j, k_j, B, s_j \leq k_j \mid C_{\max}$	-			$(4-\frac{4}{Bm})$	$O(n\log n)$	Theorem 2
$P_m \mid p\text{-}batch, r_j = r, D_j, k_j, B, s_j \leq k_j \mid C_{\max}$	-			$(3-\frac{2}{Bm})$	$O(n\log n)$	Corollary 1

1.3 Our Contribution

In this paper, we propose approximation algorithms for the PBMSM problem, which generalizes the PBMS problem by incorporating the parallelism degree of jobs. Our first approximation extends the state-of-art approximation for PBMS due to Cheng et al. [2], by simply setting each job's width to its maximum degree of parallelism. However, the original ratio analysis presented in [2,3,9] cannot be extended to derive any ratio for our algorithm. Thus, we devise a new suite of techniques for ratio analysis and show that our algorithm achieves an approximation ratio $(4 - \frac{2}{Bm})$, where B and m are, respectively, the capacities and the number of the machines. Furthermore, we refine our approximation algorithm by setting the width according to the maximum job demand and average processing capacity per processor, achieving an improved approximation ratio $(4-\frac{4}{Bm})$ while retaining the time complexity of $O(n \log n)$. The ratio proof is refined by exploiting the relationship among the maximum completion time of the algorithm, average processing capability per processor, the longest job processing time, and the maximum release time. In addition, for the special case where jobs have identical release times, our algorithm can be tuned to achieve an approximation ratio $(3-\frac{2}{Bm})$ while retaining the time complexity of $O(n \log n)$.

1.4 Organization

The rest of this paper is organized as follows: Sect. 2 presents a fast $(4 - \frac{2}{Bm})$-approximation algorithm for the PBMSM problem. Section 3 proposes an improved algorithm for the PBMSM problem. Section 4 concludes this paper.

2 A Fast $(4 - \frac{2}{Bm})$-approximation Algorithm for PBMSM

In this section, we propose an approximation algorithm with a ratio of $(4 - \frac{2}{Bm})$ for the PBMSM problem. The key idea involves setting each job's width

Algorithm 1: PBMSM ALGORITHM (H_1)

Input: machine capacity B, number of machines m, job set J with tuple $\langle r_j, D_j, k_j \rangle$ for each job $J_j \in J$.
Output: The maximum completion time.

```
Find a job with the highest demand in J, denoted as J_μ ;   // expressed as D_μ = max_{j∈J} D_j (ties broken in favor of smaller subscript).
foreach j ∈ J do
    Let s_j = k_j ;   // set the width of job J_j to its maximum degree of parallelism.
end
Schedule the job J_μ before max_{j∈J} {r_j + D_j/k_j};
Starting from the time max_{j∈J} {r_j + D_j/k_j}, schedule the remaining jobs J \ J_μ.
```

to its maximum degree of parallelism. Find a job with the highest demand in J, denoted as J_μ, and schedule J_μ before $\max_{j\in J}\{r_j + \frac{D_j}{k_j}\}$, followed by the remaining jobs scheduled after this time. The detailed algorithm is as depicted in Algorithm 1.

It remains to give the algorithm for scheduling the remaining jobs in $J \setminus J_\mu$ starting from time $\max_{i\in J}\{r_i + \frac{D_i}{k_i}\}$. We schedule the remaining jobs $J \setminus J_\mu$ by applying Algorithm A [2], also known as Algorithm LPT-FF-BS, which involves three steps: sorting jobs in non-increasing order of processing time, applying the bin packing Algorithm First-Fit to batch the re-ordered jobs, and scheduling these batches across m machines following the earliest available machine principle.

The feasibility analysis is conducted as follows: since for each $j \in J$, we have $\max_{i\in J}\{r_i + \frac{D_i}{k_i}\} \geq r_j + \frac{D_j}{k_j}$, it follows from the definition of J_μ that $\max_{i\in J}\{r_i + \frac{D_i}{k_i}\} \geq r_\mu + \frac{D_\mu}{k_\mu}$. Then, we obtain $r_\mu \leq \max_{i\in J}\{r_i + \frac{D_i}{k_i}\} - \frac{D_\mu}{k_\mu}$. Thus, we can schedule the job J_μ within the time interval $\left[\max_{i\in J}\{r_i + \frac{D_i}{k_i}\} - \frac{D_\mu}{k_\mu}, \max_{j\in J}\{r_j + \frac{D_j}{k_j}\}\right]$ on the first machine M_1.

For the sake of simplicity in exposition, the average processing capability per processor is denoted as $\frac{1}{m}P(X)$, where $\frac{1}{m}P(X) = \sum_{j\in X}\frac{D_j}{Bm}$. The optimal makespan is denoted as $C^*_{\max}(J)$ for the PBMSM problem. We then have the following two properties:

Proposition 1. $\max_{i\in J}\left\{\frac{D_i}{k_i}\right\} \leq \max_{j\in J}\left\{r_j + \frac{D_j}{k_j}\right\} \leq C^*_{\max}(J)$.

Proposition 2. $\frac{1}{m}P(J) = \frac{1}{Bm}\sum_{j\in J} D_j \leq C^*_{\max}(J)$.

To investigate the characteristics of Algorithm LPT-FF-BS, we introduce a simplified variant of the PBMSM problem that job widths are allowed to be split. We modify the second step of Algorithm LPT-FF-BS: the first job from the list is assigned to an open batch with available capacity. If the remaining capacity of the batch is insufficient to accommodate the entire job, split the job into two segments: one to fill the current batch to capacity, and the other to be allocated to a newly opened batch. All other steps are retained, resulting in the development of Algorithm FBLPT-BS.

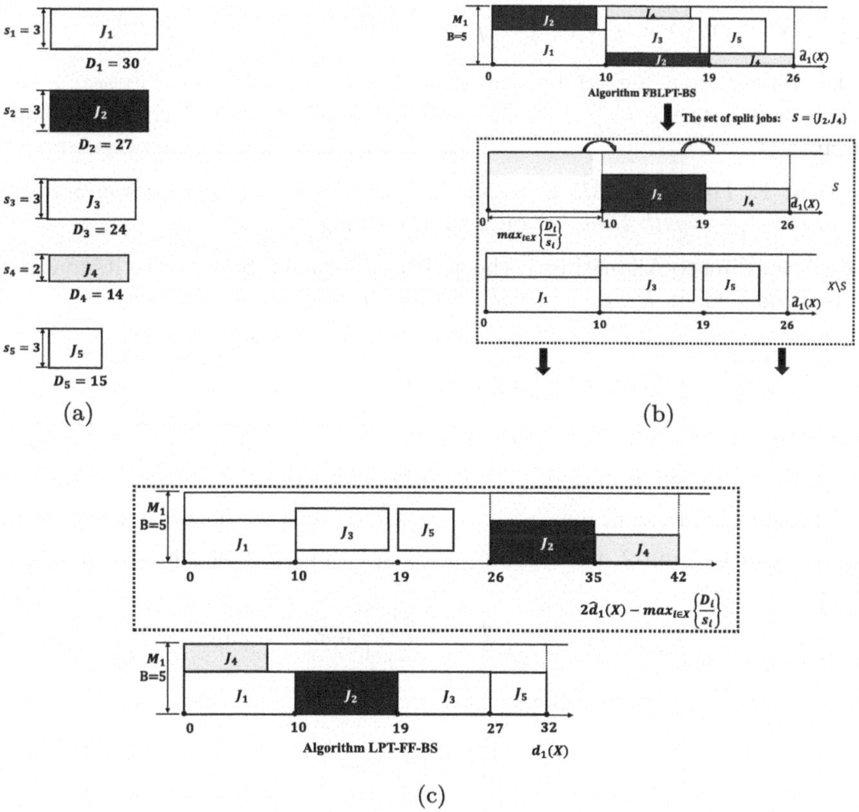

Fig. 1. Comparative analysis of processing times $d_m(X)$ and $\hat{d}_m(X)$ using Algorithms LPT-FF-BS and FBLPT-BS on a single machine. (a) Information of the job set; (b) Comparison of split and unsplit jobs using Algorithm FBLPT-BS; (c) The relationship between processing times $d_m(X)$ and $\hat{d}_m(X)$ across the algorithms.

For simplicity, if a job undergoes splitting, it will be called a split job. Given any non-empty subset X of J, let $d_m(X)$ and $\hat{d}_m(X)$ denote the processing times for scheduling the job set X on m parallel machines, starting from $\max_{j \in X}\left\{r_j + \frac{D_j}{s_j}\right\}$, applying the Algorithm LPT-FF-BS and the Algorithm FBLPT-BS, respectively. These processing times represent the durations from $\max_{j \in X}\left\{r_j + \frac{D_j}{s_j}\right\}$ to the maximum completion time of all jobs in X. Then we have the following two lemmas:

Lemma 1. $\frac{1}{m}P(X) \leq \hat{d}_m(X) \leq \frac{1}{m}P(X) + (1 - \frac{1}{Bm}) \max_{i \in X}\left\{\frac{D_i}{s_i}\right\}$.

According to Fig. 1(c), we can deduce that the value of $d_m(X)$ plus the maximum processing time of the jobs in the set X does not exceed twice the value of $\hat{d}_m(X)$. This can be expressed as $d_m(X) \le 2\hat{d}_m(X) - \max_{i\in X}\left\{\frac{D_i}{s_i}\right\}$. By Lemma 1, we have the following lemma.

Lemma 2. $\frac{1}{m}P(X) \le d_m(X) \le \frac{2}{m}P(X) + (1 - \frac{2}{Bm})\max_{i\in X}\left\{\frac{D_i}{s_i}\right\}$.

The makespan achieved by applying Algorithm 1 to the job set J on m machines is denoted as $C_{\max}^{H_1}(J)$. Based on Lemma 2, we can derive the following theorem.

Theorem 1. *For the PBMSM problem, Algorithm 1 achieves an approximation ratio of* $(4 - \frac{2}{Bm})$ *with time complexity of* $O(n \log n)$.

Proof. According to Algorithm 1, the width s_j of each job j is set to its maximum degree of parallelism k_j, i.e., $s_j = k_j$. From Proposition 1, we have:

$$\max_{i\in J}\left\{\frac{D_i}{s_i}\right\} = \max_{i\in J}\left\{\frac{D_i}{k_i}\right\} \le \max_{j\in J}\left\{r_j + \frac{D_j}{k_j}\right\} \le C^*_{\max}(J). \tag{1}$$

In Algorithm 1, the job J_μ is scheduled before $\max_{j\in J}\left\{r_j + \frac{D_j}{k_j}\right\}$, where $D_\mu = \max_{j\in J}\{D_j\}$. All remaining jobs $J \setminus J_\mu$ are scheduled by applying the Algorithm LPT-FF-BS starting from $\max_{j\in J}\left\{r_j + \frac{D_j}{k_j}\right\}$. Therefore, based on the definition of $d_m(X)$ and Lemma 2, the maximum completion time of Algorithm 1 can be expressed as:

$$C_{\max}^{H_1}(J) = \max_{j\in J}\left\{r_j + \frac{D_j}{k_j}\right\} + d_m(J \setminus J_\mu). \tag{2}$$

Since $P(J) = \sum_{j\in J}\frac{D_j}{B}$, it follows that:

$$P(J \setminus J_\mu) = P(J) - \frac{D_\mu}{B}. \tag{3}$$

It is obvious that:

$$\max_{i\in J\setminus J_\mu}\left\{\frac{D_i}{s_i}\right\} \le \max_{i\in J}\left\{\frac{D_i}{s_i}\right\}. \tag{4}$$

From (2), (3), (4), (1), Proposition 2, Lemma 2 and the condition $B \ge 2$, we can infer that

$$\begin{aligned}
C_{\max}^{H_1}(J) &= \max_{j\in J}\left\{r_j + \frac{D_j}{k_j}\right\} + d_m(J \setminus J_\mu) \\
&\le \max_{j\in J}\left\{r_j + \frac{D_j}{k_j}\right\} + \frac{2}{m}P(J \setminus J_\mu) + (1 - \frac{2}{Bm})\max_{i\in J\setminus J_\mu}\left\{\frac{D_i}{s_i}\right\} \\
&\le \max_{j\in J}\left\{r_j + \frac{D_j}{k_j}\right\} + \frac{2}{m}\left[P(J) - \frac{D_\mu}{B}\right] + (1 - \frac{2}{Bm})\max_{i\in J}\left\{\frac{D_i}{s_i}\right\} \\
&\le \max_{j\in J}\left\{r_j + \frac{D_j}{k_j}\right\} + \frac{2}{m}P(J) + (1 - \frac{2}{Bm})\max_{i\in J}\left\{\frac{D_i}{s_i}\right\} \\
&\le C^*_{\max}(J) + 2C^*_{\max}(J) + (1 - \frac{2}{Bm})C^*_{\max}(J) \\
&= (4 - \frac{2}{Bm})C^*_{\max}(J).
\end{aligned} \tag{5}$$

Algorithm 2: AN IMPROVED ALGORITHM FOR PBMSM (H_2)

Input: Machine capacity B, number of machines m, job set J with tuple $\langle r_j, D_j, k_j \rangle$ for each job $J_j \in J$.

Output: Maximum completion time.

Find a job with the highest demand in J, denoted as J_μ ; `// expressed as` $D_\mu = \max_{j \in J} D_j$ `(ties broken in favor of smaller subscript).`

$s_\mu = k_\mu$; `// Set the width of job` J_μ `to its maximum degree of parallelism.`

Schedule J_μ before $\max_{j \in J} \left\{ r_j + \frac{D_j}{k_j} \right\}$;

if $D_\mu \leq \frac{1}{m} P(J)$ **then**

  Set $s_j = 1$ for each job $J_j \in J \setminus J_\mu$;

end

else

  Set $s_j = k_j$ for each job $J_j \in J \setminus J_\mu$;

end

Starting from $\max_{j \in J} \left\{ r_j + \frac{D_j}{k_j} \right\}$, schedule the remaining set of jobs $J \setminus J_\mu$.

As demonstrated by Cheng et al. [2], the time complexity of Algorithm LPT-FF-BS is $O(n \log n)$. Given that the complexities of steps 1 and 2–4 in Algorithm 1 are each $O(n)$, and considering that step 6 of Algorithm 1 incorporates Algorithm LPT-FF-BS, the aggregate time complexity of Algorithm 1 is maintained at $O(n \log n)$. □

3 An Improved Algorithm for PBMSM

In this section, we propose an improved algorithm with a ratio of $(4 - \frac{4}{Bm})$ for PBMSM. The key idea is to initialize the width of each job based on the relationship between maximum job demand and average processing capacity per processor. Find a job with the highest demand in J, denoted as J_μ. This job J_μ is scheduled before time $\max_{j \in J} \left\{ r_j + \frac{D_j}{k_j} \right\}$, while the remaining jobs are scheduled after this time. The detailed algorithm is presented in Algorithm 2.

It remains to give the algorithm for scheduling the remaining jobs in $J \setminus J_\mu$ starting from time $\max_{i \in J} \left\{ r_i + \frac{D_i}{k_i} \right\}$. The key idea is to adjust the widths of jobs within a batch to be as small as possible while ensuring that the processing time of the created batches remains unchanged. The detailed algorithm is as depicted in Algorithm 3.

Similar to the definition of $C_{\max}^{H_1}(J)$, the makespan achieved by applying Algorithm 2 to the job set J on m machines is denoted as $C_{\max}^{H_2}(J)$. Employing the above algorithms, the following theorem with an improved approximation ratio is obtained.

Algorithm 3: SCHEDULE THE REMAINING SET OF JOBS

Input: Machine capacity B, number of machines m, job set J with tuple $\langle D_j, s_j \rangle$ for each job $J_j \in J$.
Output: Maximum completion time.

1 Reorder jobs in J in non-increasing order of processing time $p_j = \frac{D_j}{s_j}$, forming job list $\mathscr{L}$, and let batch list $\mathscr{B} = \emptyset$;
2 **foreach** $J_j \in \mathscr{L}$ **do**
3   assigned = False;
4   **foreach** *batch* $B_i \in \mathscr{B}$ **do**
        `// a pre-evaluation of the minimum inclusion condition.`
5   **if** $\lceil \frac{D_j}{P_i} \rceil + \sum_{h \in B_i} s_h \leq B$ **then**
6   Assign J_j to B_i and set s_j to $\lceil \frac{D_j}{P_i} \rceil$; `// Adjust each job width within a batch based on the batch's processing time.`
7   Update $\mathscr{L}$ by removing J_j;
8   assigned = True;
9   break ;
10   **end**
11   **end**
12   **if** *not assigned* **then**
13   Create new batch B_g and schedule J_j there with processing time $P_g = \frac{D_j}{s_j}$;
14   Update $\mathscr{B}$ by adding B_g;
15   **end**
16 **end**
17 Reorder batches in non-increasing order of processing times;
18 **while** $\mathscr{B}$ *is not empty* **do**
19   Assign first batch B_i from $\mathscr{B}$ to machine with earliest idle time;
20   Update $\mathscr{B}$ by removing B_i;
21 **end**

Theorem 2. *For the PBMSM problem, Algorithm 2 achieves an improved approximation ratio of* $(4 - \frac{4}{Bm})$ *with a time complexity of* $O(n \log n)$.

Proof. From Algorithm 3, steps 12 to 16 indicate that the processing time of a newly created batch is determined by the first job added and remains unchanged thereafter. Steps 5 to 9 indicate the adjustment of job widths in subsequent scheduling to minimize capacity usage while ensuring that their processing times do not surpass the set processing time of the batch. This approach optimizes capacity utilization and prevents the processing times of subsequent batches from exceeding those determined by the initial scheduling in Algorithm LPT-FF-BS.

In Algorithm 2, step 10 schedules the remaining jobs $J \backslash J_\mu$ using Algorithm 3 starting from time $\max_{j \in J} \left\{ r_j + \frac{D_j}{k_j} \right\}$. Consequently, the maximum completion time for Algorithm 2 is given by:

$$C_{\max}^{H_2}(J) \leq \max_{j \in J}\left\{r_j + \frac{D_j}{k_j}\right\} + d_m(J \setminus J_\mu). \tag{6}$$

Since Algorithm 2 adjusts the width of each job based on the relationship between the job's maximum demand D_μ and the average processing time per processor $\frac{1}{m}P(J)$, we will discuss the approximation ratio of this problem separately.

Case 1. If $D_\mu \leq \frac{1}{m}P(J)$:

Since the width s_j for each job J_j is a positive integer, it follows that the processing time p_j of a job does not exceed its demand D_j. Mathematically, this can be expressed as: $p_j = \frac{D_j}{s_j} \leq D_j$. By virtue of arbitrariness and Proposition 2, we can deduce:

$$\max_{i \in J}\left\{\frac{D_i}{s_i}\right\} \leq \max_{j \in J} D_j = D_\mu \leq \frac{1}{m}P(J) \leq C_{\max}^*(J). \tag{7}$$

According to Algorithm 2, the width s_j for each job j in the set J, excluding job J_μ, is set to 1. Therefore, when Algorithm FBLPT-BS is employed to schedule the job set $J \setminus J_\mu$, no jobs are split. Consequently, for the job set $J \setminus J_\mu$, the maximum completion time obtained by applying Algorithm LPT-FF-BS is equal to that obtained by applying Algorithm FBLPT-BS. Thus, it follows that: $d_m(J \setminus J_\mu) = \hat{d}_m(J \setminus J_\mu)$. Then, according to (3), (4), (7), Proposition 2, Lemma 1 and $B \geq 2$, we can deduce:

$$\begin{aligned}
d_m(J \setminus J_\mu) &= \hat{d}_m(J \setminus J_\mu) \\
&\leq \frac{1}{m}P(J \setminus J_\mu) + (1 - \frac{1}{Bm}) \max_{i \in J \setminus J_\mu}\left\{\frac{D_i}{s_i}\right\} \\
&\leq \frac{1}{m}P(J) - \frac{D_\mu}{Bm} + (1 - \frac{1}{Bm}) \max_{i \in J}\left\{\frac{D_i}{s_i}\right\} \\
&\leq \frac{1}{m}P(J) - \frac{1}{Bm} \max_{i \in J}\left\{\frac{D_i}{s_i}\right\} + (1 - \frac{1}{Bm}) \max_{i \in J}\left\{\frac{D_i}{s_i}\right\} \\
&\leq C_{\max}^*(J) + (1 - \frac{2}{Bm})C_{\max}^*(J) \\
&= (2 - \frac{2}{Bm})C_{\max}^*(J).
\end{aligned} \tag{8}$$

Furthermore, combining (6), (8) and Proposition 1, we can infer that

$$C_{\max}^{H_2}(J) \leq \max_{j \in J}\left\{r_j + \frac{D_j}{k_j}\right\} + d_m(J \setminus J_\mu) \leq (3 - \frac{2}{Bm})C_{\max}^*(J).$$

Case 2. If $D_\mu > \frac{1}{m}P(J)$:

In Algorithm 2, step 8 sets the width s_j of each job j to its maximum degree of parallelism k_j, i.e., $s_j = k_j$. Additionally, from Algorithm 3, it is observed that the processing time of the first job is maximized after reordering (i.e., $\max_{j \in J \setminus J_\mu}\left\{\frac{D_i}{k_i}\right\}$). As s_j increases, the processing time p_j of job J_j decreases.

Consequently, when $s_j = k_j$, the maximum processing time among all jobs does not exceed the optimal completion time for this problem.

Moreover, based on the algorithm analysis, the processing time for each new batch is determined by the first job added. Subsequently, the widths of the jobs scheduled thereafter are adjusted to minimize capacity usage, ensuring that their processing times do not exceed that of the batch to which they are assigned. Consequently, the processing time for all jobs does not surpass that of the first job in the job list $\mathscr{L}$. From Proposition 1, we have:

$$\max_{i\in J}\left\{\frac{D_i}{s_i}\right\} \le \max_{i\in J}\left\{\frac{D_i}{k_i}\right\} \le \max_{j\in J}\left\{r_j + \frac{D_j}{k_j}\right\} \le C^*_{\max}(J). \tag{9}$$

From (6), (3) and (4), in conjunction with Lemma 2, it can be inferred that

$$\begin{aligned} C^{H_2}_{\max}(J) &\le \max_{j\in J}\left\{r_j + \frac{D_j}{k_j}\right\} + d_m(J \setminus J_\mu) \\ &\le \max_{j\in J}\left\{r_j + \frac{D_j}{k_j}\right\} + \frac{2}{m}P(J \setminus J_\mu) + (1 - \frac{2}{Bm}) \max_{i\in J\setminus J_\mu}\left\{\frac{D_i}{s_i}\right\} \\ &\le \max_{j\in J}\left\{r_j + \frac{D_j}{k_j}\right\} + \frac{2}{m}\left[P(J) - \frac{D_\mu}{B}\right] + (1 - \frac{2}{Bm}) \max_{i\in J}\left\{\frac{D_i}{s_i}\right\}. \end{aligned}$$

Then from (9), Proposition 1, Proposition 2, and the condition $B \ge 2$, along with the assumption $D_\mu > \frac{1}{m}P(J)$, we can further deduce:

$$\begin{aligned} C^{H_2}_{\max}(J) &\le \max_{j\in J}\left\{r_j + \frac{D_j}{k_j}\right\} + \frac{2}{m}P(J) - \frac{2D_\mu}{Bm} + (1 - \frac{2}{Bm}) \max_{i\in J}\left\{\frac{D_i}{s_i}\right\} \\ &\le \max_{j\in J}\left\{r_j + \frac{D_j}{k_j}\right\} + \frac{2}{m}P(J) - \frac{2}{Bm}\frac{1}{m}P(J) + (1 - \frac{2}{Bm}) \max_{i\in J}\left\{\frac{D_i}{s_i}\right\} \\ &\le C^*_{\max}(J) + (2 - \frac{2}{Bm})C^*_{\max}(J) + (1 - \frac{2}{Bm})C^*_{\max}(J) \\ &= (4 - \frac{4}{Bm})C^*_{\max}(J). \end{aligned} \tag{10}$$

The proof of this time complexity follows a similar methodology to that outlined in the proof of Theorem 1. □

Corollary 1. *For problem $PBMSM_1$, there is an algorithm with a ratio of $(3 - \frac{2}{Bm})$ and a time complexity of $O(n \log n)$ by slightly modifying Algorithm 2.*

4 Conclusion

This paper presents a fast approximation algorithm with $O(n \log n)$ time complexity, extending existing algorithms by setting each job's width to its maximum parallelism degree. By employing innovative proof techniques, we establish an approximation ratio of $(4 - \frac{2}{Bm})$ for our algorithm. By dynamically setting the job widths using a function regarding maximum job demand and average processing capacity per processor, we improve the ratio to $(4 - \frac{4}{Bm})$ while maintaining the runtime complexity. Moreover, for jobs with identical release times, a slight adjustment yields a ratio of $(3 - \frac{2}{Bm})$. In the future, we will explore the problem against unrelated machines, generalizing the identical machines considered in the paper. Additionally, it would be interesting to extend our model to scenarios considering job rejection [10].

References

1. Chen, B., Lee, C.Y.: Logistics scheduling with batching and transportation. Eur. J. Oper. Res. **189**(3), 871–876 (2008)
2. Cheng, B., Yang, S., Hu, X., Chen, B.: Minimizing makespan and total completion time for parallel batch processing machines with non-identical job sizes. Appl. Math. Model. **36**(7), 3161–3167 (2012)
3. Dosa, G., Tan, Z., Tuza, Z., Yan, Y., Lányi, C.S.: Improved bounds for batch scheduling with nonidentical job sizes. Naval Res. Logist. (NRL) **61**(5), 351–358 (2014)
4. Fowler, J.W., Mönch, L.: A survey of scheduling with parallel batch (p-batch) processing. Eur. J. Oper. Res. **298**(1), 1–24 (2022)
5. Graham, R.L., Lawler, E.L., Lenstra, J.K., Kan, A.R.: Optimization and approximation in deterministic sequencing and scheduling: a survey. In: Annals of Discrete Mathematics, vol. 5, pp. 287–326. Elsevier (1979)
6. Guo, L., Shen, H.: Efficient approximation algorithms for the bounded flexible scheduling problem in clouds. IEEE Trans. Parallel Distrib. Syst. **28**(12), 3511–3520 (2017)
7. Jain, N., Menache, I., Naor, J., Yaniv, J.: Near-optimal scheduling mechanisms for deadline-sensitive jobs in large computing clusters. ACM Trans. Parallel Comput. (TOPC) **2**(1), 1–29 (2015)
8. Lee, C.Y., Uzsoy, R., Martin-Vega, L.A.: Efficient algorithms for scheduling semiconductor burn-in operations. Oper. Res. **40**(4), 764–775 (1992)
9. Li, S.: Makespan minimization on parallel batch processing machines with release times and job sizes. J. Softw. **7**(6), 1203–1210 (2012)
10. Ou, J., Lu, L., Zhong, X.: Parallel-batch scheduling with rejection: structural properties and approximation algorithms. Eur. J. Oper. Res. **310**(3), 1017–1032 (2023)
11. Potts, C.N., Kovalyov, M.Y.: Scheduling with batching: a review. Eur. J. Oper. Res. **120**(2), 228–249 (2000)
12. Uzsoy, R.: Scheduling a single batch processing machine with non-identical job sizes. Int. J. Prod. Res. **32**(7), 1615–1635 (1994)
13. Wu, C., Buyya, R., Ramamohanarao, K.: Cloud pricing models: taxonomy, survey, and interdisciplinary challenges. ACM Comput. Surv. (CSUR) **52**(6), 1–36 (2019)
14. Wu, X., Loiseau, P.: Efficient approximation algorithms for scheduling moldable tasks. Eur. J. Oper. Res. **310**(1), 71–83 (2023)
15. Zhang, G., Cai, X., Lee, C.Y., Wong, C.K.: Minimizing makespan on a single batch processing machine with nonidentical job sizes. Naval Res. Logist. (NRL) **48**(3), 226–240 (2001)
16. Zhang, R., Chang, P.C., Song, S., Wu, C.: A multi-objective artificial bee colony algorithm for parallel batch-processing machine scheduling in fabric dyeing processes. Knowl.-Based Syst. **116**, 114–129 (2017)

Cost-Sharing Mechanisms for the Selfish Open-End Bin Packing Problem

Zhikang Zhang(✉) and Weidong Li

School of Mathematics and Statistics, Yunnan University, Kunming, People's Republic of China
zhikang2024@163.com

Abstract. In this paper, we study games modeled after the open-end bin packing problem, where each bin has a unit cost, and each item can be viewed as a player whose goal is to choose a bin that can be accessed to minimize the cost they need to share the cost. We mainly study the cost-sharing mechanism for the maximum open-end bin packing (Max-OEBP) problem and the minimum open-end bin packing (Min-OEBP) problem. We give two lower bounds of PoA for any cost-sharing mechanism, assuming that each item lies in $\left[\frac{1}{m}, 1\right]$. Further we apply the LSB cost-sharing mechanism to both open-end bin packing models. We prove that PoA does not exceed 2 and PoS equals to 1 under the Max-OEBP model, and PoA does not exceed 4 and PoS does not exceed $\frac{71}{60}$ under the Min-OEBP model. We also show that the packing scheme output by the FFD algorithm is a Nash equilibrium under both models, which implies that we can output a stable packing scheme in polynomial time under this mechanism.

Keywords: Open-end bin packing · Cost-sharing mechanism · Price of anarchy

1 Introduction

Given a set of n items with rational sizes in $(0, 1]$, the *Open-End Bin Packing* (OEBP) problem [13,17] is to pack the items into unit-capacity bins such that the number of non-empty bins is minimized, under the condition that the size of each bin is strictly below 1 by removing one item, where the size of a bin is equal to the total size of items packed into it. A practical application of this problem based on the Hong Kong subway is given in [17].

Under the game circumstance, we are given an instance $I = \{1, 2, \ldots, n\}$ of n selfish agents and an infinite number of unit-capacity bins where each agent has an item of size $s_i \in [\frac{1}{m}, 1]$, and aims to minimize the cost-share by choosing a bin it can fit in, where $m \in \mathbb{Z}^+$. The (selfish) open-end bin packing problem [14] is to pack the items into a minimum number of bins, where a bin can be filled to a level exceeding 1, under the condition that the level of the bin is less than 1 before the last item is packed. Gai et al. [14] presented some interesting results about

S. Ghosh and Z. Zhang (Eds.): AAIM 2024, LNCS 15179, pp. 60–71, 2024.
https://doi.org/10.1007/978-981-97-7798-3_6

the existence and performance of the Nash equilibria, where a Nash equilibrium is a packing where no agent can obtain a smaller cost by unilaterally moving his item to a different bin, while other items remain in their original positions.

As mentioned in [10], the Price of Anarchy (PoA) [16] and the Price of Stability (PoS) [2] are the most important measures of the quality of the Nash equilibria reached with uncoordinated selfish agents, and defined as

$$PoA(I) = \max_{\pi \in NE(I)} \frac{n(\pi)}{OPT(I)}, PoS(I) = \min_{\pi \in NE(I)} \frac{n(\pi)}{OPT(I)},$$

where $NE(I)$ is the set of Nash equilibria of a given instance I, $n(\pi)$ is the number of used bins in the packing π and $OPT(I)$ is the optimal value of instance I. Correspondingly, the PoA (PoS) of selfish bin packing is defined as $PoA = \sup_I PoA(I)$ ($PoS = \sup_I PoS(I)$), and the asymptotic PoA (PoS) is defined as $\limsup_{I:OPT(I)\to\infty} PoA(I)$ and $\limsup_{I:OPT(I)\to\infty} PoS(I)$, respectively.

Obviously, the values of PoA and PoS are determined by the cost-sharing rules. For two different versions of selfish open-end bin packing, Gai et al. [14] proved that the (asymptotic) PoA of the proportional cost-sharing rule, under which an item pays the cost proportional to its size, is exactly 2.

The (selfish) open-end bin packing problem is closed to the selfish bin packing problem introduced by Bilò [3]. He analyzed the proportional sharing mechanism, under which an item i of size s_i in bin B has to share a cost of $\frac{s_i}{s(B)}$, where $s(B) = \sum_{h\in B} s_h$, and he proved that any feasible packing scheme under this mechanism can be adjusted to converge to a Nash Equilibrium(NE) packing in a finite number of steps, and that PoA is in [1.6,1.67]. Yu and Zhang [23] obtained a lower bound of 1.6416 and an upper bound of 1.6575 for the proportional sharing mechanism. Epstein and Kleiman [10] independently arrived at the same lower bound as the previous result and a better upper bound of 1.6428. Epstein et al. [11] considered the case of item sizes with parametric constraints and obtained a parametric upper bound on PoA. Dósa and Epstein [8] gave a new lower bound of 1.64242239616717 for the proportional sharing mechanism. Zhang and Zhang [24] used a proportional sharing mechanism for bins without large items that provides a discount for small items that stay in a bin with a large item, with a PoA of at most $\frac{22}{15}$, which is consistent with the upper bound on the PoA of the proportional sharing mechanism without large items obtained in [11]. Dósa et al. [7] proposed a weighting-based mechanism that gave a better upper bound of 1.4545 for PoA. They proved that the PoA is in the range of [1.428,1.4545], and presented an instance showing that PoA is at least $\frac{4}{3}$ if the cost-sharing mechanism does not allow items to be moved into empty bins to reduce the shared cost.

Ma et al. [19] investigated an equally shared mechanism for the bin packing game, under which all items in a bin bear the same shared cost $\frac{1}{k}$, where k is the number of items in the bin, and proved that PoA is in [1.6901,1.7]. Dósa and Epstein [9] reduced the former upper bound to below 1.7. Chen et al. [6] proposed a mechanism based on an equal sharing mechanism where a large-sized item is required to bear the full cost of the bin in which it resides, PoA is in

[1.47407,1.4748]. Nong et al. [20] gave a mechanism with $PoA \leq 1.5$ in which packing can quickly converge to a Nash equilibrium.

Wang and Zhang [22] proposed a new and simple mechanism, which they call the Local-Size-Based (LSB) cost-sharing mechanism, and show that the PoA of this mechanism is consistent with a general lower bound and that there must exist a Nash equilibrium packing that uses the minimum number of crates, i.e., $PoS = 1$, under this mechanism. Gai et al. [15] proposed a cost-sharing mechanism based on the harmonic mean with PoA in $\left[\frac{5}{3}, 1.722\right]$, after which the concept of penalty was introduced to improve the PoA to 1.5.

In addition to the classic selfish bin packing problem, other variants have been studied by scholars [1,4,5,12,21].

2 Preliminaries

Given a set of items $N = \{1, 2, \ldots, n\}$, where any item $i \in N$ has a size $s_i \in \left[\frac{1}{m}, 1\right]$. Given a batch of bins with unit cost, for any bin B, let $s(B) := \sum_{i \in B} s_i$ be the load on bin B. For the maximum open-end bin packing (Max-OEBP) problem, a bin B is valid if $s(B) - \max(B) < 1$, where $\max(B)$ be the maximum size of the item in bin B. For the minimum open-end bin packing (Min-OEBP) problem, a bin B is valid if $s(B) - \min(B) < 1$, where $\min(B)$ be the minimum size of the item in bin B. A packing scheme π is called valid if all the bins under the packing scheme π are valid. For items packed in the same bin that need to share the cost of the bin, we use $c_\pi(i)$ to denote the shared cost of item i under the packing scheme π, we have $\sum_{i \in B} c_\pi(i) = 1$.

Definition 1. If any item cannot reduce its shared cost by moving it unilaterally to another bin, then such a packing scheme is said to be a Nash equilibrium shorthanded as NE.

Wang and Zhang [22] first proposed the *Local-Size-Based* (LSB) cost-sharing mechanism, which consists of a threshold constant $\Lambda \in (0, 1]$ and a cost density function $f(x)$. A threshold constant Λ and a "cost density" function $f(x)$ will uniquely determine a cost-sharing mechanism, and given a constant Λ, the function $f(x)$ has to satisfy two conditions: 1). $f(x) \geq 0$, for any $x \in [0, 1]$; 2). $f(x)$ is monotonically decreasing on $(0, \Lambda]$ and $\int_0^\Lambda f(x)dx = 1$.

They gave a simple selection of $f(x)$ is letting $f(x) = f_o(x)$, where $f_o(x)$ is a linear function,

$$f_o(x) := \begin{cases} \frac{2}{\Lambda} - \frac{2x}{\Lambda^2}, & \text{for } 0 \leq x < \Lambda, \\ 0, & \text{for } \Lambda \leq x \leq 1. \end{cases}$$

They showed that when $\Lambda = \frac{3}{4}$, this mechanism has $PoA = \frac{4}{3}$, which achieves the best PoA that can be achieved by a cost-sharing mechanism for the selfish bin packing problem.

We apply the LSB cost-sharing mechanism to the open-end bin packing problem and use exactly the same sharing rules as the LSB cost-sharing mechanism, with the only difference that we extend the domain of definition of x to $[0, 2]$. Threshold constant Λ is in (0,1].

The items are ordered and renamed to get $s_1 \geq s_2 \geq \cdots \geq s_n$, with fixed labeling order, and if $i < j$, then i is said to be superior to j, denoted as $i \succ j$. Given a packing scheme π, for any bin B, it is useful to set $B = \{i_1, i_2, \ldots, i_l\}$, where $i_1 \succ i_2 \succ \cdots \succ i_l$, i_1 is called the bottom item and the rest of the items are called the non-bottom items, and defining $S_B^j = \sum_{h=1}^{j} s_{i_h}$ to denote the sum of the sizes of the j items with the highest priority in the bin B, we can see that $S_B^l = s(B)$. The shared cost of an item $i_j \in B$ is then defined to be

$$c_\pi(i_j) = \begin{cases} \int_0^{S_B^1} f(x)dx + \left[1 - \int_0^{s(B)} f(x)dx\right], & \text{if } j = 1, \\ \int_{S_B^{j-1}}^{S_B^j} f(x)dx, & \text{if } j > 1. \end{cases}$$

The Local-Size-Based cost-sharing mechanism has some very nice properties under the classic bin model, and these properties are still present under the open-end bin packing model as well, both in the maximum and minimum cases.

Observation 1. [22] Given a fixed LSB cost-sharing mechanism, let π be a valid packing scheme, and let B and B' be the two bins under π. Item i is moved from B to B', then the shared cost of the non-bottom items in B and B' that are superior to i remains the same before and after the movement of item i.

Observation 2. [22] For LSB cost-sharing mechanisms with the same threshold constants but different cost density functions, their Nash equilibrium packing schemes are equivalent.

3 Maximum Open-End Bin Packing Problem

For the Max-OEBP problem, we use the fact that a bin is called valid if its load is strictly less than 1 after removing the item with the largest size in the bin. In [14], they remove the last item that enters the bin so that the bin load is less than 1, and they do not allow items to move into a bin if the bin load is not less than 1. The packing scheme in Fig. 1 is NE by the definition of [14] because none of the four items can be moved to another bin, but it may not be a NE under the Max-OEBP problem we study because the packing scheme obtained after the item of size $\frac{1}{3}$ is moved to another bin is still valid in Fig. 2.

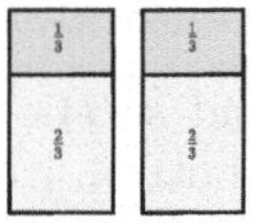

Fig. 1. An example of NE in [14].

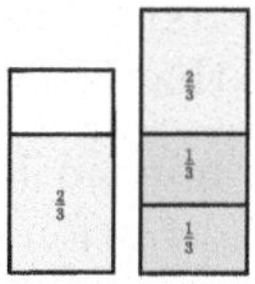

Fig. 2. The result of moving an item unilaterally.

3.1 Lower Bound for *PoA*

In this section we show that there are mechanisms to reach $PoA = PoS = 1$ when $m = 1, 2$. A lower bound under the Max-OEBP problem is given when $m \geq 3$.

Theorem 1. *If $m = 1$, then all cost-sharing mechanisms have $PoA = PoS = 1$. If $m = 2$, the proportional sharing mechanism has $PoA = PoS = 1$.*

Proof. If $m = 1$, then all items have size 1, so the only packing scheme in this case is that each item occupies a bin exclusively. This packing scheme is the NE and optimal packing scheme.

If $m = 2$, for all $i \in N$ has $s_i \in \left[\frac{1}{2}, 1\right]$, so any bin contains at most two items of which at most one is of size 1. Both the optimal packing scheme and the NE under the proportional sharing mechanism result in as many combinations as possible of an item of size 1 and an item of size less than 1 packed in a bin. This implies that any NE is the optimal packing scheme, then $PoA = PoS = 1$. □

Theorem 2. *If $m \geq 3$, then the PoA of any cost-sharing mechanism for the Max-OEBP problem is at least $\frac{3}{2} - \frac{1}{m}$.*

Proof. Let $K > 0$ be a large integer. Construct an instance containing $n(m-1)$ items of size $\frac{1}{m}$ and n items of size $\left(1 - \frac{1}{m}\right)$, where $n = 2mK$. A bin can contain at most m items of size $\frac{1}{m}$ and at most two items of size $\left(1 - \frac{1}{m}\right)$. We can construct an NE using $\frac{n(m-1)}{m} + \frac{n}{2}$ bins. Packing items of size $\left(1 - \frac{1}{m}\right)$ in pairs gets $\frac{n}{2}$ bins, and packing each m item of size $\frac{1}{m}$ in a bin gets $\frac{n(m-1)}{m}$ bins. Note that this packing scheme is NE in any cost-sharing mechanism in which Nash equilibrium exists, since all items cannot be moved to other bins or to an empty bin. Next we will show that these items can be packed into n bins, each containing one item of size $\left(1 - \frac{1}{m}\right)$ and $(m-1)$ items of size $\frac{1}{m}$. This implies that PoA is at least $\frac{\frac{n(m-1)}{m} + \frac{n}{2}}{n} = \frac{3}{2} - \frac{1}{m}$. □

3.2 LSB Cost-Sharing Mechanism

First we will analyze the conditions that need to be satisfied for an agent of an item to be motivated to change its strategy, i.e., the conditions that need to be satisfied for an item $i \in B$ to be motivated to move into B'. Let $s^h(B)$ be the sum of the sizes of the items that are superior to item h in the bin B, where $h \in N$. First, the premise that item i can move is that $s(B' \cup \{i\}) - \max(s(B' \cup \{i\})) < 1$.

Rule 1. If i is a bottom item and satisfies $s(B) < \Lambda$ and $s(B) - s_i < s(B')$, then item i has motivation to move into B' to become a bottom item.

Rule 2. If i is a bottom item and moves to be a non-bottom item in B', then item i has motivation to move.

Rule 3. If i is a non-bottom item and satisfies $s^i(B) < \Lambda$ and $s^i(B) < s^i(B')$, then item i has motivation to move into B' to become a non-bottom item.

Next we can show that the First Fit Decreasing (FFD) algorithm can output an NE for the LSB cost-sharing mechanism, which also prove the existence of Nash equilibrium for the LSB cost-sharing mechanism. The strategy of the FFD algorithm is to pack items in decreasing order of size (priority), placing items in the first bin that can be packed, i.e., guaranteeing that the bin is valid after packing, and reopening an empty bin if no such bin exists. After that an upper bound for PoA is given and it is proved that $PoS = 1$.

Theorem 3. *For the Max-OEBP problem, given any threshold constant* $\Lambda \in (0, 1]$ *and instance, the FFD algorithm can output an NE for the LSB cost-sharing mechanism.*

Proof. Let π be the packing scheme output by the FFD algorithm which uses m bins. It may be useful to set the set of bins used to be $\{B_1, B_2, \ldots, B_m\}$. According to the packing strategy of the algorithm it is known that when the kth bin is opened the load of the first $(k-1)$ bins must not be less than 1. Thus the load of the first $(m-1)$ bins is not less than 1. For $2 \leq j \leq m$, for all $i \in B_j$ has $s(B_k \cup \{i\}) - \max(B_k \cup \{i\}) \geq 1$, where $1 \leq k \leq j-1$. Thus the items in B_j cannot be moved to these bins in $\{B_1, \ldots, B_{j-1}\}$.

Next it is sufficient to show that for $1 \leq j \leq m-1$, for all $i \in B_j$ has no motivation to move into B_k, where $j+1 \leq k \leq m$. Let π' be the new packing scheme obtained by moving item $i \in B_j$ into B_k. Let b_k be the bottom item of bin B_k. If $i \succ b_k$, it will be the bottom item if it moves into B_k and $c_{\pi'}(i) = \int_0^{s_i} f(x)dx + \left[1 - \int_0^{s(B_k)+s_i} f(x)dx\right]$. When i is the bottom item in B_j, it has no motivation to move because $c_\pi(i) = \int_0^{s_i} f(x)dx \leq c_{\pi'}(i)$. When i is a non-bottom item in B_j, it also has no motivation to move because $c_\pi(i) = \int_{s^i(B_j)}^{s^i(B_j)+s_i} f(x)dx < c_{\pi'}(i)$. If $b_k \succ i$, then b_k was packed before i, which means that $s^i(B_j) \geq s^{b_k}(B_j) \geq 1 \geq \Lambda$. Therefore item i has no motivation to move because $c_\pi(i) = 0 \leq c_{\pi'}(i) = \int_{s^i(B_k)}^{s^i(B_k)+s_i} f(x)dx$. All items have no motivation to move, which implies that π is an NE. □

Lemma 1. *For the Max-OEBP problem, at most one bin in any NE has a load less than 1.*

Proof. Suppose that there exists a packing scheme π that is NE and has two bins B_k and B_h with load less than 1. Let b_k and b_h be the bottom items of B_k and B_h. It may be useful to set $b_k \succ b_h$. Let π' be the new packing scheme obtained by moving b_h into B_k. We have $c_\pi(b_h) = \int_0^{s_{b_h}} f(x)dx + \left[1 - \int_0^{s(B_h)} f(x)dx\right] > c_{\pi'}(b_h) = \int_{s^{b_h}(B_k)}^{s^{b_h}(B_k)+s_{b_h}} f(x)dx$. Thus b_h has motivation to move into B_k, which contradicts the definition of NE. □

Theorem 4. *For the Max-OEBP problem, PoA does not exceed 2.*

Proof. Let *opt* be the number of bins used in the optimal packing scheme and n^* be the number of bins used in the NE. Since the load of each bin in the optimal packing scheme does not exceed 2, we have $2 \cdot opt \geq \sum_{i=1}^{n} s_i$, and by Lemma 1, we know that $\sum_{i=1}^{n} s_i > n^* - 1$, so we have $2 \cdot opt > n^* - 1$. □

By making a simple modification to the construction in [22], it can be shown that there always exists an optimal packing scheme that is NE under the Maximum open-end bin packing model, which implies that $PoS = 1$ under this model.

Theorem 5. *For the Max-OEBP problem, there is* $PoS = 1$ *for all* $\Lambda \in (0, 1]$.

Proof. It is sufficient to show that for any instance there exists an optimal packing scheme that is NE to prove this theorem. Let $\mathcal{P}^*$ be the set of optimal packing schemes, each of which uses m^* bins, where m^* be the minimum number of bins that can be used to pack all items in N. For any $\pi \in \mathcal{P}^*$, use $\{b_1^\pi, b_2^\pi, \ldots, b_{m^*}^\pi\}$ to denote the bottom items of these m^* bins, where $b_1^\pi \succ b_2^\pi \succ \cdots \succ b_{m^*}^\pi$. For two packing schemes π, $\pi' \in \mathcal{P}^*$, either $b_1^\pi \succ b_1^{\pi'}$ or there exists $k > 1$ such that $b_j^\pi = b_j^{\pi'}$ for every $j < k$ and $b_k^\pi \succ b_k^{\pi'}$, then we say that π dominates π'. In other words, if the bottom item of π is "bigger" than the bottom item of π' in the lexicographical order, then it dominates π'. Let $\mathcal{P}_b^* \subseteq \mathcal{P}^*$ be the set of packing schemes that cannot dominate the other packing schemes in $\mathcal{P}^*$. Clearly, $\mathcal{P}_b^* \neq \emptyset$ and the packing schemes in $\mathcal{P}_b^*$ have the same bottom items, which might as well be set to $\{b_1^*, b_2^*, \ldots, b_{m^*}^*\}$.

Suppose the items in $N \backslash \{b_1^*, b_2^*, \ldots, b_{m^*}^*\}$ are $a_1 \succ a_2 \succ \cdots \succ a_{n-m^*}$. For two packing schemes π, $\pi' \in \mathcal{P}_b^*$, π is said to be lexicographically smaller than π' if there exists $l \leq n - m^*$ with $c_\pi(a_h) = c_{\pi'}(a_h)$ for every $h < l$, and $c_\pi(a_l) < c_{\pi'}(a_l)$. Since $\mathcal{P}_b^* \neq \emptyset$, there must exist a lexicographically minimal packing scheme in $\mathcal{P}_b^*$, which we define as π^*. First, there is no motivation for an item in $\{a_1, a_2, \ldots, a_{n-m^*}\}$ to move to another valid bin to reduce its shared cost, otherwise one would get a smaller packing scheme under the lexicographic order after an item move, since the shared cost of a non-bottom item that has a higher priority than it is unchanged after the move of that item, which would contradict the definition of π^*.

The next step is simply to prove that the bottom item has no motive to move. Now we claim that at most one of the m^* bins under π^* has load less than 1. Otherwise, if there exist two bins B_k and B_h with load less than 1, it may be useful to set $k < h$. Moving b_h^* into B_k gets a new packing scheme π'' with π^* dominating π'', which is a contradiction of the definition of π^*. If there exists a bin B_k with load less than 1, then that bin must be B_{m^*}, otherwise moving any item in $\{b_{k+1}^*, \ldots, b_{m^*}^*\}$ into B_k would get a packing scheme that can be dominated by π^*, contradicting the definition of π^*.

Let π_1 be the new packing scheme obtained by moving item $b_j^* \in B_j$ into B_k, where $k \neq j$. For $k > j$, the bottom item b_j^* has no motivation to move into B_k to become the bottom item, since $c_\pi(b_j^*) = \int_0^{s_{b_j^*}} f(x)dx \leq c_{\pi_1}(b_j^*) = \int_0^{s_{b_j^*}} f(x)dx + \left[1 - \int_0^{s(B_k)+s_{b_j^*}} f(x)dx\right]$. For $k < j$, the bottom item b_j^* has no motivation to move into B_k to become a non-bottom item, since the move gets a packing scheme that can be dominated by π^*, contradicting the definition of π^*. Thus all bottom items have no motivation to move, which shows that π^* is an NE. □

4 Minimum Open-End Bin Packing Problem

For the Min-OEBP problem, we use the fact that a bin is called valid if the bin load is strictly less than 1 after removing the item with the smallest size in the bin. Items can be moved provided that the packing scheme is still valid after the move. In [14] they added a rule that if a bin becomes not a valid bin after an item is moved into it, they force the smallest item in the bin, which doesn't satisfy the assumption of Min-OEBP, to move into other bin, or open a new bin. The packing scheme in Fig. 3 may not is a NE under the definition of [14] because items of size 0.3 may have motivation to move to the first bin, as shown in Fig. 4, but in the Min-OEBP model we studied such a move is not allowed, because the packing scheme obtained after the first step of the move would not be valid.

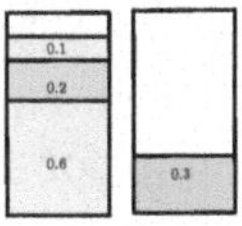

Fig. 3. An example of non-NE in [14].

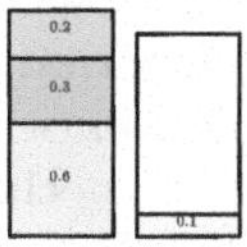

Fig. 4. The result of moving an item unilaterally.

4.1 Lower Bound for PoA

In this section, for the Min-OEBP problem we prove that $PoA = PoS = 1$ for all mechanisms for which it exist Nash equilibrium when $m = 1, 2$, and that the pricing rule is independent of the order of entry into the bin. A lower bound under the Min-OEBP problem is given when $m \geq 3$.

Theorem 6. *When $m = 1, 2$, any cost-sharing mechanism for which there exist Nash equilibria has $PoA = PoS = 1$ for the Min-OEBP problem.*

Proof. For the Min-OEBP problem, an item of size 1 in any packing scheme can be packed in one bin alone. Let N be the set of items and N_1 be the set of items of size 1. The number of bins used in the optimal packing scheme is $|N_1| + \lceil \frac{|N \setminus N_1|}{2} \rceil$. There is at most one bin in NE which has load less than 1, i.e., there is at most one bin that contains only one item. Otherwise, assuming that there exist two bins $B_k = \{b_k\}$ and $B_h = \{b_h\}$. Packing b_k and b_h into a bin will result in a strict reduction in the shared cost of at least one of the two items and no increase in the shared cost of the other item, contradicting the definition of NE. Therefore the number of bins used by any NE is also $|N_1| + \lceil \frac{|N \setminus N_1|}{2} \rceil$. This implies that any NE is the optimal packing scheme, then $PoA = PoS = 1$. □

Theorem 7. *If $m \geq 3$, then the PoA of any cost-sharing mechanism for the Min-OEBP problem is at least* $2 - \frac{4}{2+m}$.

Proof. Let $K > 0$ be a large integer. Construct an instance containing n items of sizes $\frac{1}{m}$ and $(1 - \frac{1}{m})$, where $n = 2mK$. A bin can contain at most m items of size $\frac{1}{m}$ and at most two items of size $(1 - \frac{1}{m})$. We can construct an NE using n bins. Any bin in this NE contains an item of sizes $\frac{1}{m}$ and $(1 - \frac{1}{m})$. Note that this packing scheme is NE in any cost-sharing mechanism in which Nash equilibrium exists, since all items cannot be moved to other bins or to an empty bin. Next we will show that these items can be packed into $(\frac{n}{m} + \frac{n}{2})$ bins. Packing items of size $(1 - \frac{1}{m})$ in pairs gets $\frac{n}{2}$ bins, and packing each m item of size $\frac{1}{m}$ in a bin gets $\frac{n}{m}$ bins. This implies that PoA is at least $\frac{n}{(\frac{n}{m}+\frac{n}{2})} = \frac{2m}{2+m} = 2 - \frac{4}{2+m}$. □

4.2 LSB Cost-Sharing Mechanism

There is only one difference between the movement rules of the Min-OEBP problem and the Max-OEBP problem in that the movement of item i presupposes that $s(B' \cup \{i\}) - \min(s(B' \cup \{i\})) < 1$.

First we can show that the FFD algorithm can output an NE for the LSB cost-sharing mechanism, which also prove the existence of Nash equilibrium for the LSB cost-sharing mechanism. After that an upper bound for PoA is given and it is proved that $PoS \leq \frac{71}{60}$.

Theorem 8. *For the Min-OEBP problem, given any threshold constant $\Lambda \in (0, 1]$ and instance, the FFD algorithm can output an NE for the LSB cost-sharing mechanism.*

Proof. Let π is the packing scheme output by the FFD algorithm, and let π use m bins, it may be useful to set the set of bins used to be $\{B_1, B_2, \ldots, B_m\}$, then according to the packing policy of the algorithm, the load of the first $(m-1)$ bins must not be less than 1. For the Min-OEBP problem, items in the mth bin cannot be moved to the first $(m-1)$ bins, and there is no motivation for items in the first $(m-1)$ bins to be moved unilaterally to those bins, since after moving these bins will become not valid bins.

Next we just need to show that the items in the first $(m-1)$ bins have no motivation to move to the mth bin. First, if $s(B_m) \geq 1$, then π is NE, since all items cannot be moved to other bins and the shared cost cannot be reduced by opening a new bin. If $s(B_m) < 1$, the algorithm's order of packing shows that any item in the first $(m-1)$ bins is superior to any item in B_m, and the item in the first $(m-1)$ must become the bottom item after moving to the mth bin. Let π' be the new packing scheme obtained by moving item $i \in B_j$ into B_m, where $1 \leq j \leq m-1$. First non-bottom items have no motivation to move because $c_\pi(i) = \int_{s^i(B_j)}^{s^i(B_j)+s_i} f(x)dx < \int_0^{s_i} f(x)dx \leq c_{\pi'}(i) = \int_0^{s_i} f(x)dx + \left[1 - \int_0^{s(B_m)+s_i} f(x)dx\right]$. For bottom items, since the load of the first $(m-1)$ bins is not less than 1, i.e., $s(B) \geq 1 \geq \Lambda$, this implies that the minimum shared cost of being a bottom item has already been reached. In other words, there is $c_\pi(i) = \int_0^{s_i} f(x)dx \le c_{\pi'}(i) = \int_0^{s_i} f(x)dx + \left[1 - \int_0^{s(B_m)+s_i} f(x)dx\right]$, thus these items also have no motivation to move, which implies that π is an NE. □

Lemma 2. *For the Max-OEBP problem, at most one bin in any NE has a load less than* $\frac{1}{2}$.

Proof. Suppose that there exists a packing scheme π that is NE and has two bins B_k and B_h with load less than $\frac{1}{2}$. Let b_k and b_h be the bottom items of B_k and B_h. It may be useful to set $b_k \succ b_h$. Using the same proof as Lemma 1 shows that b_h has motivation to move into B_k, which contradicts the definition of NE. □

Theorem 9. *For the Max-OEBP problem, PoA does not exceed 4.*

Proof. Using the same proof as in Theorem 4 and Lemma 2 it follows that $2 \cdot opt \ge \sum_{i=1}^{n} s_i > \frac{n^*-1}{2}$, so we have $4 \cdot opt > n^* - 1$. □

By Theorem 8 it is known that the number of bins used by the best NE for any instance will not exceed the number of bins used by the FFD algorithm, so that we can obtain an upper bound for the *PoS*.

Theorem 10. *For the Min-OEBP problem, the PoS does not exceed* $\frac{71}{60}$.

Proof. For any instance I, let $n_1(\pi)$ be the number of bins used by the best NE, m be the number of bins used by the packing scheme output by the FFD algorithm, and $OPT(I)$ be the number of bins used by the optimal packing scheme. By Theorem 8 it is known that the packing scheme outputted by the FFD algorithm is an NE, and the approximation ratio of the FFD algorithm is $\frac{71}{60}$ [18], then we have $n_1(\pi) \le m \le \frac{71}{60}OPT(I)$. This holds for any instance, thus $PoS \le \frac{71}{60}$. □

5 Conclusion

In this paper, we give two lower bounds for the cost-sharing mechanism *PoA* with parameters for the open-end bin packing problem, and also study the LSB cost-sharing mechanism under the Max-OEBP and Min-OEBP models. We also show that the FFD algorithm can be used to output a stable packing scheme in polynomial time for LSB cost-sharing mechanism. In this paper we have only given upper and lower bounds on the *PoA* of this mechanism, so future research on this mechanism can be continued to give an exact *PoA*.

Acknowledgment. This study was supported by a grant from the Recommended Exempt Graduate Student Research and Innovation Program of Yunnan University (TM-23236819).

References

1. Adar, R., Epstein, L.: Selfish bin packing with cardinality constraints. Theoret. Comput. Sci. **495**, 66–80 (2013)
2. Anshelevich, E., Dasgupta, A., Kleinberg, J., et al.: The price of stability for network design with fair cost allocation. SIAM J. Comput. **38**(4), 1602–1623 (2008)
3. Bilò, V.: On the packing of selfish items. In: IPDPS 2006, Rhodes, Greece, pp. 45–54. IEEE (2006). https://doi.org/10.1109/IPDPS.2006.1639283
4. Bilò, V., Cellinese, F., Melideo, G., et al.: Selfish colorful bin packing games. J. Comb. Optim. **40**(3), 610–635 (2020)
5. Cao, Z., Yang, X.: Selfish bin covering. Theoret. Comput. Sci. **412**(50), 7049–7058 (2011)
6. Chen, X., Nong, Q., Fang, Q.: An improved mechanism for selfish bin packing. J. Comb. Optim. **42**, 636–656 (2021)
7. Dósa, G., Kellerer, H., Tuza, Z.: Using weight decision for decreasing the price of anarchy in selfish bin packing games. Eur. J. Oper. Res. **278**(1), 160–169 (2019)
8. Dósa, G., Epstein, L.: A new lower bound on the price of anarchy of selfish bin packing. Inf. Process. Lett. **150**, 6–12 (2019)
9. Dósa, G., Epstein, L.: Quality of equilibria for selfish bin packing with cost sharing variants. Discret. Optim. **38**, 100556 (2020)
10. Epstein, L., Kleiman, E.: Selfish bin packing. Algorithmica **60**, 368–394 (2011)
11. Epstein, L., Kleiman, E., Mestre, J.: Parametric packing of selfish items and the subset sum algorithm. Algorithmica **74**, 177–207 (2016)
12. Epstein, L., Kleiman, E.: Selfish vector packing. Algorithmica **83**(9), 2952–2988 (2021)
13. Epstein, L.: Open-end bin packing: new and old analysis approaches. Discret. Appl. Math. **321**, 220–239 (2022)
14. Gai, L., Zhang, W., Luo, W., Cheng, Y.: On various open-end bin packing game. In: Du, D.-Z., Du, D., Wu, C., Xu, D. (eds.) COCOA 2021. LNCS, vol. 13135, pp. 86–95. Springer, Cham (2021). https://doi.org/10.1007/978-3-030-92681-6_8
15. Gai, L., Wu, C., Xu, D., et al.: Selfish bin packing under Harmonic mean cost sharing mechanism. Optim. Lett. **16**, 1–12 (2022)
16. Koutsoupias, E., Papadimitriou, C.: Worst-case equilibria. In: Meinel, C., Tison, S. (eds.) STACS 1999. LNCS, vol. 1563, pp. 404–413. Springer, Heidelberg (1999). https://doi.org/10.1007/3-540-49116-3_38
17. Leung, J.Y.T., Dror, M., Young, G.H.: A note on an open-end bin packing problem. J. Sched. **4**(4), 201–207 (2001)
18. Lin, M., Yang, Y., Xu, J.: On lazy bin covering and packing problems. Theoret. Comput. Sci. **411**(1), 277–284 (2010)
19. Ma, R., Dósa, G., Han, X., et al.: A note on a selfish bin packing problem. J. Global Optim. **56**(4), 1457–1462 (2013)
20. Nong, Q., Sun, T., Cheng, T.C.E., et al.: Bin packing game with a price of anarchy of $\frac{3}{2}$. J. Comb. Optim. **35**, 632–640 (2018)
21. Nong, Q., Wang, J., Gong, S., et al.: A bin packing game with cardinality constraints under the best cost rule. Discret. Math. Algorithms Appl. **11**(02), 1950022 (2019)

22. Wang, C., Zhang, G.: Best Cost-Sharing Rule Design for Selfish Bin Packing. arXiv preprint arXiv:2204.09202 (2022)
23. Yu, G., Zhang, G.: Bin packing of selfish items. In: Papadimitriou, C., Zhang, S. (eds.) WINE 2008. LNCS, vol. 5385, pp. 446–453. Springer, Heidelberg (2008). https://doi.org/10.1007/978-3-540-92185-1_50
24. Zhang, C., Zhang, G.: From packing rules to cost-sharing mechanisms. J. Comb. Optim. **44**(3), 1578–1593 (2022)

ToI-Based Data Utility Maximization for UAV-Assisted Wireless Sensor Networks

Qing Zhao[1], Zhen Li[1], Jianqiang Li[1], Jianxiong Guo[2], Xingjian Ding[1](✉), and Deying Li[3]

[1] School of Computer Science, Beijing University of Technology, Beijing, China
{zhaoqing,lijianqiang,dxj}@bjut.edu.cn, lizhen@emails.bjut.edu.cn
[2] Advanced Institute of Natural Sciences, Beijing Normal University, Zhuhai, China
jianxiongguo@bnu.edu.cn
[3] School of Information, Renmin University of China, Beijing, China
deyingli@ruc.edu.cn

Abstract. Using Unmanned Aerial Vehicles as mobile base stations is a promising way to collect data from sensor nodes, especially for large-scale wireless sensor networks. Previous works mainly focus on improving the freshness of the collected data or the energy efficiency by scheduling UAVs. Considering the fact that the sensing data in some applications is time-sensitive, that is, the value of the sensing data is based on its Timeliness of Information (ToI), which decays over time. Therefore, in this paper, we investigate the UAV Trajectory optimization problem for Maximizing the ToI-based data utility (TMT). We propose an improved deep reinforcement learning-based algorithm to address the problem, and the experience results demonstrate the effectiveness of our designs.

Keywords: Data Collection · Trajectory Optimization · Deep Reinforcement Learning · Wireless Sensor Networks

1 Introduction

Due to the characteristics of Unmanned Aerial Vehicles (UAVs) such as small size, high speed, high flexibility, and ignoring terrain, using UAVs as mobile base stations is an effective way to collect data from Wireless Sensor Networks (WSNs), especially for large-scale sensor networks. In recent years, UAV-assisted WSNs have been widely used in a variety of scenarios, such as environmental monitoring, agriculture, security, disaster rescue, and so on [1].

There are a lot of existing works investigating the UAV trajectory optimization problem, the optimization goals include maximizing the energy efficiency of UAVs [2], minimizing the data collection time [3], minimizing the Age of Information (AoI) [4], etc. Among them, the most hot topic in recent years is the AoI

This work is supported by the National Natural Science Foundation of China (with grants No. 62202016 and No. 62202055).

S. Ghosh and Z. Zhang (Eds.): AAIM 2024, LNCS 15179, pp. 72–83, 2024.
https://doi.org/10.1007/978-981-97-7798-3_7

minimization problem. The concept of AoI is first introduced by [5] to quantify the freshness of the knowledge acquired from a remote system. In WSN applications, AoI depicts the elapsed time since the data was generated until it was received by the terminal, which is an important metric in applications with high real-time requirements, such as forest fire monitoring, earthquake monitoring, etc.

However, there are few studies pay attention to the time-sensitivity of data value in WSNs. The time sensitivity of data value means that the value of the sensing data acquired by the terminal decays over time. In many WSN applications, the value of sensing data is time-sensitive. For example, in terms of agricultural temperature and humidity information monitoring, the acquired sensing data can continuously provide decision-making value for the terminal, but as the acquired data and the real-time information of the system may be increasingly different over time, the value of the acquired data surely decays. In this work, we term the time sensitivity of data value as the Timeliness of Information (ToI). Note that AoI depicts the freshness of the acquired data, while ToI reflects the value decay in the use stage after data acquisition.

In this paper, we investigate the UAV trajectory optimization problem for maximizing the ToI-based data utility in WSNs (problem TMT), where a UAV is used as a mobile base station to collect sensing data for sensor nodes. The objective of the TMT problem is to design the trajectory of the UAV to maximize the ToI-based data utility of a WSN, subject to the energy limitation of the UAV. We present a deep reinforcement learning-based algorithm that combines the Action Rejection Mechanism and the Deep Q-Network with Priority Experience Replay (ARM-PER-DQN) to solve the TMT problem. Our contributions can be summarized as follows:

- We give the model of calculating the ToI-based data utility and formulate the TVT problem. Also, we prove that the TVT problem is NP-Hard.
- We present a deep reinforcement learning-based algorithm that combines the action rejection mechanism and the deep Q-network with priority experience replay to solve our problem.
- We conduct extensive simulations to validate the performance of our designs, simulation results show the effectiveness of our algorithm.

The rest of the paper is organized as follows. In Sect. 2, we introduce the related works. In Sect. 3, we formulate the TMT problem. In Sect. 4, we describe the details of our algorithm. In Sect. 5, we evaluate the performance of our algorithm. Finally, we conclude this work in Sect. 6.

2 Related Works

In the past few years, a lot of researchers studied the data collection problem for UAV-assisted WSNs, among them the AoI minimization problem is the most hot topic and related to our work. Qin et al. [6] consider the dynamic scenarios where UAVs can perform long-term data collection tasks, their objective is to maximize the quality of service (QoS) which is based on AoI. Samir et al. [7] aim to jointly optimize the trajectories and scheduling policies of UAVs to minimize the expected weighted sum AoI. Jia et al. [8] study the data acquisition mode selection, including hovering, flying, and hybrid modes. Their objective is to minimize the average AoI of all sensor nodes. Mao et al. [9] study the AoI-optimal data collection problem for the multi-UAV-aided WSNs, they assume that the UAV could collect data from multiple sensors simultaneously, and aim to find the data collection points and flight trajectory for each UAV to minimize the maximal AoI of sensor nodes.

Some researchers consider optimizing multiple goals simultaneously for the data collection problem. Dai et al. [4] use a group of UAVs as aerial base stations to move around and collect data from multiple mobile crowdsensing users. They aimed to maximize the collected data, and geographical coverage while minimizing the AoI of all mobile users simultaneously. The optimization goals of [10] include maximizing collected data amount, geographical fairness, and minimizing the energy consumption of all UAVs. Yang et al. [11] studied a multi-objective optimization problem in a computationally friendly application scenario. Their optimization objectives include the degree of load balance, the average transmission cost, and the task slowdown.

To sum up, most of the existing works pay more attention to the freshness of the acquired data, or the frequency of data acquisition, while our work focuses on the value decay after data acquisition, which is the fundamental difference between our work and existing works.

3 Models and Problem Formulation

3.1 Network Model

We consider a wireless sensor network that contains a set of N sensors denoted as $\mathcal{S} = \{s_1, s_2, \ldots, s_N\}$. Each sensor node continuously perceives surrounding data at a pre-set frequency and only saves the latest data due to the limited storage, the data size of sensor s_i is b_i. An unmanned aerial vehicle (UAV) acts as a mobile base station to collect data from sensor nodes, the UAV is equipped with a satellite communication antenna and the collected data will be transmitted to the ground base station in real-time. Assume the UAV flies at a fixed altitude h with a constant speed v. We adopt the one-to-one transmission model in this work and consider that the UAV collects data from a sensor node by hovering above it. The energy of the battery carried by the UAV is limited (denoted by E^u), and both flying and communication consume the battery energy.

3.2 ToI-Based Data Utility

We consider the value of the sensing data to be related to its Timeliness of Information, in other words, the data is most valuable when it is just transmitted to the ground base station and then decays over time. The data of sensor nodes can be collected by the UAV multiple times, as shown in Fig. 1. In this work, we use an exponential decay model to describe the value of the sensing data over time. Given a time t, let t_i^* ($t_i^* > 0$) be the latest time when the UAV collects the sensing data of sensor node s_i before t. Note that if the UAV never collects s_i's data before t, we set $t_i^* = +\infty$. Then, at any time t, the value of the sensing data from s_i is calculated as $u_{s_i}(t) = w_i e^{-\gamma_i(t-t_i^*)}$, if $t \geq t_i^*$, otherwise, $u_{s_i}(t) = 0$, where w_i and γ_i are the value coefficient and decay coefficient of the sensing data of node s_i.

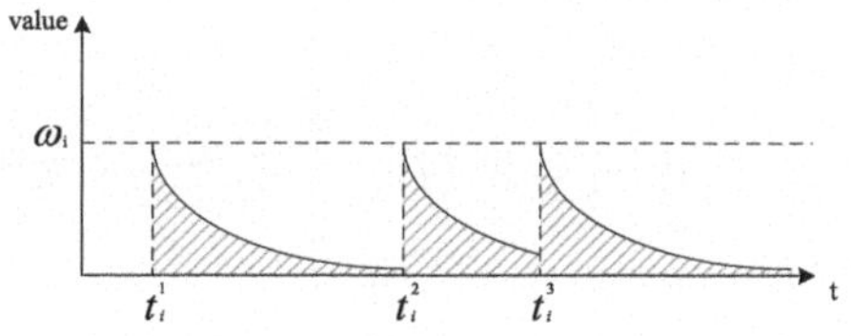

Fig. 1. The data value collected from sensor s_i over time, where t_i^1, t_i^2, and t_i^3 are the three times when the data of s_i is collected by the UAV.

As the data of sensor nodes can be collected multiple times, The utility U_{s_i} of sensing data collected from sensor node s_i can be seen as the area of the shaded region in Fig. 1. That is, $U_{s_i} = \int_{t=0}^{+\infty} u_{s_i}(t)dt$.

3.3 Problem Formulation

In this paper, we aim to find a data collection trajectory for the UAV such that the final total data utility at the ground base station can be maximized, subject to the energy constraints of the UAV (the TMT problem). The energy consumption of the UAV includes flight energy, hovering energy, and communication energy. According to [12], the propulsion power of the UAV at flying speed v can be expressed as

$$P(v) = P_\alpha\left(1+\frac{3v^2}{v_{tip}^2}\right) + P_\beta\left(\sqrt{1+\frac{v^4}{4v_0^4}} - \frac{v^2}{2v_0^2}\right)^{\frac{1}{2}} + \frac{1}{2}d_0\rho s A v^3, \qquad (1)$$

where P_α and P_β are two constants representing the blade profile power and induced power of the UAV in hovering status, respectively, v_{tip} is the tip speed of the rotor blade, v_0 is the average rotor induced velocity in hovering status, d_0, ρ, s, and A are the fuselage drag ratio, air density, rotor solidity, and rotor disc area, respectively. Then the hovering power of the UAV is $P(0) = P_\alpha + P_\beta$.

The data transmission rate (in bits/second) from a sensor node to the UAV at a distance h can be expressed as $R(h) = Blog_2\left(1+\frac{P_u\beta_0 h^{-\alpha_{loss}}}{\sigma^2}\right)$, where β_0 is the reference channel power at a reference distance 1 m, $2 \leq \alpha_{loss} \leq 4$ is the path loss exponent, B is the channel bandwidth, and σ^2 represents the white Gaussian noise power at the receiver side(the UAV) [13].

We use $T_C = \{s^0, s^1, s^2, \ldots, s^0\}$ to denote a closed data collection trajectory for the UAV, where s^0 represents the depot, and $s^i \in T_C$ denotes the *i-th* sensor node visited by the UAV. The total data utility of the trajectory T_C can be calculated as $U(T_C) = \sum_{s_i \in T_C \cap S} U_{s_i}$. The flight energy consumption of the UAV on T_C is $E_f(T_C) = P(v) \cdot \frac{D(T_C)}{v}$, where $D(T_C)$ is the distance of trajectory T_C. The hovering energy consumption and the communication energy consumption can be calculated as $E_h(T_C) = \sum_{i=1}^{|T_C|} P(0) \cdot \frac{b^i}{R(h)}$ and $E_c(T_C) = \sum_{i=1}^{|T_C|} P_c \cdot \frac{b^i}{R(h)}$, respectively.

With the models introduced above, the TMT problem can be formulated as

$$\begin{aligned} \max_{T_C} \quad & U(T_C) \\ \boldsymbol{s.t.} \quad & E_f(T_C) + E_h(T_C) + E_c(T_C) \le E^u. \end{aligned}$$

Theorem 1. *The TMT problem is NP-Hard.*

Proof. We consider a special case of the TMT problem, where the communication energy consumption of the UAV at each sensor node is 0, and consider that the communication time at each node is very short, so the hovering energy consumption of the drone can be ignored. Besides, the data value decay coefficient γ_i is sufficiently small that it is impossible for the UAV to repeatedly collect data from a sensor node to increase the ToI-based data utility. In such a special case, the TMT problem is actually to find a flying tour of the UAV to maximize the collected data utility subject to a limited path cost. Then the above special case of the TMT problem is equal to the budgeted prize-collecting traveling salesman problem [14], which is an NP-Hard problem. Therefore, the general TMT problem is also NP-Hard. □

4 A Deep RL-Based Framework for the TMT Problem

In this section, we formulate the TMT problem as a Markov Decision Process (MDP) and design a deep Reinforcement Learning (RL)-based framework to address it. To accelerate training, we implement an action rejection mechanism and prioritize experience replay. Figure 2 outlines our framework, and we'll elaborate on the design below.

4.1 Markov Decision Process

The components of the MDP we modeled are defined as follows:

State Space. Denote by p_{s_i} the process that the UAV flies to sensor node s_i from the current location and collects data from s_i. Denote s_{cur} be the current location of the UAV, then total energy cost of the UAV during p_{s_i} can be calculated as follows:

$$p_{s_i}^{cost} = P(v) \cdot \frac{d(s_{cur}, s_i)}{v} + P(0) \cdot \frac{b_i}{R(h)} + P_c \cdot \frac{b_i}{R(h)}. \tag{2}$$

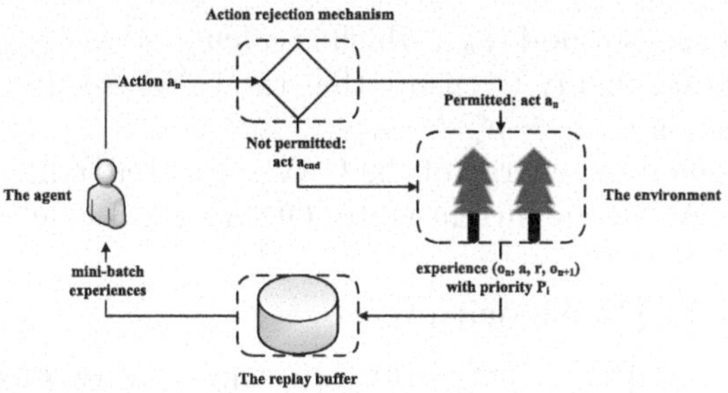

Fig. 2. The structure of the deep RL-based framework.

Denote by $p_{s_i}^{value}$ as the achieved data utility of the UAV from the process p_{s_i}, in other words, $p_{s_i}^{value}$ is the acquired data utility of the UAV if it flies to sensor node s_i and collects data at the current time. According to Fig. 1, $p_{s_i}^{value}$ is calculated as $p_{s_i}^{value} = \int_{t=t_i^*}^{t_{s_i}} u_{s_i}(t)dt$. Then the initial state space is defined as $O = \{E_r, \{< p_{s_i}^{cost}, p_{s_i}^{value} > |s_i \in \mathcal{S}\}\}$, where E_r indicates residual energy of the UAV at current state.

Action Space. The action space of the UAV is represented by $A = \{a|a = \{0, 1, \ldots, i, \ldots, N\}\}$, where $a = i$ $(i > 0)$ means the UAV will fly to s_i and collect data from it, i.e., the process p_{s_i}. The action $a = 0$ (which is also denoted by a_{end} in the following text) means that the UAV will fly back to the depot directly.

Reward Function. We define the reward of an action as the achieved data utility of the UAV by executing it. A specific action a corresponds to a process p_{s_a} as defined before, and the reward r_a is equal to $p_{s_a}^{value}$. As analyzed before, r_a can be calculated as

$$r_a = p_{s_a}^{value} = \begin{cases} \frac{w_i}{\lambda_i}, & \text{if } s_a \text{has not be visited.} \\ \int_{t=t_i^*}^{t_{s_i}} u_{s_i}(t)dt, & \text{otherwise.} \end{cases} \tag{3}$$

Note that some actions may be infeasible due to the energy limitation of the UAV, we do not consider the reward of the infeasible actions as we will use an action rejection mechanism to avoid these actions later.

4.2 Action Rejection Mechanism

Action Rejection Mechanism (ARM) is an inspection mechanism used in conjunction with our algorithm. The main function of the ARM mechanism is to judge whether an action is feasible for the UAV.

For an action a, RM first calculates the energy consumption of the UAV to execute the action a in the current state through Eq. (2), that is $p_{s_a}^{cost}$. Then calculates the flight energy cost of the UAV to return directly to the depot from the next state (from sensor node s_a), which is calculated as $e^a_{back} = P(v) \cdot \frac{d(s_a, s^0)}{v}$. The minimum residual energy to ensure that the UAV executes the action a and returns to the depot is $e^a_{safe} = p^{cost}_{s_a} + e^a_{back}$.

If the current remaining energy of the UAV is greater than e^a_{safe}, the action a is feasible for the UAV at its current state. Otherwise, the action a is infeasible.

4.3 ARM-PER-DQN Algorithm

In our proposed ARM-PER-DQN algorithm, there are two feedforward neural networks, namely the main network Q and the target network $\widehat{Q}$ [15]. These two networks have the same structure but different parameters. The parameters of the main network (denoted by θ) are used to predict the Q-value and are updated at every training step. The target network $\widehat{Q}$ calculates the target Q-value and its parameters (denoted by θ^-) will be synchronized by the main network at every T steps, where T is a hyper-parameter.

Algorithm 1. ARM-PER-DQN

Input: Mini-batch size M, step-size μ replay buffer size K_b, exponents α and β.

Initialize replay buffer, $\Delta = 0$, $p_1 = 1$
for each episode e **do**
  for each step n **do**
    Select action $a_n = \arg\max_a Q(o_n, a)$ with probability $1-\epsilon$, otherwise select a random action a_n with probability ϵ.
    Use action rejection mechanism to inspect the action a_n.
    if a_n is permitted **then**
      Execute a_n.
    else
      Execute a_{end}.
    end if
    for $i = 1 \rightarrow M$ **do**
      Compute the TD error δ_i
      Sample transition i by probability P_i using (5)
      Compute importance-sampling weight h_i in (6).
      Accumulate weight change $\Delta \Leftarrow \Delta + h_i \delta_i \Delta_\theta Q(o_n, a_n)$
    end for
    Update weights $\theta \Leftarrow \theta + \mu * \Delta$, reset $\Delta = 0$
    Synchronize θ^- with θ every T steps
  end for
end for

The training of the algorithm is based on Temporal Difference (TD) error. During the training process, the agent executes an action a, and the system state changes from the current state o_n to the next state o_{n+1}, and the agent earns a reward r_a. The Q-value estimated by the main network Q is $Q(o_n, a)$, and the TD target Q-value can be calculated as follows:

$$y(o_n, a) = r_a + \gamma \max_a \hat{Q}(o_{n+1}, a). \tag{4}$$

where $\gamma \in [0,1]$ is the discount factor, and $y(o_n, a)$ is used to replace the "ground truth". The TD error is defined as the difference between $Q(o_n, a)$ and $y(o_n, a)$, *i.e.*, $\delta_a = Q(o_n, a) - y(o_n, a)$. The mean square error between the target value and the predicted value of all state transitions as the loss function, then The loss function of all state transitions is defined as $L(\theta) = E[Q(o_n, a) - y(o_n, a)]^2$. The main network is updated through gradient descent applied to $L(\theta)$.

To maximize the value of key experiences (transitions) and accelerate training, we adopt the Priority Experience Replay strategy (PER) in our algorithm for replaying important transitions more frequently [16].

The core principle of PER is to measure the importance (prioritization) of each state transition, and sample transitions based on the importance. In detail, the proportional prioritization of a transition i is defined as $p_i = |\delta_i| + \eta$, where δ_i is the TD error, and η is a small and positive value to prevent the sampling probability from being 0. Then, the sampling probability of transition i is

$$P_i = \frac{p_i^\alpha}{\sum_{j=1}^{K_b} p_j^\alpha} \tag{5}$$

where K_b represents the size of the replay buffer, and α determines the degree of prioritization.

The biased sample distribution resulting from prioritized experience replay will affect the accuracy of the solution estimate. To address the issue of biased estimates, importance-sampling weights can be applied as suggested in [16].

$$h_i = \left(\frac{1}{K_b} \cdot \frac{1}{P_i}\right)^\beta \tag{6}$$

where β is an exponent that determines the extent of prioritization to be applied.

The details of the ARM-PER-DQN algorithm are presented in Algorithm 1.

5 Simulations

5.1 Simulation Setting

In our simulation, 50 sensor nodes are uniformly deployed in a place of 1000 m * 1000 m, and the depot is located in the bottom left corner. The battery capacity of the UAV is 6000 J, it flies at an altitude of 100 m with a maximum speed of 5 m/s. The other basic parameters are the same with the existing work [17]. With the above default parameter settings, the loss of our algorithm converges to 0 after about 50000 steps, as shown in Fig. 3, which proves the convergence of our algorithm. In the following, all the simulation results are the average of 20 runs.

We use the following three algorithms as our baselines.

DQN: It adopts the same MDP model and hyper-parameters as our algorithm. The difference is that its replay buffer does not consider the priority of different experiences.

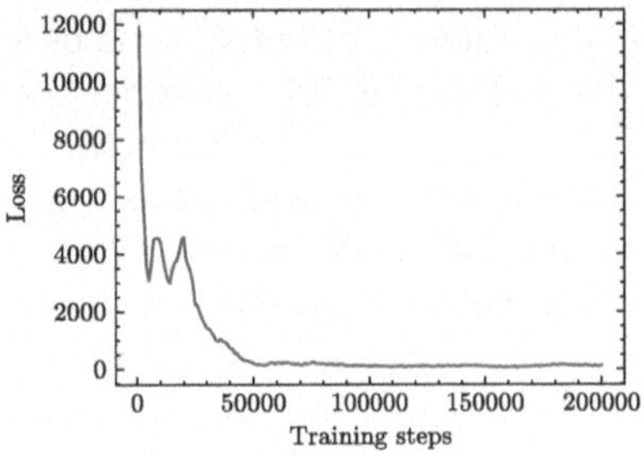

Fig. 3. The convergence process of ARM-PER-DQN.

(DVFA): In Data Value First Algorithm (DVFA), UAV always selects the sensor node that brings the maximum data value at the current time to visit.

(DVUECFA): Data Value of Unit Energy Consumption First Algorithm (DVUECFA) is similar to DVFA, the difference is that, in each step, the UAV selects the sensor node that brings the maximum data value with per unit energy consumption.

The main metric we considered is the total data utility, which is also the optimization goal of our problem. In addition, in order to compare our algorithms with baselines more comprehensively, we also include three other metrics, that is, the coverage of nodes, the flight duration of UAV, and the energy utilization rate of UAV. Note that these three metrics are not our optimization objectives. We merely aim to analyze the differences between our algorithm and the baseline algorithm more thoroughly through these metrics.

5.2 Simulation Results

Varying the Number of Sensor Nodes. As the number of nodes increases, the UAV has a wider range of options, and excluding some random factors, the reward values obtained by excellent algorithms should be on an upward trend overall. As shown in Fig. 4(a), the reward obtained by DVUECFA, ARM-PER-DQN, and DQN gradually increases as the number of nodes increases, except for DVFA which has no significant improvement. Among them, our proposed ARM-PER-DQN obtains more total reward than all other algorithms in all dimensions. Although DQN has comparable results with ARM-PER-DQN when the number of nodes is less than or equal to 50, as the number of nodes continues to increase, the reward obtained by DQN does not continue to increase as much as those obtained by ARM-PER-DQN. DVUECFA always maintains a significant gap with PER-DQN, while it keeps an upward trend. In Fig. 4(b), as the number of nodes increases, the node coverage of all the algorithms shows a decreasing trend because of the limitation of battery capacity. By comparing Fig. 4(a) with Fig. 4(c) it can be seen that ARM-PER-DQN and DQN get more reward values due to the longer flight duration, which means fuller utilization of the limited ground energy. It can be known in Fig. 4(d) that the energy utilization rate of ARM-PER-DQN and DQN has been converging to 1, which provides proof for the previous analysis. The correlation between Fig. 4(c) and Fig. 4(d) shows that there is an extremely strong dependency between the flight duration and the utilization rate of energy, which is also in line with common sense.

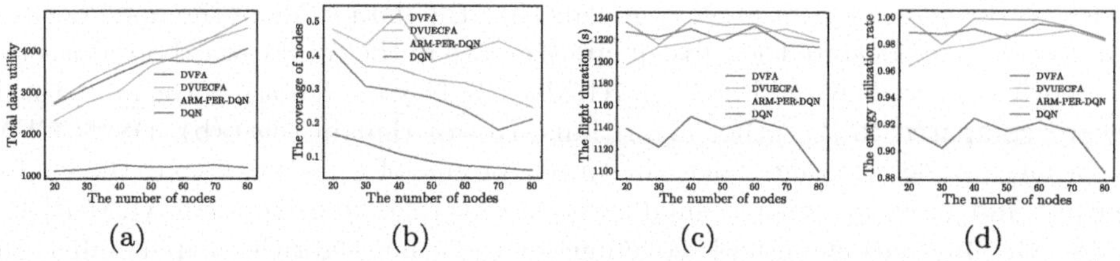

Fig. 4. The effect of the number of nodes on (a) The total reward, (b) The coverage of nodes, (c) The flight duration, and (d) The energy utilization rate

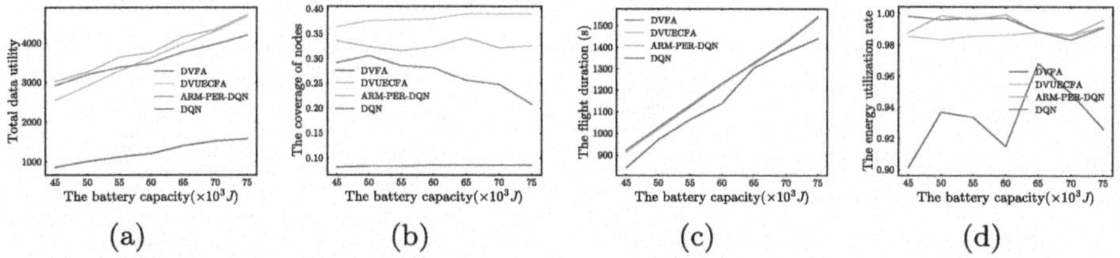

Fig. 5. The effect of the battery capacity on (a) The total reward, (b) The coverage of nodes, (c) The flight duration, and (d) The energy utilization rate

Varying the Battery Capacity of the UAV. From Fig. 5(a), it can be seen that ARM-PER-DQN outperforms the baselines on total reward. At levels with less energy (45,000 J and 50,000 J), the total reward obtained by DQN and ARM-PER-DQN is extremely close to each other. However, when the energy is more than 50,000J, the advantage of ARM-PER-DQN gradually appears, and the gap between DQN and ARM-PER-DQN is getting bigger and bigger. When the energy is 75000 J, the total reward obtained by DQN is close to that of DVUECFA, while ARM-PER-DQN still maintains a large advantage. In Fig. 5(b), notice that the node coverage of DQN starts to decrease significantly at an energy of 50,000 J, while the other algorithms are relatively stable, which indicates that DQN tends to fall into a locally optimal solution. Figure 5(c) shows that the variation in battery capacity has a small effect on the difference in the total UAV flight time of the four algorithms. Notice that in Fig. 5(d), the energy utilization rate of ARM-PER-DQN and DQN decreases when the battery capacity is more than 65,000 J, which implies that there is still space to improve the performance of these two algorithms. Overall, ARM-PER-DQN outperforms other baselines on the main metric, with an acceptable performance on the other three metrics.

Varying the Upper Bound of γ_i. γ_i is the decay coefficient of the data value defined in Sect. 3.2. As the upper limit of γ_i is raised, the average γ_i of all the nodes keeps increasing. A larger γ_i means that the data value of the node decays faster over time. Despite the subsequent decrease in the total reward obtained by all algorithms, as shown in Fig. 6(a), ARM-PER-DQN still obtains the most total reward compared to the other algorithms. Notice that in Fig. 6(b), DVUECFA has a large node coverage when the upper bound of γ_i is small, but then falls sharply and ends up with a smaller node coverage than that of ARM-PER-DQN. The fluctuations in node coverage for the other algorithms are smaller. In Fig. 6(c), it can be seen that the flight time of DVFA is strongly affected by the variation of the upper bound of γ_i, while the other three algorithms are stable. This is due to that DVFA is not able to fully utilize the limited energy in a stable way. As shown in Fig. 6(d), the variation of the upper bound of γ_i has little effect on the energy utilization rate of all the algorithms, except DVFA. Figure 6(d) also confirms the instability of the performance of DVFA.

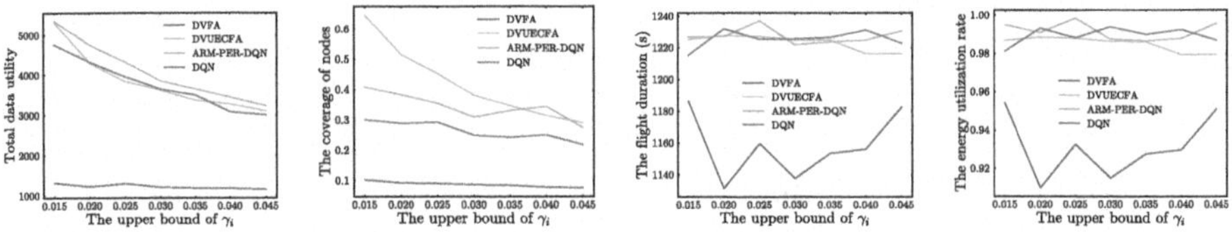

(a) (b) (c) (d)

Fig. 6. The effect of the upper bound of γ_i on (a) The total reward, (b) The coverage of nodes, (c) The flight duration, and (d) The energy utilization rate

6 Conclusion

In this paper, we model the relationship between data utility and the timeliness of information, and investigate the UAV trajectory optimization problem for maximizing the ToI-based data utility (TMT, in short). We prove that the problem to be addressed is NP-Hard, and design a deep reinforcement learning-based algorithm that combines the action rejection mechanism and the deep Q-Network with priority experience replay. At last, we evaluate the performance of our designs through extensive simulations, the results show the effectiveness of our designs.

References

1. Lai, R., et al.: Energy-efficient scheduling in UAV-assisted hierarchical wireless sensor networks. IEEE Internet Things J. (2024)
2. Liu, J., Zhang, H.: Height-fixed UAV enabled energy-efficient data collection in RIS-aided wireless sensor networks. IEEE Trans. Wireless Commun. **22**(11), 7452–7463 (2023)
3. Luo, C., Satpute, M.N., Li, D., Wang, Y., Chen, W., Wu, W.: Fine-grained trajectory optimization of multiple UAVs for efficient data gathering from WSNs. IEEE/ACM Trans. Networking **29**(1), 162–175 (2020)
4. Dai, Z., et al.: Aoi-minimal UAV crowdsensing by model-based graph convolutional reinforcement learning. In: IEEE INFOCOM 2022-IEEE Conference on Computer Communications, pp. 1029–1038. IEEE (2022)
5. Kaul, S., Gruteser, M., Rai, V., Kenney, J.: Minimizing age of information in vehicular networks. In: 2011 8th Annual IEEE Communications Society Conference on Sensor, Mesh and Ad Hoc Communications and Networks, pp. 350–358. IEEE (2011)
6. Qin, Z., Zhang, X., Zhang, X., Lu, B., Liu, Z., Guo, L.: The UAV trajectory optimization for data collection from time-constrained IoT devices: a hierarchical deep Q-network approach. Appl. Sci. **12**(5), 2546 (2022)
7. Samir, M., Assi, C., Sharafeddine, S., Ebrahimi, D., Ghrayeb, A.: Age of information aware trajectory planning of UAVs in intelligent transportation systems: a deep learning approach. IEEE Trans. Veh. Technol. **69**(11), 12382–12395 (2020)
8. Jia, Z., Qin, X., Wang, Z., Liu, B.: Age-based path planning and data acquisition in UAV-assisted IoT networks. In: 2019 IEEE International Conference on Communications Workshops (ICC Workshops), pp. 1–6. IEEE (2019)
9. Mao, C., Liu, J., Xie, L.: Multi-UAV aided data collection for age minimization in wireless sensor networks. In: 2020 International Conference on Wireless Communications and Signal Processing (WCSP), pp. 80–85. IEEE (2020)
10. Dai, Z., Liu, C.H., Han, R., Wang, G., Leung, K., Tang, J.: Delay-sensitive energy-efficient UAV crowdsensing by deep reinforcement learning. IEEE Trans. Mob. Comput. (2021)
11. Yang, L., Yao, H., Wang, J., Jiang, C., Benslimane, A., Liu, Y.: Multi-UAV-enabled load-balance mobile-edge computing for IoT networks. IEEE Internet Things J. **7**(8), 6898–6908 (2020)
12. Zeng, Y., Xu, J., Zhang, R.: Energy minimization for wireless communication with rotary-wing UAV. IEEE Trans. Wireless Commun. **18**(4), 2329–2345 (2019)
13. Zeng, Y., Zhang, R.: Energy-efficient UAV communication with trajectory optimization. IEEE Trans. Wireless Commun. **16**(6), 3747–3760 (2017)
14. Paul, A., Freund, D., Ferber, A., Shmoys, D.B., Williamson, D.P.: Budgeted prize-collecting traveling salesman and minimum spanning tree problems. Math. Oper. Res. **45**(2), 576–590 (2020)
15. Mnih, V., et al.: Human-level control through deep reinforcement learning. Nature **518**(7540), 529–533 (2015)
16. Schaul, T., Quan, J., Antonoglou, I., Silver, D.: Prioritized experience replay. arXiv preprint arXiv:1511.05952 (2015)
17. Luo, C., Liu, N., Hou, Y., Hong, Y., Chen, Z., Li, D.: Trajectory optimization of laser-charged UAV to minimize the average age of information for wireless rechargeable sensor network. Theoret. Comput. Sci. **945**, 113680 (2023)

Parity-Constrained Weighted k-Center

Xinlan Xia[1], Lu Han[1(✉)], and Lili Mei[2]

[1] School of Science, Beijing University of Posts and Telecommunications, Beijing, People's Republic of China
{xiaxinlan,hl}@bupt.edu.cn

[2] School of Cyberspace, Hangzhou Dianzi University, Hangzhou, People's Republic of China
meilili@zju.edu.cn

Abstract. This paper studies the parity-constrained weighted k-center (PARW k-center) problem. In the PARW k-center problem, we are given a set of vertices in a metric space with distances and a non-negative budget k. Additionally, each vertex is associated with a non-negative weight and an odd or even parity requirement. The objective is to select a subset of vertices as open centers and assign each vertex to an open center so as to minimize the maximum distance of any vertex to its assigned center, but subject to that the total weight of all open centers is no more than k, and that the number of vertices assigned to each open center meets its parity requirement.

As our main contribution, we present a 10-approximation algorithm for the PARW k-center problem, which consists of three main phases. The first phase utilizes the concept of maximal independent set to initially determine a solution (i.e., determine a set of open centers and assignments of all the vertices) whose total weight of open centers does not exceed the budget k. By pairing open centers that do not meet their parity requirements under the initial assignments, the second and third phases respectively modify the open centers and assignments, leading to a feasible solution.

Keywords: approximation algorithms · parity-constrained · k-Center

1 Introduction

The k-center problem is one of the most fundamental and classical NP-hard problems in the domain of combinatorial optimization [4–6,11,14,21], and it has been well-studied due to its broad applications in machine learning, communication networks, and management science [3,12,17,19]. In the k-center problem, we are given a set of vertices in a metric space with distances and a non-negative integer k. The aim is to select at most k vertices as open centers and assign each vertex

Lu Han is supported by Beijing Municipal Natural Science Foundation (No. Z220004), National Natural Science Foundation of China (No. 12371321). Lili Mei is supported by National Natural Science Foundation of China (No. 12201594).

S. Ghosh and Z. Zhang (Eds.): AAIM 2024, LNCS 15179, pp. 84–93, 2024.
https://doi.org/10.1007/978-981-97-7798-3_8

to an open center in such a way that the maximum distance between any vertex and its assigned open center is minimized. Gonzalez [8] and Hochbaum and Shmoys [12] gave the first 2-approximation algorithm for the k-center problem, and this approximation ratio is tight under the assumption that P $\neq$ NP [12]. For an optimization problem, an α-approximation algorithm is a polynomial time algorithm that, for each instance of this problem, produces a solution whose objective value is within a factor of α of the optimal value, and we call α the approximation ratio of the algorithm. Many variants of the k-center problem and their corresponding approximation algorithms have been proposed in the past years, including the fair k-center problem that can avoid the biased results in data summarization [17], the capacitated k-center problem which applied in the location of emergency facilities [15], etc. [4,14].

The weighted k-center (W k-center) problem is one of the most meaningful variants of the k-center problem, and Hochbaum and Shmoys [13] design a 3-approximation algorithm for it. Compared with the classical k-center problem, the W k-center problem additionally gives a non-negative weight to each vertex, and the goal is to select vertices as open centers and assign each vertex to an open center, such that the total weight of all open centers is no more than k, and the maximum distance between any vertex and its assigned open center is minimized. Bajpai et al. [2] also named the W k-center problem as the knapsack center problem to reflect the knapsack kind constraint on the open centers, and An et al. [1] called the W k-center problem as the budgeted center problem.

Recently, Kim et al. [16] proposed another interesting variant of the k-center problem named the parity-constrained k-center (PAR k-center) problem and gave its first 6-approximation algorithm. In the PAR k-center problem, an odd or even parity requirement is given for each vertex. If a vertex is selected as an open center, the number of vertices assigned to it must meet its parity requirement. Parity plays an important role in combinatorial optimization areas and has several meaningful applications [7,9,10,18,20]. To the best of our knowledge, very few works concentrate on the parity-constrained version of the clustering problems. Besides the PAR k-center problem, Kim et al. [16] also proposed the parity-constrained facility location (PAR FL) problem and gave its $(3\rho + 2)$-approximation algorithm, where ρ is the currently best approximation ratio for the uncapacitated facility location problem. In the PAR FL problem, we are given sets of facilities and clients, an opening cost and an odd or even parity requirement for each facility, and an assignment cost for each facility-client pair. Assume that the assignment costs are non-negative, symmetric, and satisfy the triangle inequality. The goal is to select a subset of facilities to open and assign each client to an open facility, so as to minimize the total cost of opening facilities and assigning clients, but subject to that the number of clients assigned to each open facility meets its parity requirement.

The status quo that the research on the parity-constrained version of the clustering problems is very limited stimulates us to study a more general version of the PAR k-center problem, namely the parity-constrained weighted k-center (PARW k-center) problem. In the PARW k-center problem, a set of vertices in a metric space with distances and a non-negative budget k are given. Each vertex has a non-negative weight and an odd or even parity requirement. The goal is to select a subset of vertices as open centers and assign each vertex to an open center, so as to minimize the maximum distance of any vertex to its assigned center, but subject to that the total weight of all open centers is no more than the budget k, and that the number of vertices assigned to each open center meets its parity requirement. A real-life scenario of the PARW k-center problem is related to grouping participants for different activities. Consider that we have a certain budget to launch some activities at a number of locations. Due to the type of its venue, each location has its own launching cost and parity of the number of participants required for launching an activity. Each participant must join an activity, and the cost of joining an activity is related to the locations of both the participant and the activity. We are concerned about where to launch activities while staying within budget so that the cost of each participant is reasonable.

As our main contribution, we propose the PARW k-center problem and design its first 10-approximation algorithm. Our algorithm has three main phases. Based on the work of Hochbaum and Shmoys [13] on the W k-center problem, in the first phase, we utilize the concept of the maximal independent set to initially determine a solution (i.e., determine a set of open centers and assignments of all the vertices) whose total weight does not exceed the budget k. Note that the initial solution may not satisfy the parity requirements of some open centers. Inspired by the idea of the pairing of Kim et al. [16] in dealing with the PAR k-center, we pair up all the open centers whose parity requirements are not satisfied under the initial assignments. Then, according to the open centers of each pair, we respectively modify the open centers and assignments in the second and third phases to obtain a feasible solution. The biggest challenge in designing our algorithm is how to carefully carry out a suitable pairing process, since the process of Kim et al. [16] only focus on a simple cardinality constraint, and directly using their process will likely lead to a solution violating the budget.

The rest of our paper is organized as follows. Section 2 defines the PARW k-center. Section 3 provides our approximation algorithm along with the analysis.

2 Preliminaries

Here is the formal description of the PARW k-center problem. In a PARW k-center instance $\mathcal{I}$, we are given a non-negative budget k, a set of vertices V in a metric space (V, d) with a distance function $d : V \times V \rightarrow \mathbb{R}_{\geq 0}$. We assume that $d(i,i) = 0$, $d(i,j) = d(j,i)$ for any vertices $i, j \in V$, and that $d(i,j) + d(j,t) \leq d(i,t)$ for any vertices $i, j, t \in V$. Each vertex $i \in V$ is associated with a non-negative weight $w_i \geq 0$ and an odd or even parity requirement represented by a parity constraint function $\pi : V \rightarrow \{\text{odd, even}\}$. Denote by (S, σ) a solution of the instance $\mathcal{I}$, where $S \subseteq V$ is a set of open centers selected from all the vertices, and σ is a mapping of assignments from each vertex $j \in V$ to some center in S. The solution (S, σ) is feasible if the following two types of constraints are satisfied:

- Budget Constraint: The sum of the weights of all open centers does not exceed the budget k, i.e.,

$$\sum_{i \in S} w_i \leq k.$$

- Parity Constraints: For each open center $i \in S$, the parity of the number of vertices assigned to i is the same as its parity requirement, i.e., for each $i \in S$, we have that

$$|\{j \in V : \sigma(j) = i\}| \text{ is in accord with } \pi(i).$$

The goal is to find a feasible solution whose maximum distance of any vertex to its assigned center is minimized.

For a given PARW k-center instance $\mathcal{I} = (V, d, k, \{w_i\}_{i \in V}, \pi)$, denote by (S^*, σ^*) its optimal solution and OPT the corresponding optimal value, i.e.,

$$\text{OPT} = \max_{j \in V} d(j, \sigma^*(j)).$$

It is worth noting that, since there are at most polynomial pair of vertices in the given instance, and the optimal value is one of the distances of the pairs, we can assume that the optimal value of OPT is known.

For a graph $G = (V, E)$ and a pair of vertices $i, j \in V$, denote by $d_G(i, j)$ the number of edges on the shortest path between i and j, and denote by $d_G(i, S)$ the minimum number of edges on the shortest paths between i and all the vertices in S. Let $G^p = (V(G^p), E(G^p))$ be a vertex-induced graph of graph $G = (V, E)$, where $V(G^p) := V$, and for each pair of vertices $i, j \in V$, $(i, j) \in E(G^p)$ if and only if $d_G(i, j) \leq p$. For each vertex $i \in V$, let $N_G(i)$ consist by any vertex $j \in V$ satisfying $d_G(i, j) = 1$. For a non-negative integer L, define $[L] = \{1, 2, ..., L\}$.

3 Approximation Algorithm and Analysis

This section presents a constant-factor approximation algorithm, consisting of three main phases, for the PARW k-center problem. In the first phase, for any given PARW k-center instance $\mathcal{I}$, we construct its initial solution $(S_{\text{ini}}, \sigma_{\text{ini}})$ whose budget constraint is satisfied. Note that the parity constraints of the solution $(S_{\text{ini}}, \sigma_{\text{ini}})$ may not be satisfied. Therefore, in the second and third phases, we respectively modify the open centers of S_{ini} and the assignment of σ_{ini} to finally obtain a solution satisfying both the budget and parity constraints. Due to space limitations, the proofs of all the essential lemmas are removed but will appear in a full version of this paper.

3.1 The Phase of Constructing an Initial Solution

The formal description of the first phase is given in the Algorithm 1. For any given PARW k-center instance $\mathcal{I} = (V, d, k, \{w_i\}_{i \in V}, \pi)$, the first phase pays attention to each vertex pair whose distance is no more than the optimal value of OPT, so as to construct a certain number of connected components $G_1, G_2, ..., G_L$ based on the given set of vertices V. For each connected component $G_l = (V_l, E_l)$ where $l \in [L]$, we decide which of its vertices are selected as centers and how all vertices in it are assigned. Combine the sets of selected centers and the assignments in all the connected components finally yields a solution $(S_{\text{ini}}, \sigma_{\text{ini}})$ satisfying the budget constraint for the instance $\mathcal{I}$.

Although the solution $(S_{\text{ini}}, \sigma_{\text{ini}})$ may not meet some centers' parity constraints, it has some good properties as shown in Lemma 1, which allow us to derive a feasible solution after some reasonable modifications.

Lemma 1. *For the initial solution $(S_{\text{ini}}, \sigma_{\text{ini}})$ obtained by Algorithm 1, we have that*

(i) Solution $(S_{\text{ini}}, \sigma_{\text{ini}})$ satisfies the budget constraint, i.e., $\sum_{i \in S_{\text{ini}}} w_i \leq k$,
(ii) The objective value of the solution $(S_{\text{ini}}, \sigma_{\text{ini}})$ is bounded by 3OPT.

In order to prove Lemma 1, the following Lemma 2 is essential. Lemma 2 demonstrates the relationships between the centers selected by the Algorithm 1 and centers selected in the optimal solution. For each connected component $G_l = (V_l, E_l)$ where $l \in [L]$, assume that vertices in V_l are partitioned into m_l clusters $\{\mathcal{C}_1, ..., \mathcal{C}_{m_l}\}$ in the optimal solution, and denote by $S_l^* = \{s^*(1), s^*(2), ..., s^*(m_l)\}$ the selected centers in the optimal solution under each connected component G_l. Clearly, the maximum distance between any two vertices in the same cluster is no more than 2OPT. Consider that D_l is a maximal independent set of G_l^2, then the distance between every two vertices in set D_l is no less than 2OPT. It can be observed that each cluster contains at most one vertex in D_l. For each center $s^*(i)$ opened in the optimal solution, let i be the vertex in D_l contained in cluster $\mathcal{C}_i$ (if it exists), and $s(i)$ be the open center selected by the algorithm according to i. Since $s(i)$ and $s^*(i)$ are both adjacent to i in G_l, it can be demonstrated from step 2.3 that $w_{s(i)} \leq w_{s^*(i)}$ holds.

Lemma 2. *In each connected component $G_l = (V_l, E_l)$ where $l \in [L]$, we have that $w_{s(i)} \leq w_{s^*(i)}$ for each $i \in D_l$.*

3.2 The Phase of Modifying the Open Centers

In the second phase, we modify the set of open centers in the initial solution to obtain a modified center set S. For the solution $(S_l^{\text{ini}}, \sigma_l^{\text{ini}})$ of each connected component G_l, we divide the modification of the open centers set into five cases in the Algorithm 2.

Lemmas 3 and 4 show that the budget constraint holds under each case.

Lemma 3. *For* cases (A) *and* (B), *we have that* $\sum_{i \in S_l} w_i \leq \sum_{i \in S_l^*} w_i$ *for each* $l \in [L]$.

Lemma 4. *For* cases (C), (D), *and* (E), *we have that* $\sum_{i \in S_l} w_i \leq \sum_{i \in S_l^*} w_i$ *for each* $l \in [L]$.

Algorithm 1:. Construct an initial solution.

Input: A PARW k-center instance $\mathcal{I}$ and its optimal value of OPT.
Output: A solution $(S_{\text{ini}}, \sigma_{\text{ini}})$ and a group of solutions and trees.
Step 1 Construct several connected components.
For a given PARW k-center instance $\mathcal{I} = (V, d, k, \{w_i\}_{i \in V}, \pi)$, construct a graph $G = (V, E)$, where $E := \{(i,j) : i, j \in V, d(i,j) \leq \text{OPT}\}$. Let L be the number of connected components in G, and $G_l = (V_l, E_l)$ the lth connected component with $l \in [L]$, and V_l and E_l the vertex and edge sets of G_l.
Step 2 For each connected component G_l, pick up a center set S_l^{ini}.
Step 2.1 Initialization
Set $D_l := \emptyset$, $V_l^{\text{un}} := V$, and $\sigma_l^{\text{ini}}(j) := \emptyset$ for each $j \in V$. Let $\mathcal{T}_l = (V(\mathcal{T}_l), E(\mathcal{T}_l))$ be a tree in this connected component and initialize $V(\mathcal{T}_l) := \emptyset$, $E(\mathcal{T}_l) := \emptyset$.
Step 2.2 Select a maximal independent set in G_l^2
Arbitrarily pick a vertex $i \in V$, and update $D_l := \{i\}$, $V(\mathcal{T}_l) := \{i\}$. Set $\sigma_l^{\text{ini}}(j) := i$, for each $j \in N_{G_l^2}(i) \cap V_l^{\text{un}}$. Update $V_l^{\text{un}} := V_l^{\text{un}} \backslash N_{G_l^2}(i)$.
while $V_l^{\text{un}} \neq \emptyset$ **do**
Select a vertex $i \in V_l$ such that $d_{G_l}(i, D_l) = 3$. Update $D_l := D_l \cup \{i\}$, and $V(\mathcal{T}_l) := V(\mathcal{T}_l) \cup \{i\}$, $E(\mathcal{T}_l) := E(\mathcal{T}_l) \cup (i,j)$, where j is a vertex from $V(\mathcal{T}_l)$ such that $d_{G_l}(i,j) = 3$. Update $\sigma_l^{\text{ini}}(j) := i$, for each $j \in N_{G_l^2}(i) \cap V_l^{\text{un}}$. Update $V_l^{\text{un}} := V_l^{\text{un}} \backslash N_{G_l^2}(i)$.
Step 2.3 Select a set of open centers S_l^{ini}
Initialize $S_l^{\text{ini}} := \emptyset$.
for each $i \in D_l$ **do**
Set $s(i) = \arg\min_{t \in N_{G_l}(i)} w_t$. Update $S_l^{\text{ini}} := S_l^{\text{ini}} \cup \{s(i)\}$, $\sigma_l^{\text{ini}}(i) := s(i)$.
Step 3 Construct a solution $(S_{\text{ini}}, \sigma_{\text{ini}})$.
For each $j \in V$, let $i = \sigma_l^{\text{ini}}(j)$ and update $\sigma_l^{\text{ini}} := s(i)$. For each $i \in S_l^{\text{ini}}$, update $\sigma_l^{\text{ini}}(i) := i$. Set $S_{\text{ini}} := \bigcup_{l \in [L]} S_l^{\text{ini}}$, and $\sigma_{\text{ini}}(j) := \sigma_l^{\text{ini}}(j)$ for every $j \in V$, where $l \in [L]$. For each $i \in V(\mathcal{T}_l)$, replace i by $s(i)$. For each edge $(i,j) \in E(\mathcal{T}_l)$, replace it by $(s(i), s(j))$.
Output the solution $(S_{\text{ini}}, \sigma_{\text{ini}})$, a group of solution $(S_l^{\text{ini}}, \sigma_l^{\text{ini}})$ and trees.

Algorithm 2:. Modify the open centers.

Input: PARW k-center instance $\mathcal{I} = (V, d, k, \{w_i\}_{i\in V}, \pi)$, a group of solutions $(S_l^{\text{ini}}, \sigma_l^{\text{ini}})$ and trees $\mathcal{T}_l$, where $l \in [L]$.

Output: A set of open centers S for the PARW k-center instance $\mathcal{I}$.

Step 1 For each connected component $G_l = (V_l, E_l)$, select a set of open centers S_l, where $l \in [L]$.

for each solution $(S_l^{\text{ini}}, \sigma_l^{\text{ini}})$ with $l \in [l]$ **do**

Define S_l^{inv} as a set consisting of centers in connected component G_l, for which the parity constraint is not satisfied. $\mathcal{O}_l$ denotes the set of vertices in V_l with odd-constraint. If $S_l^{\text{inv}} = \emptyset$, update $S_l := S_l^{\text{inv}}$, otherwise we consider the following cases.

Case (A). $|S_l^{\text{ini}}| = 1$, and there exists a vertex $t \in V_l$ such that the parity of $|V_l|$ is same as π_t.
In this case, if $|V_l|$ is odd, let vertex $j := \arg\min_{i\in V_l\cap\mathcal{O}_l} w_i$ and update $S_l := \{j\}$; otherwise, let j denotes the current opened center, $j_1 := \arg\min_{i\in V_l\setminus\mathcal{O}_l} w_i$, $j_2 := \arg\min_{i\in V_l\cap\mathcal{O}_l} w_i$, update $S_l := \{j_1\}$ if $w_{j_1} \le w_j + w_{j_2}$; if $w_{j_1} > w_j + w_{j_2}$, update $S_l := \{j, j_2\}$.

Case (B). $|S_l^{\text{ini}}| = 1$, and there doesn't exist any vertex $t \in V_l$ such that the parity of $|V_l|$ is same as π_t.
In this case, let $j := \arg\min_{i\in V_l} w_i$, update $S_l := S_l^{\text{ini}} \cup \{j\}$.

Case (C). $|S_l^{\text{ini}}| \ge 2$, and $|S_l^{\text{inv}}|$ is even.
In this case, update $S_l := S_l^{\text{ini}}$.

Case (D). $|S_l^{\text{ini}}| \ge 2$, $|S_l^{\text{inv}}|$ is odd, and $\mathcal{O}_l \cap S_l^{\text{ini}} \ne \emptyset$.
In this case, arbitrarily choose a vertex $j \in S_l^{\text{ini}} \cap \mathcal{O}_l$ and update $S_l := S_l^{\text{ini}} \setminus \{j\}$.

Case (E). $|S_l^{\text{ini}}| \ge 2$, $|S_l^{\text{inv}}|$ is odd, and $\mathcal{O}_l \cap S_l^{\text{ini}} = \emptyset$.
In this case, let $r := \arg\min_{i\in V_l\cap\mathcal{O}_l} w_i$, and $j := \arg\max_{i\in S_l^{\text{ini}}} w_i$. Update $S_l := S_l^{\text{ini}} \cup \{r\} \setminus \{j\}$.

Step 2 Obtain a set of final open centers.

Set $S := \bigcup_{l\in[L]} S_l$.

Output the open center set S of PARW k-center instance $\mathcal{I}$.

Algorithm 3:. Modify the assignments.

Input: PARW k-center instance $\mathcal{I} = (V, d, k, \{w_i\}_{i \in V}, \pi)$, a set of open centers S, the group of solutions $(S_l^{\text{can}}, \sigma_l^{\text{can}})$ and trees $\mathcal{T}_l$, and a set of open centers S_l, where $l \in [L]$.

Output: A feasible solution (S, σ) for the PARW k-center instance $\mathcal{I}$.

Step 1 For each connected component $G_l = (V_l, E_l)$, modify the solution $(S_l^{\text{ini}}, \sigma_l^{\text{ini}})$, where $l \in [L]$.

for each solution $(S_l^{\text{ini}}, \sigma_l^{\text{ini}})$ with $l \in [l]$ **do**

Define S_l^{inv} as a set of vertices consisting of centers in solution $(S_l^{\text{ini}}, \sigma_l^{\text{ini}})$ in which the parity constraint is not satisfied. $\mathcal{O}_l$ denotes the set of vertices in V_l with odd-constraint. If $S_l^{\text{inv}} = \emptyset$, update $\sigma_l(j) := \sigma_l^{\text{ini}}(j)$ for each $j \in V_l$, otherwise we consider the following cases.

Case (A). $|S_l^{\text{ini}}| = 1$, and there exists a vertex $t \in V_l$ such that the parity of $|V_l|$ is same as $\pi(t)$.

In this case, if $|S_l| = 1$, denote i as the only opened center in S_l, and update $\sigma_l(j) := i$ for each $j \in V_l$; If $|S_l| = 2$, denote i as the newly opened center in the second phase, update $\sigma_l(i) := i$, and update $\sigma_l(j) := \sigma_l^{\text{ini}}(j)$ for each $j \in V_l \backslash \{i\}$.

Case (B). $|S_l^{\text{ini}}| = 1$, and there doesn't exist any vertex $t \in V_l$ such that the parity of $|V_l|$ is same as π_t.

In this case, denote i as the newly opened center in the second phase, update $\sigma_l(i) := i$, and update $\sigma_l(j) := \sigma_l^{\text{ini}}(j)$ for each $j \in V_l \backslash \{i\}$.

Case (C). $|S_l^{\text{ini}}| \geq 2$, and $|S_l^{\text{inv}}|$ is even.

In this case, we proceed following two substeps,

Substep 1: Find reassigning paths on tree $\mathcal{T}_l$

Call Algorithm 4 for details.

Substep 2: Modify the assignment.

for each reassigning path $p = (s_1, s_2, ..., s_n)$ **do**

Reassign vertex s_i to s_{i+1}, and update $\sigma_l^{\text{ini}}(s_i) := s_{i+1}$ for each $i = 1, ..., n-1$.

Update $\sigma_l(j) := \sigma_l^{\text{ini}}(j)$ for each $j \in V_l$.

Case (D). $|S_l^{\text{ini}}| \geq 2$, $|S_l^{\text{inv}}|$ is odd, and $\mathcal{O}_l \cap S_l^{\text{ini}} \neq \emptyset$.

In this case, denote i as the vertex that close in the second phase, and arbitrarily choose a vertex $i_c \in \mathcal{T}_l$ such that $d_{G_l}(i, i_c) = 3$. For each vertex j assigned to i in solution $(S_l^{\text{ini}}, \sigma_l^{\text{ini}})$, update $\sigma_l^{\text{ini}}(j) := i_c$. Execute substep 1 and 2 same as case C.

Case (E). $|S_l^{\text{ini}}| \geq 2$, $|S_l^{\text{inv}}|$ is odd, and $\mathcal{O}_l \cap S_l^{\text{ini}} = \emptyset$.

In this case, denote r as the newly opened center in the second phase, and denote r_p as the
vertex such that $\sigma_l^{\text{ini}}(r) = r_p$. Update $V(\mathcal{T}_l) := V(\mathcal{T}_l) \cup \{r\}$ and $E(\mathcal{T}_l) := E(\mathcal{T}_l) \cup (r, r_p)$. denote i as the vertex that is closed in the second phase, arbitrarily choose a vertex $i_c \in \mathcal{T}_l$ such that $d_{G_l}(i, i_c)$ $= 3$. For each vertex j assigned to i in solution $(S_l^{\text{ini}}, \sigma_l^{\text{ini}})$, update $\sigma_l^{\text{ini}}(j) := i_c$. Execute substep 1 and 2 same as case (C).

Step 2 Construct a solution satisfying the parity constraints.

Set $S := \bigcup_{l \in [L]} S_l$ and $\sigma(j) := \sigma_l(j)$ for any $j \in V$, where $l \in [L]$.

Output the solution (S, σ).

3.3 The Phase of Modifying the Assignments

In the third phase, we modify the assignments of each connected solution G_l and obtain a solution (S, σ) that satisfies both the budget and parity constraints. As can be seen in the Algorithm 3, we still consider 5 cases to deal with the parity constraints. The Algorithm 4 is a sub-algorithm of the Algorithm 3 in the modification of cases (C), (D), and (E), and this idea of modification is first presented by Kim et al. [16].

Algorithm 4:. Find reassigning paths on tree $\mathcal{T}_l$ [16].

Input: A tree $\mathcal{T}_l = (V(\mathcal{T}_l), E(\mathcal{T}_l))$, a vertex set $S_l^{\text{inv}} \subseteq V(\mathcal{T}_l)$ consists of centers whose parity constraint is unsatisfied.

Output: A group of reassigning paths.

For the given tree $\mathcal{T}_l$ consists of a subset of vertices from V_l, denote $\mathcal{T}_l(u)$ as a subtree that is rooted at $u \in V(\mathcal{T}_l)$, denote $\mathcal{T}_l^2$ as the 2-th power graph of $\mathcal{T}_l$. Reassigning paths are node-disjoint paths on $\mathcal{T}_l^2$ between invalid facilities, and exposed paths are node-disjoint paths on $\mathcal{T}_l^2$ with one and only one of its endpoints invalid.

for each subtree $\mathcal{T}_l(u)$ **do**
  if u is one of $\mathcal{T}_l$'s leaves **then**
    If u's parity constraint isn't satisfied, identify $p = (u)$ as an exposed path within $\mathcal{T}_l$.
  else
    Collect the exposed paths from each subtree $\mathcal{T}_l(c)$ of $\mathcal{T}_l(u)$, where c is the child of u. Pair each two of the exposed paths arbitrarily by connecting the valid endpoints. Extend the remaining exposed path(if any) to u to form a new exposed path.
  Update u to be its father node in $\mathcal{T}_l$.

Lemmas 5 and 6 demonstrate the performance of the Algorithm 3 under different cases.

Lemma 5. *For* cases (A) *and* (B), *we have that* $d(i, \sigma(i)) \leq 6\text{OPT}$ *for each* $i \in V_l$.

Lemma 6. *For* cases (C), (D), *and* (E), *we have that* $d(i, \sigma(i)) \leq 10\text{OPT}$ *for each* $i \in V_l$.

Combining Lemmas 5 and 6 yields our main result.

Theorem 1. *There exists a* 10*-approximation algorithm for the PARW* k*-center problem.*

References

1. An, H.C., Bhaskara, A., Chekuri, C., Gupta, S., Madan, V., Svensson, O.: Centrality of trees for capacitated k-center. Math. Program. **154**(1), 29–53 (2015)
2. Bajpai, T., Chakrabarty, D., Chekuri, C., Negahbani, M.: Revisiting priority k-center: fairness and outliers. In: Proceedings of the International Colloquium on Automata, Languages, and Programming, No. 21, pp. 21:1–21:20 (2021)
3. Bateni, M., Esfandiari, H., Fischer, M., Mirrokni, V.: Extreme k-center clustering. In: Proceedings of the AAAI Conference on Artificial Intelligence, vol. 35, pp. 3941–3949 (2021)
4. Chakrabarty, D., Goyal, P., Krishnaswamy, R.: The non-uniform k-center problem. ACM Trans. Algorithms **16**(4), 1–19 (2020)
5. Chen, X., Ji, S., Wu, C., Xu, Y., Yang, Y.: An approximation algorithm for diversity-aware fair k-supplier problem. Theor. Comput. Sci. 114305 (2023)
6. Chen, X., Xu, D., Xu, Y., Zhang, Y.: Parameterized approximation algorithms for sum of radii clustering and variants. In: Proceedings of the AAAI Conference on Artificial Intelligence, vol. 38, no. 18, pp. 20666–20673 (2024)
7. Frank, A., Király, Z.: Graph orientations with edge-connection and parity constraints. Combinatorica **22**, 47–70 (2002)
8. Gonzalez, T.F.: Clustering to minimize the maximum intercluster distance. Theoret. Comput. Sci. **38**, 293–306 (1985)
9. Grötschel, M., Lovász, L., Schrijver, A.: The ellipsoid method and its consequences in combinatorial optimization. Combinatorica **1**, 169–197 (1981)
10. Grötschel, M., Lovász, L., Schrijver, A.: Corrigendum to our paper "the ellipsoid method and its consequences in combinatorial optimization". Combinatorica **4**(4), 291–295 (1984)
11. Han, L., Xu, D., Xu, Y., Yang, P.: Approximation algorithms for the individually fair k-center with outliers. J. Global Optim. **87**(2), 603–618 (2023)
12. Hochbaum, D.S., Shmoys, D.B.: A best possible heuristic for the k-center problem. Math. Oper. Res. **10**(2), 180–184 (1985)
13. Hochbaum, D.S., Shmoys, D.B.: A unified approach to approximation algorithms for bottleneck problems. J. ACM **33**(3), 533–550 (1986)
14. Khuller, S., Pless, R., Sussmann, Y.J.: Fault tolerant k-center problems. Theoret. Comput. Sci. **242**(1–2), 237–245 (2000)
15. Khuller, S., Sussmann, Y.J.: The capacitated k-center problem. SIAM J. Discret. Math. **13**(3), 403–418 (2000)
16. Kim, K., Shin, Y., An, H.C.: Constant-factor approximation algorithms for parity-constrained facility location and k-center. Algorithmica **85**(7), 1883–1911 (2023)
17. Kleindessner, M., Awasthi, P., Morgenstern, J.: Fair k-center clustering for data summarization. In: Proceedings of the International Conference on Machine Learning, pp. 3448–3457 (2019)
18. Lanthaler, M., Lechner, W.: Minimal constraints in the parity formulation of optimization problems. New J. Phys. **23**(8), 083039 (2021)
19. Lim, A., Rodrigues, B., Wang, F., Xu, Z.: k-center problems with minimum coverage. Theoret. Comput. Sci. **332**(1–3), 1–17 (2005)
20. Tong, P., Lawler, E.L., Vazirani, V.V.: Solving the weighted parity problem for gammoids by reduction to graphic matching. In: Progress in Combinatorial Optimization, pp. 363–374 (1984)
21. Xu, Y., Chau, V., Wu, C., Zhang, Y., Zissimopoulos, V., Zou, Y.: A semi brute-force search approach for (balanced) clustering. Algorithmica **86**(1), 130–146 (2024)

Optimal Bridge, Twin Bridges and Beyond: Inserting Edges into a Road Network to Minimize the Constrained Diameters

Zhidan Feng[1], Henning Fernau[1], and Binhai Zhu[2(✉)]

[1] University of Trier, FB 4 – Informatikwissenschaften, 54286 Trier, Germany
zhidanfeng@mail.sdu.edu.cn, fernau@uni-trier.de

[2] Gianforte School of Computing, Montana State University, Bozeman, MT 59717, USA
bhz@montana.edu

Abstract. Given a road network modelled as a planar straight-line graph $G = (V, E)$ with $|V| = n$, let $(u, v) \in V \times V$, the shortest path (distance) between u, v is denoted as $\delta_G(u, v)$. Let $D(G) = \max_{(u,v)} \delta_G(u, v)$, for $(u, v) \in V \times V$, which is called the diameter of G. Given a disconnected road network modelled as two disjoint trees T_1 and T_2, this paper first aims at inserting one or two edges (bridges) between them to minimize the (constrained) diameter $D(T_1 \cup T_2 \cup I_j)$ going through the inserted edges, where $I_j, j = 1, 2$, is the set of inserted edges with $|I_1| = 1$ and $|I_2| = 2$. The corresponding problems are called the *optimal bridge* and *twin bridges* problems. Since when more than one edge are inserted between two trees the resulting graph becomes more complex, for the general network G we consider the problem of inserting a minimum of k edges such that the shortest distances $\delta_G(u_i, v_i)$ between a given set of m pairs of points $P = \{(u_i, v_i) \mid u_i, v_i \in V, i \in [m]\}$ are all decreased.

The main results of this paper are summarized as follows:

- We show that the optimal bridge problem can be solved in $O(n^2)$ time and that a variation of it has a near-quadratic lower bound unless SETH fails. The proof also implies that the famous 3-SUM problem does have a near-quadratic lower bound for large integers, e.g., each of the n input integers has $\Omega(\log n)$ decimal digits. We then give a simple factor-2 $O(n \log n)$-time approximation algorithm for the optimal bridge problem.
- We present an $O(n^4)$-time algorithm to solve the twin bridges problem, exploiting some new property not existing in the optimal bridge problem.
- For the general problem of inserting k edges to reduce the (graph) distances between m given pairs of points, we show that the problem is NP-complete — even if the given and resulting graphs are planar and the inserted edges must have a bounded length.

Keywords: Road networks · Optimal bridge · Geometric algorithms · Fine-grained complexity · NP-completeness

S. Ghosh and Z. Zhang (Eds.): AAIM 2024, LNCS 15179, pp. 94–108, 2024.
https://doi.org/10.1007/978-981-97-7798-3_9

1 Introduction

Geometric spanners have received a lot of attention since 1986 [8]. In the majority of the literature, the problem is to construct a geometric (or metric) graph G from scratch such that the maximum stretch factor (or dilation) of G is minimized (or approximated). Recently, a breakthrough result of Gudmundsson and Wong proved that by inserting k edges greedily into a connected metric graph M, the dilation of the resulting graph can be approximated with a factor of $O(k)$ [10].

In this paper, we consider a similar problem of inserting edges to an existing road network G (modelled as a planar straight-line graph of n vertices) such that the diameter of the resulting graph is minimized/reduced. We start from the basic problem of inserting one and two edges into two disjoint trees T_1 and T_2 to minimize the constrained diameter, going through the inserted edges, of the resulting graphs, which are called the *optimal bridge* and *twin bridges* problems. These problems could appear in some scenario, for example, in the recent work by Higashikawa et al., where the vertices of G could be in red and black and a path must be planned with no red-red edge (i.e., no edges whose vertices are both red) [3]. Naturally, such a path could be planned by deleting all red-red edges in G, possibly resulting in disjoint components (disjoint trees or a forest when G is a tree). Then, to ensure connectivity and minimizing/reducing the dilation some non-red-red edges much be inserted — even though how to do that is left open.

Coming back to the optimal bridge and twin bridges problems, we present $O(n^2)$ and $O(n^4)$ time algorithms to solve them. For the former, we show that a variation of it, the one-bridge decision problem, has a near-quadratic lower bound unless the Strong Exponential Time Hypothesis (SETH) fails. Our method modifies the traditional one by Williams [19] and it can be used to prove that the famous 3-SUM problem has a near-quadratic lower bound when the n input integers are large, i.e., each has $\Omega(\log n)$ digits. Then, a simple $O(n \log n)$ time approximation algorithm is designed achieving a factor of two (the factor is tight). Consequently, this approximation algorithm can be used as a subroutine to approximate the more general problem of connecting $k+1$ trees with k edges into a tree T so as to minimize the diameter of T, with a factor of 4 and a running time of $O(n \log n)$. For the twin bridges problem, not surprisingly, when two edges are added between T_1 and T_2, the resulting graph is not a tree anymore. Hence we need to consider a new scenario that never occurs in the optimal bridge problem.

On the other hand, from a pure geometric setting, the optimal bridge problem is well-studied when the input is a pair of polygons. The problem was first studied by Cai, Xu and Zhu in 1999 for the case where the input is a pair of convex polygons P and Q, with a total of n vertices [7]. (In this case, the problem is to construct a connecting segment pq between P, Q such that $p \in P$ and $q \in Q$ and $\max_{x \in P} |xp| + |pq| + \max_{y \in Q} |qy|$ is minimized. Here, $|xp|$ is the Euclidean distance between points x and p.) The problem was solved in $O(n^2 \log n)$ time and was subsequently improved to $O(n)$ by using geometric properties [4,12,16]. When the polygons P, Q are not necessarily convex, the problem is to find a connecting segment pq between P, Q such that $p \in P$ and $q \in Q$ to minimize $\max_{x \in P} \delta_P(x,p) + |pq| + \max_{y \in Q} \delta_Q(q,y)$. (Here $\delta_P(x,p)$ is the geodesic distance between x and p in polygon P.) For this problem, there are two versions: (1) p, q are visible to each other, and (2) p, q do not have to be visible to each other. For the first case, the problem can be solved in $O(n^2)$ time by Kim and Shin [12], improved to $O(n \log^3 n)$ time by Tan [17]. For the second case, Bhosle and Gonzalez gave an $O(n^2 \log n)$ time algorithm [5] (and they solved a variation when p, q could be connected by one or more segments in the same amount of time [6]). When P, Q are rectilinear and p, q are connected by one or two rectilinear segments, Wang gave an optimal $O(n)$ time algorithm [18].

Also in the geometric setting, when P, Q are both convex, a simple $O(n)$ time greedy 2-approximation algorithm was given by Cai, Xu and Zhu [7]. The algorithm is simply to connect two closest points between P and Q. Slightly later, Ahn, Cheong and Shin [1] gave a factor-$\sqrt{2}$ approximation by connecting the centers of P and Q (i.e., the centers of the circumscribing circles of P and Q), which can be used to obtain a factor-$2\sqrt{2}$ approximation to connect $k+1$ convex polygons by inserting k edges to minimizing the (geodesic) diameter. In our case, we show that the greedy algorithm by Cai, Xu and Zhu can be adapted to the road network setting to achieve a factor-2 approximation and the factor is tight. The algorithm can also be applied to obtain a factor-4 approximation to connect $k+1$ trees into a tree T by inserting k edges so as to minimize the diameter of T.

As discussed earlier, when two or more edges are added between T_1 and T_2, the resulting graph is not a tree anymore. Hence, algorithms must make use of some new properties. We prove such properties and present an $O(n^4)$ time algorithm to solve the problem.

On the other hand, it is noted that when two or more edges are inserted to connect T_1 and T_2 it might be possible that it cannot reduce the diameter further — regardless of if the second edge is inserted between T_1 and T_2 or between the vertices in $T_i, i \in \{1,2\}$. (An example is when T_1 and T_2 are two disjoint segments on the same line.) Hence, the problem here is more on inserting edges to a connected graph to minimize/reduce the diameter, closer to the setting by Gudmundsson and Wong [10].

Finally, we investigate the complexity of the following problem: given a connected road network $G = (V, E)$ and a set of pairs $P = \{(u,v) \mid u, v \in V\}$, insert k edges such that the distances $\delta_G(u,v)$, with $(u,v) \in P$, can all be reduced. We prove that this problem is NP-complete by a reduction from Vertex Cover on planar 2-connected cubic graphs [13].

Due to space constraints, some details are left out but can be found in the full version posted on arXiv.

2 Preliminaries

We define a road network as a (weighted) planar straight-line graph $G = (V, E)$. Each vertex (node) $p \in V$ is a point in the plane with coordinates $p = (x(p), y(p))$. The weight between two points p, q is their Euclidean distance which is denoted as $|pq|$, and is equal to $\sqrt{(x(p) - x(q))^2 + (y(p) - y(q))^2}$. Given $u, v \in V$, the shortest path between them on G is denoted as $\delta_G(u, v)$. (We also slightly abuse the notation to use $\delta_G(u, v)$ as the length of the corresponding shortest path.) The diameter of G is defined as $\max_{u,v \in V} \delta_G(u, v)$.

A tree $T = (V(T), E(T))$ is a connected road network with no cycle. As well-known facts, between $u, v \in V(T)$ the shortest path $\delta_T(u, v)$ is unique. Also, if $|V(T)| = n$, then the shortest path tree from any node u to all other nodes can be computed in $O(n)$ time by running the breadth-first search (BFS) algorithm starting at u. Consequently, the all-pair shortest paths problem on T can be solved in $O(n^2)$ time; moreover, when this information is stored as a table, the length of $\delta_T(u, v)$ can be returned in $O(1)$ time. In fact, $\max_{x \in V(T)} \delta_T(u, x)$, the maximum distance from $u \in V(T)$ to a (leaf) node x, can be computed in $O(n)$ time by looking at the row/column maximum; moreover, once stored, this distance can be queried in $O(1)$ time. Throughout the paper, we assume such a table on all-pair shortest path distances is associated with the corresponding tree.

3 The Optimal Bridge Problem

The optimal bridge problem is defined as follows: given two tree road networks (trees for short) T_1 and T_2, add an edge (p, q), with $p \in V(T_1)$, $q \in V(T_2)$, such that the *(constrained) diameter* through pq of the resulting graph $T_1 \cup T_2 \cup \{(p, q)\}$, i.e., the maximum of $\delta_{T_1}(x, p) + |pq| + \delta_{T_2}(q, y)$ with $x \in V(T_1)$ and $y \in V(T_2)$, is minimized.

An $O(n^2)$ time algorithm is easy to design to solve the optimal bridge problem:

1. For each pair, $p \in V(T_1)$, $q \in V(T_2)$, record
$$\max_{x \in V(T_1)} \delta_{T_1}(x, p) + |pq| + \max_{y \in V(T_2)} \delta_{T_2}(q, y).$$
2. Return the minimum distance recorded and then compute the corresponding path between x and y.

It is intriguing to know if this $O(n^2)$ time bound can be further improved. For the (pure) geometric version where two polygons with vertices n are given and the bridge must be a segment pq where p is visible from q, an $O(n \log^3 n)$ time algorithm is known [17].

We define the *one-bridge* problem as the decision version of the optimal bridge: given constants C_1 and C_2, determine if a bridge pq exists such that $|pq| = C_1$ and its *specific* solution value $\delta_{T_1}(x, p) + |pq| + \delta_{T_2}(q, y) = C_2$, where

x is a leaf in T_1 and y is a leaf in T_2. Clearly, one-bridge is a variation of the optimal bridge problem. We show next that the one-bridge problem cannot be solved in $O(n^{2-\epsilon})$ time unless the SETH (Strong Exponential Time Hypothesis) fails. Thus, even though this does not give a similar lower bound for the optimal bridge problem, it does give some evidence that it is probably hard to improve the $O(n^2)$-time bound for the optimal bridge problem.

3.1 A Near-Quadratic Lower Bound for the One-Bridge Problem

We prove the conditional lower bound by considering a variation of Williams' Orthogonal Vector (OV) problem [19], which we call *Complementary Orthogonal Vectors* problem (COV for short).

Given a binary (0/1) vector with d dimensions, $v = (v_1, v_2, \ldots, v_d)$, define (with $\bar{0} = 1$ and $\bar{1} = 0$) $\bar{v} = (\bar{v}_1, \bar{v}_2, \ldots, \bar{v}_d)$. We say that v and $\bar{v}$ are *complementary* to each other. Clearly, v and $\bar{v}$ are orthogonal, i.e., the dot product satisfies $v \cdot \bar{v} = 0$. Roughly, in Williams' case, we given a k-SAT instance ϕ' with a set of n variables $A \cup B$, with $|A|$ and $|B|$ being (neglecting rounding) of the same size $n/2$. Then two sets of binary vectors in d-dimension, A' and B', which correspond to how a partial assignment of A and B satisfy the d clauses in ϕ', are constructed. Here the j-th component of a d-vector $u \in A'$ is 0 if the partial assignment of A can satisfy the j-th clause in ϕ', and 1 otherwise. (A vector in B' is similarly defined.) Then, Williams' problem is to find if there are binary vectors $u \in A'$ and $v \in B'$ such that they are orthogonal, i.e., $u \cdot v = 0$. In our *Complementary Orthogonal Vectors* (COV) problem, which could be of independent interest, we have two sets of *ternary* vectors A and B such that each entry in a vector of A and B is an element of $\{0, 1, 2\}$, and we want to find a pair of binary vectors $u \in A$ and $v \in B$ such that they are complementary.

To modify William's proof, we make the following changes: (1) Use One-in-three SAT instead of k-SAT; (2) Modify the definition for constructing the sets of d-vectors.

Let ϕ be an One-in-three SAT instance composed of n variables and m disjunctive clauses where the i-th clause F_i contains three literals and is in the form of $(x_{i,1} \vee x_{i,2} \vee x_{i,3})$. The problem is to determine, for $i = 1..m$, exactly one of the three literals in each clause F_i, i.e., $x_{i,1}, x_{i,2}$ and $x_{i,3}$, which is assigned TRUE. One-in-three SAT is a well-known NP-complete problem with $m = \Omega(n)$ [15].

We arbitrarily partition the variables in ϕ into two equal-sized parts V_A and V_B (we can assume that n is even, though it does not really matter for the result). Each of the m-vectors $u \in A$ is determined by an assignment α_A of V_A (i.e., a partial assignment of the variables in ϕ), where

$$u = (u_1, u_2, \cdots, u_m),$$

and

$$u_i = \begin{cases} 0 & \text{if } F_i \text{ is satisfied with exactly one TRUE literal by } \alpha_A, \\ 1 & \text{if } F_i \text{ is not satisfied by } \alpha_A, \\ 2 & \text{if } F_i \text{ is satisfied with at least two TRUE literals by } \alpha_A. \end{cases} \quad (1)$$

Similar to (1), we could define an m-vector $v \in B$ determined by an assignment α_B of V_B. In a preprocessing, we could remove all non-binary vectors from A and B, but this does not affect the lower bound proof (for the worst case). Hence we will stick with the ternary vectors as input for COV.

Then, similar to Williams' idea, we can claim that ϕ has a valid truth assignment if and only if there are vectors $u \in A$ and $v \in B$ which are complementary (i.e., COV has a solution). As there are $2^{n/2}$ assignments for V_A and V_B respectively, the above reduction takes $2^{n/2} \cdot O(m)$ time. If COV could be computed in $O(N^{2-\epsilon})$ time, where N is the input size for COV, One-in-three SAT could be solved in $2^{n-n\epsilon/2} \cdot O(m^{2-\epsilon})$ time — which would contract the SETH. We hence have the following theorem.

Theorem 1. *The Complementary Orthogonal Vectors problem with input size N cannot be solved in $O(N^{2-\epsilon})$ unless the SETH fails.*

We next reduce COV to the one-bridge problem as follows: Let the input for COV be two sets A and B, each containing n_1 and n_2 (0/1/2)-vectors in m-dimensional space, respectively, with $n_1 + n_2 = n$, Hence the total size of COV is $N = mn$. For each m-vector $u \in A$, with $u = (u_1, u_2, \cdots, u_m)$, we construct a path L_u of segment lengths ℓ_i, i.e., $L_u = (\ell_1, \ell_2, \cdots, \ell_m)$, where

$$\ell_i = \begin{cases} \frac{1}{3^{i-1}} & \text{if } u_i = 0, \\ 0 & \text{if } u_i = 1, \\ 4 & \text{if } u_i = 2. \end{cases} \tag{2}$$

Here ℓ_i is the length of the i-th segment of L_u. (The fact that $\ell_i = 0$ means that we could have duplicated geometric points on L_u.) Similarly, given an m-vector $v \in B$, with $v = (v_1, v_2, \cdots, v_m)$, we can construct a path L_v of segment lengths ℓ'_i, i.e., $L_v = (\ell'_1, \ell'_2, \cdots, \ell'_m)$, with

$$\ell'_i = \begin{cases} \frac{1}{3^{i-1}} & \text{if } v_i = 0, \\ 0 & \text{if } v_i = 1, \\ 4 & \text{if } v_i = 2. \end{cases} \tag{3}$$

Let $C_1 = 0$ and $C_2 = C + 2$ with

$$C = 1 + \left(\frac{1}{3}\right) + \left(\frac{1}{3}\right)^2 + \cdots + \left(\frac{1}{3}\right)^{m-1} = \frac{3}{2}\left(1 - \left(\frac{1}{3}\right)^m\right). \tag{4}$$

Then we put these straight paths L_u's not containing segment lengths of four downward along the Y-axis starting at $(0, C)$. We also add another node $(0, C+1)$ and connect it to $(0, C)$. For L_u's which contain a segment of length 4, we just convert each of them into a path centered at $(0, C)$, in a star fashion. This would give us T_1. Similarly, we put the straight path L_v's not containing a segment of length 4, each corresponding to a binary vector $v \in B$, upward and starting at $(0, 0)$ and we also add another node $(0, -1)$ and connect it to $(0, 0)$. For L_v's

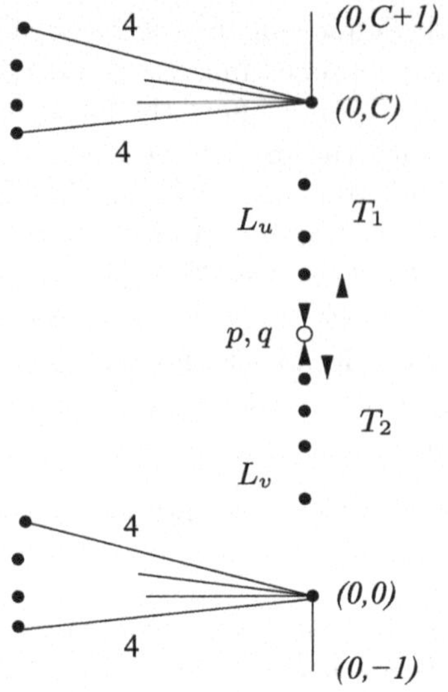

Fig. 1. An example for the reduction from COV to the one-bridge decision problem. The vertical paths, which should all be on the Y-axis, are drawn for a better visualization.

containing a segment of length 4, we put them in a star centered at $(0,0)$. This gives us the second tree T_2. See Fig. 1 for an example.

This reduction obviously takes $O(N) = O(mn)$ time, where the total number of points in T_1 and T_2 is $O(N)$. Finally, we claim that COV has a solution if and only if between T_1 and T_2 there is a bridge pq with length $C_1 = 0$ and with a specific solution value of $C_2 = C + 2$. We prove this "iff" relation next, focusing on the "if-part" as the "only-if" part is trivial. We first prove the following lemma regarding the geometric sequence $(1, \frac{1}{3}, \frac{1}{3^2}, \cdots, \frac{1}{3^{m-1}})$. Note that as $C < \frac{3}{2}$, $C_2 < 3.5$. Hence, we can ignore any path with a segment length of 4 in our proof, as they would incur a specific solution of value at least 4, which is greater than C_2.

Lemma 1. *For* $0 \le i < m-1$, $\left(\frac{1}{3}\right)^i > \sum_{j=i+1..m-1} \left(\frac{1}{3}\right)^j$.

Proof. By calculation,

$$\sum_{j=i+1..m-1} \left(\frac{1}{3}\right)^j = \left(\frac{1}{3}\right)^{i+1} \cdot \frac{3}{2}\left(1 - \left(\frac{1}{3}\right)^{m-i-1}\right) < \frac{1}{2} \cdot \left(\frac{1}{3}\right)^i < \left(\frac{1}{3}\right)^i.$$

□

We then prove the "if-part" by proving its contrapositive: if COV does not have a solution, then between T_1 and T_2 there does not exist a bridge pq with length $C_1 = 0$ and with a specific solution value of $C_2 = C + 2$. To finish this proof, we need the following two lemmas.

Lemma 2. *Let $u \in A, v \in B$; moreover, $u = (u_1, u_2, \cdots, u_m)$ and $v = (v_1, v_2, \cdots, v_m)$ with $u_i, v_i \in \{0, 1\}$. If there exists an i with $u_i = v_i = 0$ and for all $k < i$ exactly one of u_k and v_k is equal to zero, then the sum of segment lengths in L_u and L_v is greater than C.*

Proof. Here i is the smallest index such that $u_i = v_i = 0$ and for all $k < i$ exactly one of u_k and v_k is equal to zero. Then, by definition, both $\ell_i = \left(\frac{1}{3}\right)^{i-1}$ and $\ell'_i = \left(\frac{1}{3}\right)^{i-1}$ would contribute to the sum of segment lengths in L_u and L_v. Following Lemma 1,

$$\left(\frac{1}{3}\right)^{i-1} > \sum_{j=i..m-1} \left(\frac{1}{3}\right)^{j}.$$

In other words, the value of this additional copy of $\left(\frac{1}{3}\right)^{i-1}$ is greater than the sum of all possible shorter segment lengths, each appearing exactly once, which is necessary to achieve a sum of C. Hence there is no bridge pq with a length of zero connecting T_1 and T_2, and the lemma is proven. □

Lemma 3. *Let $u \in A, v \in B$; moreover, $u = (u_1, u_2, \cdots, u_m)$ and $v = (v_1, v_2, \cdots, v_m)$ with $u_i, v_i \in \{0, 1\}$. If there exists an i with $u_i = v_i = 1$ and if, for all $k < i$, exactly one of u_k and v_k is equal to zero, then the sum of segment lengths in L_u and L_v is smaller than C.*

Proof. Symmetrically, let i be the smallest index such that $u_i = v_i = 1$ and, for all $k < i$, exactly one of u_k and v_k is equal to zero. Again, by definition, both $\ell_i = \left(\frac{1}{3}\right)^{i-1}$ and $\ell'_i = \left(\frac{1}{3}\right)^{i-1}$ would not contribute to the sum of segment lengths in L_u and L_v. Again, by the proof of Lemma 1,

$$\frac{1}{2} \cdot \left(\frac{1}{3}\right)^{i-1} > \sum_{j=i..m-1} \left(\frac{1}{3}\right)^{j}.$$

In other words, for all possible segment lengths shorter than $\left(\frac{1}{3}\right)^{i-1}$, each possibly appearing twice in L_u as well as in L_v, their sum would not make up for $\left(\frac{1}{3}\right)^{i-1}$ which is needed to have a sum of C for the segments in L_u and L_v. Again, there cannot be a bridge pq with a length of zero connecting T_1 and T_2. □

Theorem 2. *Given two trees with a total of N vertices and non-negative constants C_1 and C_2, the one-bridge decision problem cannot be solved in $O(N^{2-\epsilon})$ time unless the SETH fails. This holds even if $C_1 = 0$.*

Proof. As claimed earlier, COV has a solution if and only if between T_1 and T_2 there is a bridge pq with length $C_1 = 0$ and with a specific solution value of $C_2 = C + 2$. We finish to prove this "iff" relation, focusing on the "if-part" as the "only-if" is trivial.

Suppose that there is a bridge pq between T_1 and T_2 with length $C_1 = 0$ and with a specific solution value of $C_2 = C + 2$. Since $C < 1.5$ by Eq. (4), we have

$C+2 = C_2 < 3.5$. Hence, the only possible specific solution value C_2 in the one-bridge instance is the (tree) distance between $(0, C+1)$ and $(0, -1)$, as all non-vertical segments have length 4. As $C_1 = 0$ (i.e., there is a bridge of length zero between T_1 and T_2), the sum of segment lengths in L_u and L_v, which are paths in T_1 and T_2 respectively, must be exactly C. Following Lemma 2 and Lemma 3, for that to happen, each component in the geometric sequence $(1, \frac{1}{3}, \frac{1}{3^2}, \cdots, \frac{1}{3^{m-1}})$ must appear exactly once in L_u and L_v. Recall that the total number of points in T_1 and T_2 is $O(N)$. Then, if the one-bridge problem could be solved in $O(N^{2-\epsilon})$ time, the Complementary Orthogonal Vectors problem would also be solvable in $O(N^{2-\epsilon})$ time. But this is a contradiction to Theorem 1, asumming SETH. □

3.2 Implication to the 3-SUM Conjecture

The 3-SUM problem is defined as follows: given a set S of n integers, decide if there are $a, b, c \in S$ such that $a + b + c = 0$. The problem was initially posed by Gajentaan and Overmars [9], who conjectured that the problem has a lower bound of $\Omega(n^2)$. Then the problem was solved in $o(n^2)$ time by Baran et al. [2] (later even for real numbers, see [11]). Hence the new conjecture for 3-SUM is by Patrascu, which is that 3-SUM has a lower bound of $\Omega(n^{2-o(1)})$ [14]. We show that when the integers in S are large, i.e., each one has $\Omega(\log n)$ decimal digits, then the 3-SUM conjecture is true (in fact, this almost matches the original conjecture by Gajentaan and Overmars).

Theorem 3. *The 3-SUM problem with N large integers, each with $\Omega(\log N)$ decimal digits, cannot be solved in $O(N^{2-\epsilon})$ time unless the SETH fails.*

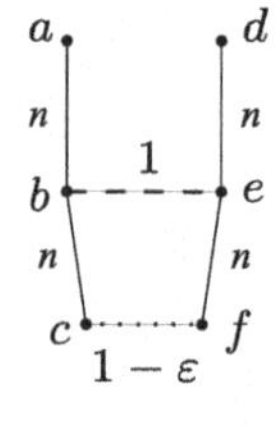

Fig. 2. An example where two paths (a, b, c) and (d, e, f) need to be connected into a spanning tree. The edges be and cf are positioned by rotating a segment of length n at b and e slightly to achieve that $|cf| = 1 - \epsilon$. Adding (b, e) would result in an optimal diameter of $2n + 1$; with the greedy method of adding the shortest edge between the two paths, (c, f) is added to achieve a diameter of $4n + 1 - \varepsilon$. (The distances in the figure are not measured geometrically.)

3.3 A Simple Approximation

Even though we cannot solve the one-bridge problem in $o(n^2)$ time (neither we could for the optimal bridge problem so far), a subquadratic approximation algorithm for the optimal bridge problem can be easily designed as follows.

1. Compute the closest pair (p, q), where $p \in V(T_1)$ and $q \in V(T_2)$.
2. Return $\max_{x \in V(T_1)} \delta_{T_1}(x, p) + |pq| + \max_{y \in V(T_2)} \delta_{T_2}(q, y)$ and the corresponding path between x and y.

The closest pair between $V(T_1)$ and $V(T_2)$ can be computed in $O(n \log n)$ time using a standard method, say, Voronoi diagrams. Note that $\delta_{T_1}(x, p)$ (resp. $\delta_{T_2}(q, y)$) can be computed in linear time by running BFS starting at p (resp. q) on the tree T_1 (resp. T_2); i.e., we do not need the $O(n^2)$ time preprocessing as for the exact algorithm.

Theorem 4. *The optimal bridge problem can be solved in $O(n^2)$ time and can be approximated in $O(n \log n)$ time with a factor of 2 (and the factor is tight).*

Proof. The running times are straightforward. For the approximation factor claim, we argue as follows. Let OPT be the optimal solution value, (p^*, q^*) be the optimal bridge and (p, q) be the approximate solution computed. Also, let $x^* \in V(T_1)$ and $y^* \in V(T_2)$ be chosen such that $\delta_{T_1}(x, p^*) + |p^*q^*| + \delta_{T_2}(q^*, y)$ is maximized. Then,

$$\max_{x \in V(T_1)} \delta_{T_1}(x, p) \leq D(V(T_1)),$$

where $D(V(T_1))$ is the diameter of the tree T_1. Similarly, $\max_{y \in V(T_2)} \delta_{T_2}(q, y) \leq D(V(T_2))$. On the other hand, we have $\text{OPT} = \delta_{T_1}(x^*, p^*) + |p^*q^*| + \delta_{T_2}(q^*, y^*)$. Also, there are some $\zeta^*, \eta^* \in V(T_1)$ such that, by the choice of x^*,

$$D(V(T_1)) = \delta_{T_1}(\zeta^*, \eta^*) \leq \delta_{T_1}(\zeta^*, p^*) + \delta_{T_1}(p^*, \eta^*) \leq 2 \cdot \delta_{T_1}(x^*, p^*).$$

Then, $\delta_{T_1}(x^*, p^*) \geq D(V(T_1))/2$ and (similarly) $\delta_{T_2}(q^*, y^*) \geq D(V(T_2))/2$. Consequently, the approximation solution value APP satisfies

$$\begin{aligned} \text{APP} &= \max_{x \in V(T_1)} \delta_{T_1}(x, p) + |pq| + \max_{y \in V(T_2)} \delta_{T_2}(q, y) \\ &\leq D(V(T_1)) + |pq| + D(V(T_2)) \\ &\leq 2\delta_{T_1}(x^*, p^*) + |p^*q^*| + 2\delta_{T_2}(q^*, y^*) \\ &\leq 2 \cdot \text{OPT}. \end{aligned}$$

To see the tightness of the approximation factor, we refer to Fig. 2. In this figure, the approximate bridge is (c, f), giving a solution value of $4n + 1 - \varepsilon$. The optimal bridge is (b, e), resulting a solution value of $2n + 1$. Clearly, we have $\lim_{n \to \infty} \frac{4n+1-\varepsilon}{2n+1} = 2$. □

We note that for the geometric version when two convex polygons are given, the approximation algorithm can make use of the centers of the polygons; hence the factor can be improved to $\sqrt{2}$; see [1]. Nonetheless, if we need to connect $k + 1$ tree road networks using k edges to minimize the diameter of the resulting tree, combining our approximation with a method to [1] can be used to obtain a factor of 4.

Corollary 1. *There is a 4-approximate algorithm in $O(kn \log n)$ time to connect $k + 1$ trees into a tree T by using k edges to minimize the diameter of T.*

4 The Twin Bridges Problem

In this section, we focus on the twin bridges problem of two road networks which are given as disjoint (geometric) trees. Given two disjoint trees T_1 and T_2 with a total of n vertices (which are points in the plane), we aim to add pairs (p_1, q_1) and (p_2, q_2) with $p_i \in V(T_1), q_i \in V(T_2), i \in \{1, 2\}$, and $p_1 \neq p_2, q_1 \neq q_2$, as new edges/bridges into T_1 and T_2 to form a road network $T'' = T_1 \cup T_2 \cup \{(p_1, q_1), (p_2, q_2)\}$, where the goal is to minimize the maximum distance between nodes of T'' whose corresponding path must pass through one or two of these bridges. We loosely call this minimum distance the *constrained* diameter of T''. See Fig. 3 for an example.[1] First of all, notice that T'' is not a tree anymore. The constrained diameter of T'' can appear in four cases:

1. It is the maximum of $\delta_{T_1}(x, p_1) + |p_1 q_1| + \delta_{T_2}(q_1, y)$, with $x \in V(T_1)$ and $y \in V(T_2)$;
2. It is the maximum of $\delta_{T_1}(x, p_2) + |p_2 q_2| + \delta_{T_2}(q_2, y)$, with $x \in V(T_1)$ and $y \in V(T_2)$;
3. It is the maximum of $\delta_{T_1}(x, p_1) + |p_1 q_1| + \delta_{T_2}(q_1, q_2) + |q_2 p_2| + \delta_{T_1}(p_2, z)$, with $x, z \in V(T_1)$; and
4. It is the maximum of $\delta_{T_2}(y, q_1) + |q_1 p_1| + \delta_{T_1}(p_1, p_2) + |p_2 q_2| + \delta_{T_2}(q_2, w)$, with $y, w \in V(T_2)$.

Note that Case 1 and Case 2 are similar to the optimal bridge problem in some way; in fact, it is a matter of partitioning T_i into $T_{i,1}$ and $T_{i,2} (i = 1, 2)$ by deleting an edge in T_i such the two pairs $(T_{i,k}, T_{j,l})$, with $\{i, j\} = \{1, 2\}$ and $k, l \in \{1, 2\}$, form instances of the optimal bridge problem — with the resulting optimal bridges $p_1 q_1$ and $p_2 q_2$ respectively. (See also the next lemma for the details.) On the other hand, Case 3 and Case 4 are to improve the diameter

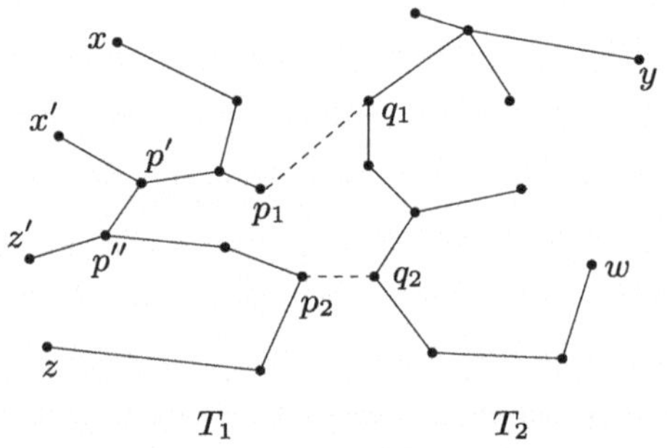

Fig. 3. An illustration for the twin bridges problem.

[1] Note that since T_1 and T_2 are initially disconnected, at least one edge/bridge between T_1 and T_2 must be added to make a connected tree. Consequently, after the first bridge is added, if we do not require the second edge to be a bridge between T_1 and T_2, then adding the second edge in T_1 or T_2 is a trivial problem. This case is not included in our twin bridges problem.

of T_1 (resp. T_2) using both the bridges p_1q_1 and p_2q_2. (In the optimal bridge problem, it is impossible to improve the diameter of T_1 and T_2 using the bridge pq. The reason is that pq is a cut edge between T_1 and T_2, and going from T_1 through pq will lead one to T_2 or vice versa.)

We refer to Fig. 3, and, assuming a proper partitioning for T_1 and T_2 is known as above and the pairing of $(T_{1,k}, T_{2,l})$ and $(T_{1,l}, T_{2,k})$ is fixed. Then for $x, p_1 \in T_{1,k}$, $p_2 \in T_{2,l}$, and $y, q_1 \in T_{2,l}$ and $q_2 \in T_{2,l}$ (or vice versa), with $\{k, l\} = \{1, 2\}$, we define

$$\begin{aligned} f_1(x, y, p_1, q_1, p_2, q_2) &:= \delta_{T_{1,k}}(x, p_1) + |p_1q_1| + \delta_{T_{2,l}}(q_1, y)\,, \\ f_2(x, y, p_1, q_1, p_2, q_2) &:= \delta_{T_{1,l}}(x, p_2) + |p_2q_2| + \delta_{T_{2,k}}(q_2, y)\,, \text{ and} \\ f(x, y, p_1, q_1, p_2, q_2) &:= \min\{f_1(x, y, p_1, q_1, p_2, q_2), f_2(x, y, p_1, q_1, p_2, q_2)\}\,. \end{aligned}$$

Clearly, to compute $f(-)$, the crucial part is to find the partition of T_1 and T_2. We show next that Case 1 and Case 2 can be solved in $O(n^4)$ time, and the following lemma is proved first.

Lemma 4. *Suppose that the optimal solution for the twin bridges problem occurs in Case 1 or Case 2, there must be two edges $e_i \in E(T_i)$, with $i \in \{1, 2\}$, such that each $T_i - \{e_i\}$ results in a pair of trees $(T_{i,1}, T_{i,2})$ and the problem is reduced to solving two optimal bridge problems with input instances $(T_{1,1}, T_{2,1})$ and $(T_{2,1}, T_{2,2})$, or vice versa.*

Proof. We refer to Fig. 3. Suppose that p_1q_1 and p_2q_2, $p_i \in V(T_1), q_i \in V(T_2)$ for $i \in \{1, 2\}$, form the optimal solution for the twin bridges problem under Case 1 or Case 2. Also suppose that the optimal solution is formed by $x \in V(T_1), y \in V(T_2)$, through the bridge p_1q_1; and $z \in V(T_1), w \in V(T_2)$, through the bridge p_2q_2. Then in $T_1 \cup T_2 \cup \{p_1q_1\} \cup \{p_2q_2\}$ there is a cycle $C^* = (p_1, \cdots, p_2, q_2, \cdots, q_1)$, which is 2-connected. Let $e_1 = (p', p'')$ be an edge on C^* but not on either of the constrained diameters between x, y and z, w, and also on C^*, $\delta_{C^*}(p', p_2) \geq \delta_{C^*}(p'', p_1)$ (or $\delta_{T_1}(p'', p_2) \geq \delta_{T_1}(p', p_1)$). Then (p', p'') can be deleted to decompose T_1 into $T_{1,1}$ (containing p_1) and $T_{1,2}$ (containing p_2). The reason we could delete e_1 is that for a point x' in $T_{1,1}$ which is closer to p' than p'', we have

$$\begin{aligned} \delta_{T_1}(x', p_2) &= \delta_{T_{1,1}}(x', p') + \delta_{C^*}(p', p_2) \\ &\geq \delta_{T_{1,1}}(x', p') + \delta_{C^*}(p'', p_1) \\ &= \delta_{T_{1,1}}(x', p') + |p'p''| + \delta_{T_{1,1}}(p', p_1) \\ &\geq \delta_{T_{1,1}}(x', p') + \delta_{T_{1,1}}(p', p_1) \\ &= \delta_{T_{1,1}}(x', p_1). \end{aligned}$$

In other words, deleting e_1 would not affect the (optimal) diameter from a point in T_1 through the bridge p_1q_1. Similarly, if (p', p'') is on C^* but not on either of the (optimal) constrained diameters, and we have $\delta_{C^*}(p', p_2) \leq \delta_{C^*}(p'', p_1)$

(or $\delta_{T_1}(p'', p_2) \le \delta_{T_1}(p', p_1)$), then (p', p'') can be deleted such that for a point z' in $T_{1,2}$ which is closer to p'' than p', we have $\delta_{T_1}(z', p_1) \ge \delta_{T_{1,2}}(z', p_2)$. Symmetrically, T_2 can also be decomposed into $T_{2,1}, T_{2,2}$ by deleting some edge $e_2 \in E(T_2)$. □

With this lemma, it is easy to solve Case 1 and Case 2 in $O(n^4)$ time. We enumerate the $O(n^2)$ pairs of edges (e_1, e_2) with $e_i \in E(T_i), i \in \{1, 2\}$. For each pair, we delete them from T_1 and T_2. Then for the two resulting pairs of trees, one from T_1 and the other from T_2, we solve the optimal bridge problem in $O(n^2)$ time.

It is noted, surprisingly, that the two optimal bridges p_1q_1 and p_2q_2 could intersect in some cases, contrary to many geometric problems. In Fig. 4, we show such an example, $\triangle bcd$ is initially an equilateral triangle with da being the height of the edge bc, we then move a slightly out of the triangle and moving c toward b to have $|bd| > |bc| = |cd| = |ad|$. We also set $|ax| = |by| = |cz| = |dw| = \varepsilon$. If the two intersecting bridges bc and ad are chosen, the optimal solution value is $|bc| + 2\varepsilon$; while if we choose the non-intersecting bridges ac and and bd, the solution value is $|bd| + 2\varepsilon$, which is larger. Therefore, when in practical applications the two bridges must not intersect, we might need to seek sub-optimal solutions.

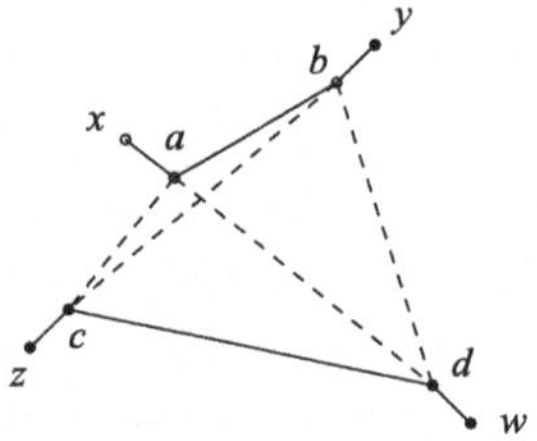

Fig. 4. An example for the twin bridges problem in which the two optimal bridges must intersect. The two input trees are paths: $T_1 = (x, a, b, y)$ and $T_2 = (z, c, d, w)$.

It remains to handle Case 3 and 4 now. We focus on Case 3 here as Case 4 can be similarly handled. By definition, in Case 3 the optimal solution is the maximum of $\delta_{T_1}(x, p_1) + |p_1q_1| + \delta_{T_2}(q_1, q_2) + |q_2p_2| + \delta_{T_1}(p_2, z)$, with $x, z \in V(T_1)$. This can be solved as follows. First, let $\delta_{T_1}(x, z)$ be the diameter of T_1; naturally, that implies x, z are leaves of T_1. Then the problem is to identify two vertices on $\delta_{T_1}(x, z)$, say p_1 and p_2, and also two vertices on T_2, say q_1 and q_2, such that

$$g(p_1, q_1, p_2, q_2) := \delta_{T_1}(p_1, p_2) - \{|p_1q_1| + \delta_{T_2}(q_1, q_2) + |q_2p_2|\}$$

is maximized. Clearly, this can be solved by enumerating all 4-tuples (p_1, q_1, p_2, q_2), which incur a cycle $\langle p_1, q_1, \delta_{T_2}(q_1, q_2), q_2, p_2, \delta_{T_1}(p_2, p_1)\rangle$, where $\delta_{T_1}(p_1, p_2)$ is a subpath on the diameter $\delta_{T_1}(x, z)$. Once (p_1, q_1, p_2, q_2) is given, $g(p_1, q_1, p_2, q_2)$ can be computed in $O(1)$ time. Hence all we need to do is to select the maximum of these $O(n^4)$ values for Case 3 (and symmetrically for Case 4). By evaluating the recorded values for all recorded solutions in Case 1 and 2 as well, we therefore have the following theorem.

Theorem 5. *The twin bridges problem can be solved in $O(n^4)$ time.*

5 RDBP is NP-complete

Here we consider the more general problem of inserting k edges in a planar road network such that for a given set of pairs of vertices we would like to reduce the shortest path distance between each pair. We call this problem Reducing Distances Between Pairs (RDBP), and we have the following theorem.

Theorem 6. *RDBP is* NP*-complete on a given planar graph with degree at most four and the inserted edges have bounded lengths.*

Our reduction is from Vertex Cover which is known to be NP-complete even for planar 2-connected cubic graphs [13]. Two interesting features of the reduction are that (1) the graph obtained by adding the optimum choice of edges is still planar and that (2) the resulting graph has maximum degree of at most five. Both features needed not be true according to our problem definition. In the sense of dealing with crossings, it is hence not necessary to 'build bridges' in this model.

6 Concluding Remarks

There are several open problems from this paper. For example, for the optimal bridge problem itself, can a near quadratic lower bound be proved? Or, can we prove SETH-based lower bounds for any of the other problems dealt with in this paper? Also, for RDBP, can we show the NP-completeness of its natural variation, i.e., P is not given and the goal is to minimize the diameter of the resulting graph G'? Furthermore, RDBP can be considered on graph classes different from planar graphs. As indicated in the discussions above, we could also require that the graph obtained after adding the edges belongs to the considered graph class (or that this is not necessarily the case).

References

1. Hee-Kap Ahn, Otfried Cheong, and Chan-Su Shin. Building bridges between convex regions. *Computational Geometry*, 25(1-2):161–170, 2003.
2. Ilya Baran, Erik D. Demaine, and Mihai Puatracscu. Subquadratic algorithms for 3SUM. *Algorithmica*, 50(4):584–596, 2008.
3. Sergey Bereg, Yuya Higashikawa, Naoki Katoh, Junichi Teruyama, Yuki Tokuni, and Binhai Zhu. Red-black spanners for mixed-charging vehicular networks. In *Proceedings of the 29th Annual Computing and Combinatorics Conference, COCOON, Part I*, volume 14422 of *Lecture Notes in Computer Science*, pages 220–231. Springer, 2023.

4. Binay K. Bhattacharya and Robert Benkoczi. On computing the optimal bridge between two convex polygons. *Information Processing Letters*, 79(5):215–221, 2001.
5. Amit M. Bhosle and Teofilo F. Gonzalez. Exact and approximation algorithms for finding an optimal bridge connecting two simple polygons. *International Journal on Computational Geometry and Applications*, 15(6):609–630, 2005.
6. Amit M. Bhosle and Teofilo F. Gonzalez. Finding optimal geodesic bridges between two simple polygons. In *Proceedings of the 23rd Annual Canadian Conference on Computational Geometry, CCCG*, pages 1–6, 2011.
7. Leizhen Cai, Yin-Feng Xu, and Binhai Zhu. Computing the optimal bridge between two convex polygons. *Information Processing Letters*, 69(3):127–130, 1999.
8. Paul Chew. There is a planar graph almost as good as the complete graph. In Alok Aggarwal, editor, *Proceedings of the Second Annual ACM SIGACT/SIGGRAPH Symposium on Computational Geometry*, pages 169–177. ACM, 1986.
9. Anka Gajentaan and Mark H. Overmars. On a class of $O(n^2)$ problems in computational geometry. *Computational Geometry*, 5:165–185, 1995.
10. Joachim Gudmundsson and Sampson Wong. Improving the dilation of a metric graph by adding edges. *ACM Transactions on Algorithms*, 18(3):20:1–20:20, 2022.
11. Allan Grønlund Jørgensen and Seth Pettie. Threesomes, degenerates, and love triangles. In *55th IEEE Annual Symposium on Foundations of Computer Science, FOCS*, pages 621–630. IEEE Computer Society, 2014.
12. Sung Kwon Kim and Chan-Su Shin. Computing the optimal bridge between two polygons. *Theory of Computing Systems*, 34(4):337–352, 2001.
13. Bojan Mohar. Face covers and the genus problem for apex graphs. *Journal of Combinatorial Theory, Series B*, 82(1):102–117, 2001.
14. Mihai Patrascu. Towards polynomial lower bounds for dynamic problems. In Leonard J. Schulman, editor, *Proceedings of the 42nd ACM Symposium on Theory of Computing, STOC*, pages 603–610. ACM, 2010.
15. Thomas J. Schaefer. The complexity of satisfiability problems. In Richard J. Lipton, Walter A. Burkhard, Walter J. Savitch, Emily P. Friedman, and Alfred V. Aho, editors, *Proceedings of the 10th Annual ACM Symposium on Theory of Computing, STOC*, pages 216–226. ACM, 1978.
16. Xuehou Tan. On optimal bridges between two convex regions. *Information Processing Letters*, 76(4-6):163–168, 2000.
17. Xuehou Tan. Finding an optimal bridge between two polygons. *International Journal on Computational Geometry and Applications*, 12(3):249–262, 2002.
18. D. P. Wang. An optimal algorithm for constructing an optimal bridge between two simple rectilinear polygons. *Information Processing Letters*, 79(5):229–236, 2001.
19. Ryan Williams. A new algorithm for optimal 2-constraint satisfaction and its implications. *Theoretical Computer Science*, 348(2-3):357–365, 2005.

Dynamic DR-Submodular Maximization with Linear Costs over the Integer Lattice

Yuanyuan Qiang and Bin Liu(✉)

School of Mathematical Sciences, Ocean University of China, Qingdao 266100, People's Republic of China
binliu@ouc.edu.cn

Abstract. Submodular maximization is to maximize a submodular function under specific constraints, which plays a crucial role in both theoretical combinatorial optimization and practical machine learning. Submodular functions are defined as a class of set functions that exhibit the property of diminishing returns. The domain of submodular functions can be extended from sets to the integer lattice, with numerous ongoing studies. This paper focuses on designing dynamic algorithms for the problem of maximizing a non-negative monotone DR-submodular function f minus a non-negative linear cost function c, both defined over the integer lattice. Under a box constraint, our dynamic algorithm achieves a $(\frac{1}{2}, 1)$-approximation ratio with amortized time complexity of $O(\frac{r^* ||\boldsymbol{b}||_1 \log ||\boldsymbol{b}||_1 \log ||\boldsymbol{b}||_\infty}{\epsilon^2})$, where the amortized time refers to the number of oracle queries of f per update. Significantly, our dynamic algorithms are capable of handling the scenarios where multiple identical elements are simultaneously inserted or deleted at each time. This paper provides a comprehensive framework that extends beyond traditional set-based approaches, offering insights into the behavior and performance of dynamic algorithms in the context of multisets.

Keywords: Submodular maximization · Lattice submodular · Dynamic algorithm

1 Introduction

In combinatorial optimization, numerous well-known objectives possess a crucial property referred to as submodularity. Submodularity is characterized by the decreasing marginal value as additional elements are selected. Formally, assume E is a ground set, $F : 2^E \rightarrow \mathbb{R}$ is *submodular* if it satisfies $F(S \cup \{e\}) - f(S) \geq F(T \cup \{e\}) - f(T)$ for any $S \subseteq T \subseteq E$ and $e \in E \backslash T$, which is equivalent to $F(A) + F(B) \geq F(A \cup B) + F(A \cap B)$ for any $A, B \subseteq E$. F is *monotone*(*non-decreasing*) if $F(S) \leq F(T)$ for any $S \subseteq T \subseteq E$. Therefore, many combinatorial optimization problems, such as max cut in directed or undirected graphs, specific constraint satisfaction problems, and maximum facility location problems, can all be characterized as submodular maximization problems. Moreover, in recent

S. Ghosh and Z. Zhang (Eds.): AAIM 2024, LNCS 15179, pp. 109–121, 2024.
https://doi.org/10.1007/978-981-97-7798-3_10

decades, submodular maximization has found extensive and diverse applications, such as data mining and categorical feature compression in machine learning. However, the majority of submodular maximization problems are classified as NP-hard. This implies that the optimal solutions cannot be obtained through polynomial algorithms unless $P = NP$.

For a finite ground set $E = \{e_1, e_2, \ldots, e_n\}$, we say that a function $f : \mathbb{Z}_+^n \to \mathbb{R}$ is (*lattice*) *submodular* if $f(\boldsymbol{x}) + f(\boldsymbol{y}) \geq f(\boldsymbol{x} \vee \boldsymbol{y}) + f(\boldsymbol{x} \wedge \boldsymbol{y})$ for all $\boldsymbol{x}, \boldsymbol{y} \in \mathbb{Z}^n$, where $\boldsymbol{x} \vee \boldsymbol{y}$ and $\boldsymbol{x} \wedge \boldsymbol{y}$ denote the coordinate-wise maximum and minimum, respectively. Unlike submodular set functions, submodularity defined over the integer lattice is not equivalent to the diminishing return property. We say that $f : \mathbb{Z}_+^n \to \mathbb{R}$ is *diminishing return submodular* (*DR-submodular*) if $f(\boldsymbol{x} + \boldsymbol{\chi}_e) - f(\boldsymbol{x}) \geq f(\boldsymbol{y} + \boldsymbol{\chi}_e) - f(\boldsymbol{y})$ for any $\boldsymbol{x}, \boldsymbol{y} \in \mathbb{R}$ with $\boldsymbol{x} \leq \boldsymbol{y}$ and $e \in E$, where $\boldsymbol{\chi}_e$ is a unit n-dimensional vector with $\boldsymbol{x}(e) = 1$ and 0 elsewhere. In this paper, to balance benefits and costs, we focus on the maximization problems where the objective function is of the form $f - c$, where f is a non-negative monotone DR-submodular function and c is a non-negative linear function, as demonstrated below.

$$\max_{\boldsymbol{x} \in \mathbb{Z}_+^n} f(\boldsymbol{x}) - c(\boldsymbol{x}), \quad \text{subject to } \boldsymbol{x} \in \Gamma,$$

where the objective $f - c$ may no longer be non-negative and monotone. Assume all identical elements form a group. In the dynamic setting of this problem, there are many applications, such as group dynamic influence maximization with costs([10,14]). In a social network, new groups join and existing groups drop out over time. Using the classical algorithm for each network change is time-consuming. However, the dynamic model efficiently handles this scenario.

In this paper, for the above problem, we consider the case when Γ is a box constraint in the dynamic setting. Since the objective function is defined over the multisets, we extend the classical dynamic models where only one element is allowed to be inserted or deleted at each time to a model where multiple identical elements are allowed to be inserted or deleted at each time.

Organization. The rest of this paper is organized as follows. In Sect. 2, we provide the results already obtained related to our work. In Sect. 3, we give some definitions and notations on submodular maximization over the integer lattice. Section 4 designs and analyzes our dynamic algorithm for the box constrained dynamic DR-submodular maximization with linear costs. Section 5 then gives our conclusion.

2 Related Work

Submodular maximization has been broadly studied for the past few decades. In this section, we will only introduce some of the most relevant available results related to our work.

2.1 Set Submodular Maximization with Linear Costs

For the set submodular maximization problems where the objective function is a non-negative monotone submodular function F subtracted by a non-negative linear function C, some results have been obtained. It is evident that $F - C$ remains submodular but loses the properties of non-negativity and monotonicity. And thus unconstrained submodular maximization with linear costs is even NP-hard. Moreover, in this scenario, there is no multiplicative approximation guarantee possible in polynomial time ([6,19]). However, previous research such as Feldman [8], Du et al. [3] and Sviridenko et al. [24] has shown that the approximation guarantee of an algorithm $\mathcal{A}$ can be characterized by the inequality $F(G)-C(G) \geq \alpha F(O)-\beta C(O)$, where G represents the output solution of the algorithm, O signifies an optimal solution, and the inequality holds for some $0 \leq \alpha, \beta \leq 1$. If the inequality holds for all instances, we refer to the algorithm $\mathcal{A}$ as a (α, β)-*bicriteria approximation* algorithm.

For the problem of unconstrained monotone submodular maximization with linear costs, in an offline setting, Harshaw et al. [13] obtained the best bicriteria approximation ratio of $(1 - \frac{1}{e}, 1)$ by using the distorted greedy algorithm. Additionally, in the streaming setting, Nikolakaki et al. [18] gave a novel approach achieving a $(\frac{1}{2}, 1)$-approximate ratio for the unconstrained case and a $(\frac{3-\sqrt{5}}{2}, 1)$-approximate ratio for the cardinality constrained case, respectively. Instead of maximizing $F - C$ directly, their algorithms maximize a scaled function $F - s \cdot C$, where $s \geq 1$ is an absolute constant in their streaming approximate algorithms. This innovative method allows for more effective approximations under different constraints and settings.

2.2 Lattice Submodular Maximization

The study of submodular maximization over the integer lattice originates from distinct contexts. Fujishige [9] explored submodular functions over distributive lattices and related polyhedra. Murota [17] examined a subclass of submodular functions over the integer lattice, addressing their maximization using discrete convex analysis. For specific lattice submodular maximization problems, Soma et al. [20] provided a $(1 - \frac{1}{e})$-approximation algorithm for monotone lattice submodular functions under a knapsack constraint. Gottschalk and Peis [11] presented a $\frac{1}{3}$-approximation algorithm for maximizing lattice submodular functions over a bounded integer lattice. Soma and Yoshida [23] developed polynomial-time $(1 - \frac{1}{e} - \epsilon)$-approximation algorithms for DR-submodular functions with cardinality, polymatroid, and knapsack constraints. For the unconstrained non-monotone case, Soma and Yoshida [22] offered a $\frac{1}{2+\epsilon}$-approximation algorithm with a runtime of $O(\frac{n \log^2 B}{\epsilon})$, where n is the ground set size and B is the maximum coordinate value. Gu et al. [12] proposed a $\frac{1}{2}$-approximation algorithm with a runtime of $O(n \log B)$.

Note that Ene and Nguyen [5] give a generic reduction from DR-submodular setting to submodular setting. Using this reduction, one can translate the results for unconstrained and constrained submodular maximization to the DR-submodular setting for many types of constraints in a unified manner. However, this paper focuses more on the definition of multisets, considering dynamic algorithms for the DR-submodular maximization problems. It addresses dynamic scenarios involving the insertion or deletion of multiple identical elements, which cannot be handled through simple reduction.

2.3 Dynamic Submodular Maximization

In a dynamic submodular maximization model, "dynamic" denotes changes in the ground set over time. The classical dynamic model assumes a sufficiently large ground set with an initially defined submodular objective function. Starting with an empty ground set, elements are inserted or deleted over time, creating new submodular maximization problems at each step. The goal is to provide a feasible solution with a certain approximation guarantee at every step while minimizing update time, referred to as "amortized time."

For dynamic monotone submodular maximization with a cardinality constraint k and ground set size n, Monemizadeh [16] and Lattanzi et al. [15] independently proposed dynamic algorithms. Lattanzi et al. [15] designed a randomized algorithm with $O(\frac{\log^6 k \log^2 n}{\epsilon^6})$ amortized update time and a $\frac{1}{2+\epsilon}$-approximation. Monemizadeh [16] developed a dynamic algorithm achieving the same approximation with $O(\frac{k^2 \log^3 n}{\epsilon^2})$ amortized update time. For maximizing a monotone submodular function over the matroid constraint, Dütting et al. [4] proposed a randomized algorithm with $\widetilde{O}(m^2)$ amortized update time and a $\frac{1}{4}$-approximation, where m is the rank of the matroid. Banihashem et al. [2] later improved this to a $\frac{1}{4+\epsilon}$-approximation with $O(m \log m \log^3 \frac{m}{\epsilon})$ amortized update time, and reduced Monemizadeh [16]'s time complexity to $O(\frac{k \log^2 k}{\epsilon})$ while maintaining the same approximation ratio.

We highlight the differences between streaming and dynamic models. Streaming algorithms only allow element insertion, while dynamic algorithms permit both insertion and deletion. In streaming settings ([1,7,25]), solutions are computed using a maintained sketch, with space complexity as a key measurement. In dynamic models ([15,16]), elements arrive sequentially, with a focus on amortized time rather than space complexity, and all elements are retained unless deleted.

3 Preliminaries

Throughout the paper, denote $E = \{e_1, e_2, \cdots, e_n\}$ as the ground set with n elements in total. Let $\boldsymbol{x} \in \mathbb{Z}_+^E$ be a n-dimensional vector where coordinate i corresponding to element $e_i \in E$ and $\boldsymbol{x}(e_i)$ be the component of coordinate $e_i (e_i \in E)$ of $\boldsymbol{x}$. Let $\{\boldsymbol{x}\}$ be a multiset where the element e_i appears $\boldsymbol{x}(e_i)$ times, and $\boldsymbol{x}(E) = |\{\boldsymbol{x}\}| = \sum_{e_i} \boldsymbol{x}(e_i)$. Denote $\boldsymbol{\chi}_e$ as a unit vector where $\boldsymbol{x}_e = 1$ and 0 elsewhere. For any $\boldsymbol{x}$ and $\boldsymbol{y}$, $\boldsymbol{x} \leq \boldsymbol{y}$ refers to $\boldsymbol{x}(e_i) \leq \boldsymbol{y}(e_i)$ for each $e_i \in E$, and let $f(\boldsymbol{y} \mid \boldsymbol{x}) = f(\boldsymbol{x} + \boldsymbol{y}) - f(\boldsymbol{x})$.

In this paper, we consider the following problem in the dynamic setting, where $\mathbb{Z}_+^E = \{\boldsymbol{x} \mid \boldsymbol{x} = (x_1, x_2, \cdots, x_{|E|}), x_i \in \mathbb{Z}_+, 1 \le i \le |E|\}$.

Problem P. Given a ground set E and a upper bound $\boldsymbol{b}$, for a monotone DR-submodular function $f : \mathbb{Z}_+^E \to \mathbb{R}_+$ and a linear function $c : \mathbb{Z}_+^E \to \mathbb{R}_+$,

$$\max_{\boldsymbol{x} \in \mathbb{Z}_+^E} f(\boldsymbol{x}) - c(\boldsymbol{x}), \quad \text{subject to } \boldsymbol{x} \in \mathcal{B} = \{\boldsymbol{x} \mid \boldsymbol{x} \le \boldsymbol{b}, \boldsymbol{b} \in \mathbb{Z}_+^E\}.$$

In Problem P, $\mathcal{B}$ is called a box constraint. The objective function $f(\boldsymbol{x}) - c(\boldsymbol{x})$ can be verified that it is still a DR-submodular function. Since for any $\boldsymbol{x}, \boldsymbol{y} \in \mathbb{Z}_+^E$ with $\boldsymbol{x} \le \boldsymbol{y}$, the following inequalities hold. $[f(\boldsymbol{x}+\boldsymbol{\chi}_e) - c(\boldsymbol{x}+\boldsymbol{\chi}_e)] - [f(\boldsymbol{x}) - c(\boldsymbol{x})] = f(\boldsymbol{x}+\boldsymbol{\chi}_e) - f(\boldsymbol{x}) - c(\boldsymbol{y}+\boldsymbol{\chi}_e) + c(\boldsymbol{y}) \ge f(\boldsymbol{y}+\boldsymbol{\chi}_e) - f(\boldsymbol{y}) - c(\boldsymbol{y}+\boldsymbol{\chi}_e) + c(\boldsymbol{y}) = [f(\boldsymbol{y}+\boldsymbol{\chi}_e) - c(\boldsymbol{y}+\boldsymbol{\chi}_e)] - [f(\boldsymbol{y}) - c(\boldsymbol{y})]$, where the first inequality holds for the DR-submodularity of f.

4 The Dynamic Algorithms and Analysis

In this section, we propose a dynamic approximate algorithm for Problem P. Note that Problem P is equivalent to the following problem.

Problem P'. Given a ground set E, for a monotone DR-submodular function $f : \mathbb{Z}_+^E \to \mathbb{R}_+$ and a linear function $c : \mathbb{Z}_+^E \to \mathbb{R}_+$,

$$\max_{\boldsymbol{x} \in \mathbb{Z}_+^E} f(\boldsymbol{x}) - c(\boldsymbol{x}), \quad \text{subject to } \{\boldsymbol{x}\} \subseteq \{\boldsymbol{b}\}.$$

We first design an offline loop algorithm, Algorithm 1, which serves as a frequent subroutine in other algorithms. This algorithm employs a combination of greedy methods and binary search to select elements and put them into the output solution. On the other hand, it filters out those elements that do little to increase the value of the objective function. For Problem P', we then develop a $(\frac{1}{2}, 1)$-approximation algorithm, Algorithm 2, for a specified ground set $\boldsymbol{z}_0 = \boldsymbol{b}$. Based on Algorithm 2, we create insertion and deletion algorithms capable of handling multiple simultaneous insertions and deletions of identical elements.

For element e_i, assume that the number of this element is inserted or deleted in the dynamic algorithm is r_i. However, for simplicity, the number of identical elements inserted or deleted in the insertion or deletion algorithms is uniformly denoted by r. The essence of insertion algorithm is to get an approximate solution of current problem by making full use of the information from the previous operation. Indeed, the algorithm obtains an output solution by comparing the cases where these r_i elements e_i were included in the previous offline algorithm to the cases where they were not included. Additionally, we adopt a sampling technique to reduce the time complexity. Similarly, we design the deletion algorithm that can handle multiple simultaneous deletions of identical elements. Combining the existing algorithms, we get the final dynamic Algorithm 5 for Problem P. Note that δ in the algorithms is a parameter related to probability, which is set to $1/\epsilon^2$ in this paper. According to Lemma 1, ϵ cannot approach infinity.

Algorithm 1. Offline Loop Algorithm: OLA(i, $\boldsymbol{x}_j$, $\boldsymbol{z}_j$ for $j \in \{0, 1, \ldots, i\}$)

Input: i and $\boldsymbol{x}_j$, $\boldsymbol{z}_j$ for $j \in \{0, 1, \ldots, i\}$
Output: $\boldsymbol{x}$, i^* and $\boldsymbol{x}_i, \boldsymbol{z}_i$ for $i \in \{0, 1, \ldots, i^*\}$
1: Let $s = \delta \log(||\boldsymbol{b}||_1)$, $\widetilde{F} = f - 2c$
2: **while** $\boldsymbol{z}_i \neq \boldsymbol{0}$ **do**
3: $\quad i \leftarrow i + 1$, $\boldsymbol{x}_i = \boldsymbol{x}_{i-1}$
4: $\quad$ Let $\{\boldsymbol{y}_i\} \subseteq \{\boldsymbol{z}_{i-1}\}$ and $\{\boldsymbol{y}_i\}$ be a multiset of elements sampled with the probability $p = \min\{\frac{s}{|\{\boldsymbol{z}_{i-1}\}|}, 1\}$ from $\{\boldsymbol{z}_{i-1}\}$
5: $\quad$ **for** $e \in \{e_j : \boldsymbol{y}_i(e_j) > 0\}$ **do**
6: $\qquad l^* \leftarrow$ Binary-Search$(0, \min\{\boldsymbol{y}_i(e) - \boldsymbol{x}_i(e), \boldsymbol{b}(e) - \boldsymbol{x}_i(e)\}, 0, \widetilde{F})$
7: $\qquad \boldsymbol{x}_i \leftarrow \boldsymbol{x}_i + l^* \boldsymbol{\chi}_e$
8: $\quad$ **end for**
9: $\quad \boldsymbol{z}_i = \boldsymbol{z}_{i-1} - \boldsymbol{y}_i$
10: $\quad$ **for** $e \in \{e_j : [(\boldsymbol{z}_{i-1} - \boldsymbol{y}_i) \vee \boldsymbol{0}](e_j) > 0\}$ **do**
11: $\qquad$ **if** $\widetilde{F}(\boldsymbol{\chi}_e \mid \boldsymbol{x}_i) \leq 0$ **then**
12: $\qquad\quad \boldsymbol{z}_i \leftarrow \boldsymbol{z}_i - \boldsymbol{z}_i(e)\boldsymbol{\chi}_e$
13: $\qquad$ **end if**
14: $\quad$ **end for**
15: **end while**
16: $i^* \leftarrow$ last i after leaving the while loop
17: $\boldsymbol{x} \leftarrow \boldsymbol{x}^*$
18: **return** $\boldsymbol{x}$, i^* and $\boldsymbol{x}_i, \boldsymbol{z}_i$ for $i \in \{0, 1, \ldots, i^*\}$.

In Algorithm 1, suppose that e is the element being processed in line 5. The binary search subroutine is defined as Binary-Search$(\alpha, \beta, \theta, f) = \max\{\alpha \leq l \leq \beta : f(l\boldsymbol{\chi}_e \mid \boldsymbol{x}) > \theta\}$. Next, we analyze the approximation guarantee of Algorithm 2 in Theorem 1.

Theorem 1. *Algorithm 2 returns a solution $\boldsymbol{x}$ satisfying*

$$f(\boldsymbol{x}) - c(\boldsymbol{x}) \geq \frac{1}{2} f(\boldsymbol{x}^*) - c(\boldsymbol{x}^*),$$

where $\boldsymbol{x}^*$ is an optimal solution of Problem P.

Proof. We first show that for each $e^* \in \{\boldsymbol{x}^*\} \backslash \{\boldsymbol{x}\}$,

$$\widetilde{F}(\boldsymbol{\chi}_{e^*} \mid \boldsymbol{x}) \leq 0. \tag{1}$$

Now consider the while loop iterations in the algorithm. On the one hand, if there exists iteration i^0, $i^0 \leq i^*$, such that $\boldsymbol{y}_{i^0}(e^*) > 0$, denote

$$l_{e*}^{i^0} = \max\left\{0 \leq l \leq \min\{\boldsymbol{y}_{i^0}(e^*) - \boldsymbol{x}_{e*}^{i^0}(e^*), \boldsymbol{b}(e) - \boldsymbol{x}_{e*}^{i^0}(e^*)\} : \widetilde{F}(l\boldsymbol{\chi}_{e^*} \mid \boldsymbol{x}_{e*}^{i^0}(e^*) > 0\right\},$$

where $\boldsymbol{x}_{e*}^{i^0}$ is the current $\boldsymbol{x}$ when e^* is in line 5 of the i^0th while loop in Algorithm 1. By contradiction, if $\widetilde{F}(\boldsymbol{\chi}_{e^*} \mid \boldsymbol{x}) > 0$, it follows that

$$\begin{aligned}\widetilde{F}((l_{e^*}^{i^0}+1)\boldsymbol{\chi}_{e^*} \mid \boldsymbol{x}_{e^*}^{i^0}) &= \widetilde{F}(\boldsymbol{x}_{e^*}^{i^0}+(l_{e^*}^{i^0}+1)\boldsymbol{\chi}_{e^*}) - \widetilde{F}(\boldsymbol{x}_{e^*}^{i^0}) \\ &= [\widetilde{F}(\boldsymbol{x}_{e^*}^{i^0}+(l_{e^*}^{i^0}+1)\boldsymbol{\chi}_{e^*}) - \widetilde{F}(\boldsymbol{x}_{e^*}^{i^0}+l_{e^*}^{i^0}\boldsymbol{\chi}_{e^*})] \\ &+ [\widetilde{F}(\boldsymbol{x}_{e^*}^{i^0}+l_{e^*}^{i^0}\boldsymbol{\chi}_{e^*}) - \widetilde{F}(\boldsymbol{x}_{e^*}^{i^0})] \\ &\geq [\widetilde{F}(\boldsymbol{x}+\boldsymbol{\chi}_{e^*}) - \widetilde{F}(\boldsymbol{x})] + [\widetilde{F}(\boldsymbol{x}_{e^*}^{i^0}+l_{e^*}^{i^0}\boldsymbol{\chi}_{e^*}) - \widetilde{F}(\boldsymbol{x}_{e^*}^{i^0})] \\ &= \widetilde{F}(\boldsymbol{\chi}_{e^*} \mid \boldsymbol{x}) + \widetilde{F}(l_{e^*}^{i^0}\boldsymbol{\chi}_{e^*} \mid \boldsymbol{x}_{e^*}^{i^0}) \\ &> 0,\end{aligned}$$

which contradicts with the definition of $l_{e^*}^{i^0}$. Thus Eq. (1) holds. On the other hand, if for any $i \leq i^*$, $\boldsymbol{y}_i(e^*) = 0$. Since the while loop in Algorithm 1 iterates from a complete set $\{\boldsymbol{z}_0\}$ to a empty set $\{\boldsymbol{z}_{i^*}\}$ and by the filtering process of Algorithm 1 (the second for loop), it is obvious that the element e^* will be sieved out, and thus (1) holds. Therefore, it is obtained from the above

$$\sum_{e^* \in \{\boldsymbol{x}^*\} \setminus \{\boldsymbol{x}\}} \widetilde{F}(\boldsymbol{\chi}_{e^*} \mid \boldsymbol{x}) \leq 0.$$

Let $\boldsymbol{x}^* = (x_1^*, x_2^* ..., x_n^*)$, $\boldsymbol{x} = (x_1, x_2, ..., x_n)$ and $\boldsymbol{\chi}_i$ be a $\{0,1\}$-vector where the i_{th} component is 1 and 0 elsewhere. Without loss of generality, assume that $\boldsymbol{x}^* \vee \boldsymbol{x} = (x_1^*, ..., x_m^*, x_{m+1}, ..., x_n) = (x_1 + (x_1^* - x_1), ..., x_m + (x_m^* - x_m), x_{m+1}, ..., x_n)$, and then we can obtain the following inequalities due to the DR-submodularity of $\widetilde{F}$.

$$\widetilde{F}(\boldsymbol{x}^* \vee \boldsymbol{x}) - \widetilde{F}(\boldsymbol{x}) \leq \sum_{e^* \in \{\boldsymbol{x}^*\} \setminus \{\boldsymbol{x}\}} \widetilde{F}(\boldsymbol{\chi}_{e^*} \mid \boldsymbol{x}),$$

which is equivalent to

$$\begin{aligned}\sum_{e^* \in \{\boldsymbol{x}^*\} \setminus \{\boldsymbol{x}\}} \widetilde{F}(\boldsymbol{\chi}_{e^*} \mid \boldsymbol{x}) &\geq \widetilde{F}(\boldsymbol{x}^* \vee \boldsymbol{x}) - \widetilde{F}(\boldsymbol{x}) \\ &= f(\boldsymbol{x}^* \vee \boldsymbol{x}) - 2c(\boldsymbol{x}^* \vee \boldsymbol{x}) - f(\boldsymbol{x}) + 2c(\boldsymbol{x}) \\ &= [f(\boldsymbol{x}^* \vee \boldsymbol{x}) - f(\boldsymbol{x})] - 2[c(\boldsymbol{x}^* \vee \boldsymbol{x}) - c(\boldsymbol{x})] \\ &\geq f(\boldsymbol{x}^*) - f(\boldsymbol{x}) - 2c(\boldsymbol{x}^*),\end{aligned}$$

where the last inequality is obtained from the monotonicity of f and the linearity of c. Then it follows

$$f(\boldsymbol{x}^*) - f(\boldsymbol{x}) - 2c(\boldsymbol{x}^*) \leq 0. \tag{2}$$

By the method of adding elements in Algorithm 1, we have

$$\widetilde{F}(\boldsymbol{x}) = \sum_{e \in \{e' \mid \boldsymbol{y}_i(e') > 0\}} \widetilde{F}(l_e^i \boldsymbol{\chi}_e | \boldsymbol{x}_e^i) > 0.$$

Thus $f(\boldsymbol{x}) - 2c(\boldsymbol{x}) \geq 0$, which is equivalent to $c(\boldsymbol{x}) \leq \frac{1}{2}f(\boldsymbol{x})$. And then following inequalities holds.

$$f(\boldsymbol{x}) - c(\boldsymbol{x}) \geq \frac{1}{2}f(\boldsymbol{x}) \geq \frac{1}{2}[f(\boldsymbol{x}^*) - 2c(\boldsymbol{x}^*)] = \frac{1}{2}f(\boldsymbol{x}^*) - c(\boldsymbol{x}^*),$$

where the second inequality holds for Eq. (2). Therefore, the theorem follows. □

Algorithm 2. Offline Algorithm for Problem P: OAP

Input: monotone submodular function f, $\boldsymbol{b}$
Output: $\boldsymbol{x}$, i^*,and $\boldsymbol{x}_i, \boldsymbol{z}_i$ for $0 \leq i \leq i^*$
1: Let $\boldsymbol{x}_0 = 0$, $\boldsymbol{z}_0 = \boldsymbol{b}$
2: Invoke OLA(0, $\boldsymbol{x}_0$, $\boldsymbol{z}_0$)
3: **return** $\boldsymbol{x}$, i^*,and $\boldsymbol{x}_i, \boldsymbol{z}_i$ for $0 \leq i \leq i^*$.

Now we turn to analyze the query complexity of Algorithm 2. We first bound the number of iterations of the outer while loop by using the results of the following Lemma 1. This lemma provides a critical upper bound that significantly influences our subsequent analysis. With this bound established, we proceed to examine the inner computational steps, ensuring each phase of the algorithm is accounted for. And then the final query complexity of Algorithm 2 can be obtained in Lemma 2.

Lemma 1. *At each while loop i of Algorithm 2, with probability at least* $1 - \frac{1}{||\boldsymbol{b}||_1^\delta}$, *we have* $\boldsymbol{x}_i(E) \geq \boldsymbol{x}_{i-1}(E) + 1$.

Proof. Now let us consider the beginning of iteration i in Algorithm 2's while loop. It can be obtained from the process of Algorithm 2 that for each $e \in \{\boldsymbol{z}_{i-1}\}$,

$$f(\chi_e \mid \boldsymbol{x}_{i-1}) > 0.$$

Denote A as the event that there is at least one element in the multiple set $\{\boldsymbol{y}_i\}$, then we can get

$$\Pr[\boldsymbol{x}_i(E) \geq \boldsymbol{x}_{i-1}(E) + 1] = \Pr[A].$$

By Algorithm 2, we can also obtain

$$\Pr[A] = 1 - [1 - \frac{\delta \log ||\boldsymbol{b}||_1}{\boldsymbol{z}_{i-1}(E)}]^{\boldsymbol{z}_{i-1}(E)} \geq 1 - \mathrm{e}^{-\frac{\delta \log ||\boldsymbol{b}||_1}{\boldsymbol{z}_{i-1}(E)} \cdot \boldsymbol{z}_{i-1}(E)} = 1 - \frac{1}{||\boldsymbol{b}||_1^\delta},$$

where the first inequality holds for $1 + x \leq \mathrm{e}^x$ for any $x \in \mathbb{R}$. □

Based on Lemma 1, the following corollary is easily derived, providing an upper bound on the number of iterations for the outer while loop in Algorithm 2. Note that $||\boldsymbol{b}||_1$ is the sum of all components of $\boldsymbol{b}$, which is indeed the number of elements in multiset $\{\boldsymbol{b}\}$.

Corollary 1. *In Algorithm 2, the number of iterations of the while loop i^* satisfies: $i^* \leq O(||\boldsymbol{b}||_1)$.*

Lemma 2. *The query complexity in Algorithm 2 is $O(n||\boldsymbol{b}||_1 \log ||\boldsymbol{b}||_\infty)$.*

Proof. In Algorithm 2, the query complexity in the first for loop is $O(n \log ||\boldsymbol{b}||_\infty)$ and in the second for loop is $O(n)$. By Corollary 1, we have the query complexity in Algorithm 2 is at least $O(n||\boldsymbol{b}||_1 \log ||\boldsymbol{b}||_\infty)$. □

Algorithm 3. Insertion Algorithm for P: IAP(e, r, i^*, $\boldsymbol{x}_i, \boldsymbol{z}_i$ for $0 \leq i \leq i^*$)

Input: Insert r elements e, and $i^*, \boldsymbol{x}_i, \boldsymbol{z}_i$ for $0 \leq i \leq i^*$
Output: $\boldsymbol{x}$, $\boldsymbol{x}_i, \boldsymbol{z}_i$ for $0 \leq i \leq i^*$

$\boldsymbol{z}_0 \leftarrow \boldsymbol{z}_0 + r\chi_e$
for $i = 1$ to i^* **do**
  $\pi \leftarrow$ a random variable equal to 1 with probability $p' = \min\{1 - \frac{C_s^{\boldsymbol{z}_{i-1}(E)-r}}{C_s^{\boldsymbol{z}_{i-1}(E)}}, 1\}$ and 0 otherwise
  if $\pi = 1$ **then**
    Invoke OLA($i-1$, $\boldsymbol{x}_j$, $\boldsymbol{z}_j$ for $j \in \{0, 1, \ldots, i-1\}$)
    Break
  else
    if $\widetilde{F}(\chi_e \mid \boldsymbol{x}_i) > 0$ **then**
      $\boldsymbol{z}_i \leftarrow \boldsymbol{z}_i + r\chi_e$
    else
      Break
    end if
  end if
end for
return $\boldsymbol{x}$, $\boldsymbol{x}_i, \boldsymbol{z}_i$ for $0 \leq i \leq i^*$.

We have now obtained the query complexity of the offline Algorithm 2. In the insertion algorithm, when r elements e are inserted, they are first added to the ground set $\boldsymbol{z}_0$. In each iteration i of the for loop, if at least one element e is sampled with probability p', the process restarts from iteration i. If $\widetilde{F}(\chi_e \mid \boldsymbol{x}_i) > 0$, the r elements e are added to $\boldsymbol{z}_i$. If no element e is sampled or $\widetilde{F}(\chi_e \mid \boldsymbol{x}_i) \leq 0$, the algorithm break the for loop and ends. Next, we will analyze the query complexity of the insertion algorithm IAP(e, r, i^*, $\boldsymbol{x}_i, \boldsymbol{z}_i$ for $0 \leq i \leq i^*$) using the probability techniques and the results of Lemma 2.

Lemma 3. *Let the value of δ is $1/\epsilon^2$, the query complexity in the insertion algorithm IAP(e, r, i^*, $\boldsymbol{x}_i, \boldsymbol{z}_i$ for $0 < i < i^*$) is $O(\frac{r||\boldsymbol{b}||_1 \log ||\boldsymbol{b}||_1 \log ||\boldsymbol{b}||_\infty}{\epsilon^2})$.*

Using a proof method similar to that of Lemma 3, the proof process of Lemma 4 can be obtained. In the deletion algorithm , when $r(r < \boldsymbol{z}_0(e))$ elements e are deleted, they are first removed from the ground set $\{\boldsymbol{z}_0\}$. If an element e exists in $\boldsymbol{x}_j$ for $j \in \{0, 1, \ldots, i^*\}$, the process restarts from iteration j. If no element e remains in $\boldsymbol{z}_i$ for $i \in \{0, 1, \ldots, i^*\}$, meaning all r elements e are filtered out, the algorithm terminates.

Lemma 4. *Let the value of δ is $1/\epsilon^2$, the query complexity in the deletion algorithm DAP(e, r, i^*, $\boldsymbol{x}_i, \boldsymbol{z}_i$ for $0 \le i \le i^*$) is $O(\frac{r\|\boldsymbol{b}\|_1 \log \|\boldsymbol{b}\|_1 \log \|\boldsymbol{b}\|_\infty}{\epsilon^2})$.*

Algorithm 4. Deletion Algorithm for P: DAP(e, r, i^*, $\boldsymbol{x}_i, \boldsymbol{z}_i$ for $0 \le i \le i^*$)

Input: Delete $r(r < \boldsymbol{z}_0(e))$ elements e, and $i^*, \boldsymbol{x}_i, \boldsymbol{z}_i$ for $0 \le i \le i^*$
Output: $\boldsymbol{x}$, $\boldsymbol{x}_i, \boldsymbol{z}_i$ for $0 \le i \le i^*$

```
1: z_0 ← z_0 − r χ_e
2: for i = 1 to i* do
3:    if x_i(e) > z_0(e) − r then
4:       Invoke OLA(i − 1, x_j, z_j for j ∈ {0, 1, ..., i − 1})
5:       Break
6:    else if z_i(e) > z_0(e) − r − x_i(e) then
7:       z_i ← z_i − [z_i(e) − z_0(e) + x_i(e) + r] χ_e
8:    end if
9: end for
10: return x, x_i, y_i, z_i for 0 ≤ i ≤ i*.
```

Finally, we establish the following theorem for the problem of the box constrained dynamic DR-submodular maximization with linear costs over the integer lattice.

Theorem 2. *Dynamic Insertion/Deletion Algorithm for Problem P (Algorithm 5) can achieve a $(\frac{1}{2}, 1)$-approximation ratio and $O(\frac{r^*\|\boldsymbol{b}\|_1 \log \|\boldsymbol{b}\|_1 \log \|\boldsymbol{b}\|_\infty}{\epsilon^2})$ amortized time, where ϵ is a parameter related to the probability and r^* is the average number of inserted and deleted elements at each time.*

Algorithm 5. Dynamic Insertion/Deletion Algorithm for Problem P

Input: A sequences $\{\text{Update}(e_1, r_1), \cdots, \text{Update}(e_m, r_m)\}$, where Update$(e_t, r_t)$ indicates insert or delete $r_{i=t}$ elements e_t
Output: $\boldsymbol{x}^t$ at any time $t \in [m]$

```
1: z_0 ← 0
2: for each Update(e_t, r_t) do
3:    if Update(e_t, r_t) is Insert(e_t, r_t) then
4:       Invoke IAP(e_t, r_t, i*, x_i, z_i for 0 ≤ i ≤ i*)
5:    else if Update(e_t, r_t) is Delete(e_t, r_t) then
6:       Invoke DAP(e_t, r_t, i*, x_i, z_i for 0 ≤ i ≤ i*)
7:    end if
8: end for
9: return x^t at any time t ∈ [m].
```

In this section, for the problem of box constrained dynamic DR-submodular maximization with linear costs over the integer lattice, we first design an offline Algorithm 2. Then we show that it can achieve a $(\frac{1}{2}, 1)$-approximation with the query complexity of $O(n||\boldsymbol{b}||_1 \log ||\boldsymbol{b}||_\infty)$. Based on the offline algorithm, we then propose an insertion algorithm which allows multiple elements to be inserted simultaneously, Algorithm 3. Similarly, we also propose a deletion algorithm which allows multiple elements to be deleted simultaneously, Algorithm 4. At last, we obtain our dynamic Algorithm 5. In Theorem 1, we show that it can achieve a $(\frac{1}{2}, 1)$-approximation with the amortized time of $O(\frac{r^*||\boldsymbol{b}||_1 \log ||\boldsymbol{b}||_1 \log ||\boldsymbol{b}||_\infty}{\epsilon^2})$, where ϵ is a parameter related to the probability and r^* is the average number of inserted and deleted elements at each time.

5 Conclusion

This paper studies Problem P, which is equivalent to Problem P'. Specifically, Problem P is to maximize the difference between a non-negative monotone DR-submodular function and a linear function over the integer lattice with a box constraint. This work introduces and analyzes dynamic models for the problem, marking the first consideration of dynamic algorithms in these contexts. Unlike the general dynamic model, this paper considers a dynamic scenario where types of elements come one by one in a streaming manner, rather than the individual element. At some points in time, some type of elements may need to be added to or deleted from the current ground set simultaneously. For Problem P, the dynamic Algorithm 5 provides an approximate solution with the $(\frac{1}{2}, 1)$-approximation ratio at each time t. In addition, the amortized query time for the algorithms is also analyzed in this paper.

Acknowledgement. This work was supported in part by the National Natural Science Foundation of China (11971447).

Disclosure of Interests. The authors declare that they have no conflict of interest.

References

1. Badanidiyuru, A., Mirzasoleiman, B., Karbasi, A., Krause, A.: Streaming submodular maximization: massive data summarization on the fly. In: Proceedings of the 20th ACM SIGKDD International Conference on Knowledge Discovery and Data Mining, pp. 671–680. ACM press, New York, New York, USA (2014)
2. Banihashem, K., Biabani, L., Goudarzi, S., Hajiaghayi, M., Jabbarzade, P., Monemizadeh, M.: Dynamic algorithms for matroid submodular maximization. In: Proceedings of the 35th Annual ACM-SIAM Symposium on Discrete Algorithms, pp. 3485–3533. SIAM press, Alexandria, Virginia, USA (2024)
3. Du, D., Li, Y., Xiu, N., Xu, D.: Simultaneous approximation of multi-criteria submodular function maximization. J. Oper. Res. Soc. China **2**(3), 271–290 (2014)
4. Dütting, P., Fusco, F., Lattanzi, S., Norouzi-Fard, A., Zadimoghaddam, M.: Fully dynamic submodular maximization over matroids. In: Proceedings of the 40th International Conference on Machine Learning, pp. 8821–8835. PMLR press, Honolulu, Hawaii, USA (2023)

5. Ene, A., Nguyen, H.L.: A reduction for optimizing lattice submodular functions with diminishing returns. arXiv:1606.08362, arXiv preprint (2016)
6. Feige, U.: A threshold of $\ln n$ for approximating set cover (preliminary version). In: Proceedings of the 28th Annual ACM Symposium on Theory of Computing, pp. 314–318. ACM press, Philadelphia, Pennsylvania, USA (1996)
7. Feldman, M., Norouzi-Fard, A., Svensson, O., Zenklusen, R.: The one-way communication complexity of submodular maximization with applications to streaming and robustness. In: Proceedings of the 52nd Annual ACM SIGACT Symposium on Theory of Computing, pp. 1363–1374. ACM press, Chicago, Illinois, USA (2020)
8. Feldman, M.: Guess free maximization of submodular and linear sums. Algorithmica **83**(3), 853–878 (2021)
9. Fujishige, S.: Submodular Functions and Optimization, 2nd edn. Elsevier, Amsterdam (2005)
10. Gehlot, H., Sundaram, S., Ukkusuri, S.V.: Algorithms for influence maximization in socio-physical networks. J. Comb. Optim. **45**(1), 1 (2023)
11. Gottschalk, C., Peis, B.: Submodular function maximization on the bounded integer lattice. In: Proceedings of the 13th International Workshop on Approximation and Online Algorithms, pp. 133–144. Springer press, Patras, Greece (2015)
12. Gu, S., Shi, G., Wu, W., Lu, C.: A fast double greedy algorithm for non-monotone DR-submodular function maximization. Discrete Math. Algorithms Appl. **12**(1), 2050007:1–2050007:11 (2020)
13. Harshaw, C., Feldman, M., Ward, J., Karbasi, A.: Submodular Maximization beyond Non-negativity: Guarantees, Fast Algorithms, and Applications. In: Proceedings of the 36th International Conference on Machine Learning, pp. 2634–2643. PMLR press, Long Beach, California, USA (2019)
14. Kempe, D., Kleinberg, J., Tardos, É.: Maximizing the spread of influence through a social network. In: Proceedings of the 9th ACM SIGKDD International Conference on Knowledge Discovery and Data Mining, pp. 137–146. ACM press, Washington, D.C., USA (2003)
15. Lattanzi, S., Mitrović, S., Norouzi-Fard, A., Tarnawski, J. M., Zadimoghaddam, M.: Fully dynamic algorithm for constrained submodular optimization. In: Proceedings of the 34th Advances in Neural Information Processing Systems, pp. 12923–12933. Neurips proceedings press, virtual (2020)
16. Monemizadeh, M.: Dynamic submodular maximization. In: Proceedings of the 34th Advances in Neural Information Processing Systems, pp. 9806–9817. Neurips proceedings press, virtual (2020)
17. Murota, K.: Discrete convex analysis. Math. Program. **83**, 313–371 (1998)
18. Nikolakaki, S.M., Ene, A., Terzi, E.: An efficient framework for balancing submodularity and cost. In: Proceedings of the 27th ACM SIGKDD Conference on Knowledge Discovery and Data Mining, pp. 1256–1266. ACM press, virtual (2021)
19. Papadimitriou, C., Yannakakis, M.: Optimization, approximation, and complexity classes. In: Proceedings of the 20th Annual ACM Symposium on Theory of Computing, pp. 229–234. ACM press, Chicago, Illinois, USA (1988)
20. Soma, T., Kakimura, N., Inaba, K., Kawarabayashi, K. I.: Optimal budget allocation: theoretical guarantee and efficient algorithm. In: Proceedings of the 31st International Conference on Machine Learning, pp. 351–359. JMLR press, Beijing, China (2014)
21. Soma, T., Yoshida, Y.: A generalization of submodular cover via the diminishing return property on the integer lattice. In: Proceedings of the 29th Advances in Neural Information Processing Systems, pp. 847–855. Neurips proceedings press, Montreal, Quebec, Canada (2015)

22. Soma, T., Yoshida, Y.: Non-monotone DR-submodular function maximization. In: Proceedings of the 31st AAAI Conference on Artificial Intelligence, pp. 898–904. AAAI press, San Francisco, California, USA (2017)
23. Soma, T., Yoshida, Y.: Maximizing monotone submodular functions over the integer lattice. Math. Program. **172**(1–2), 539–563 (2018)
24. Sviridenko, M., Vondrák, J., Ward, J.: Optimal approximation for submodular and supermodular optimization with bounded curvature. Math. Oper. Res. **42**(4), 1197–1218 (2017)
25. Yang, R., Xu, D., Guo, L., Zhang, D.: Sequence submodular maximization meets streaming. J. Comb. Optim. **41**(1), 43–55 (2021)

A Distributed Approximation Algorithm for the Total Dominating Set Problem

Limin Wang[1], Zhao Zhang[2], Donglei Du[3], Yaping Mao[4,5], and Xiaoyan Zhang[6(✉)]

[1] School of Mathematical Sciences, Henan Institute of Science and Technology, Xinxiang 453003, China
wanglimin0721@163.com
[2] College of Mathematics and Computer Sciences, Zhejiang Normal University, Zhejiang 321004, China
[3] Faculty of Management, University of New Brunswick, Fredericton E3B 5A3, New Brunswick, Canada
[4] School of Mathematics and Statistics, Qinghai Normal University, Qinghai 810008, China
[5] Academy of Plateau Science and Sustainability, Xining, Qinghai 810008, China
[6] School of Mathematical Science and Institute of Mathematics Key Laboratory of Ministry of Education Numerical Simulation of Large Scale , Complex Systems, Nanjing Normal University, Jiangsu 210023, China
zhangxiaoyan@njnu.edu.cn

Abstract. Given a network graph $G = (V, E)$, a subset $T \subseteq V$ is said to be a total dominating set (TDS) if every $v \in V$ is adjacent to at least one node in T. In this paper, we present a distributed algorithm for the TDS problem via the LP relaxation techniques. For a positive integer k and maximum degree Δ, the proposed algorithm outputs a fractional total dominating set of expected size $O(k\Delta^{\frac{2}{k}})|TDS_{OPT}|$. The distributed algorithm runs in $O(k^2)$ communication rounds, and the algorithm use messages of size $O(\log \Delta)$. The fractional solution is then rounded to obtain an integer total dominating set for the original problem.

Keywords: Total dominating set · Distributed · Approximation algorithm · Rounding algorithm

1 Introduction

Domination is one of classical and interesting topics in graph theory. It has a wide range of applications in theoretical and real-world problems, such as complex ecosystems [4], coding theory [8], electrical networks [18], monitoring communication [23], among others. We refer to [1,16,17] for more theoretical results and practical applications.

A subset $D \subseteq V$ of nodes of G is a *dominating set* of G if each node in $V \setminus D$ is adjacent to at least one node in D. Dominating sets that satisfy certain specific

S. Ghosh and Z. Zhang (Eds.): AAIM 2024, LNCS 15179, pp. 122–132, 2024.
https://doi.org/10.1007/978-981-97-7798-3_11

properties, such as weakly-connected and connected, are often used in routing protocols of ad hoc networks [5,11]. The nodes in a dominating set represent the servers, which provide acceptable level of services in a network. However, when a server crashes, the provided service will be compromised. A total dominating set provides added fault tolerance because every server is adjacent to at least one other server, which serves as a backup in case of server crash. Since every node is dominated by another node in the total dominating set, this non-differentiation of domination property is suitable for peer-to-peer networks [7]. As a variation of dominating set, the concept of total dominating set was introduced in [10]. In a graph $G = (V, E)$, a subset $T \subseteq V$ is said to be a *total dominating set* if each $v \in V$ is adjacent to at least one node in T. The minimum total dominating set (MTDS) problem is to find a total dominating set of the minimum size.

In this paper, we propose a distributed algorithm for this MTDS problem. Unlike the established distributed algorithms, such as those the algorithms in [2,3] that are concerned with special graphs (planar graphs with no 3-cycles or 4-cycles), this work focuses on general graphs.

1.1 Related Work

Total domination of graphs, introduced in [10], has been under extensive study in graph theory [16,17,20,21]. In [13], Garey et al. showed that it is NP-hard to compute an MTDS in general graphs, and is also MAX SNP-hard [24,27,28]. For some special graph classes, some complexity results are summarized in [20].

The first linear time algorithm was designed to compute the total domination number of a tree [26]. A simple heuristic algorithm was given in [19] to find a TDS of size at most $n(1+\ln \delta)/\ln \delta$, where n is the number of nodes in G, and δ is the minimum degree of G. Chlebík and Chlebíková [9] presented an $(H_\Delta - 1/2)$-approximation algorithm for the TDS problem, where $H_i = \sum_{h=1}^{i} 1/h$ is the i-th harmonic number. They also proved that there exist constants $a > 0$ and $b \geq 3$ such that for each $\Delta \geq b$ it is NP-hard to approximate MTDS within a factor $(\ln \Delta - a \ln \ln \Delta)$ in bipartite graphs.

Zhu [30] proposed a greedy algorithm for MTDS with approximation ratio $\ln(\Delta - 0.5) + 1.5$, where Δ is the maximum degree, and showed that some approximation hardness for the MTDS problem on some special graph classes. Specifically, Zhu [30] provided a lower bound on approximability for MTDS in bounded degree graphs and regular graphs: it is NP-hard to approximate MTDS within factor $391/390$ for 3-bounded degree graphs, within factor $100/99$ for 4-bounded degree graphs, within factor $53/52$ for 5-bounded degree graphs, within factor $51/50$ for 6-bounded degree graphs, within factor $681/680$ for 3-regular graphs, within factor $250/249$ for 4-regular graphs and within factor $169/168$ for 5-regular graphs. Schaudt and Schrader [29] proved that it is NP-hard to find a TDS T such that the induced subgraph $G[T]$ satisfies some specified properties.

Alipour et al. [2] designed a distributed algorithm for the MTDS problem in a planar graph without 3-cycles, where the approximation ratio and communication rounds are 16 and $O(1)$, respectively. In a planar graph without 4-cycles, Alipour and Jafari [3] proposed another distributed algorithm for the problem, where the ratio is 9, and the communication rounds is also $O(1)$. A total dominating set T is called *minimal* when no proper subset of T is a total dominating set of G. Goddard et al. [14] proposed a self-stabilizing algorithm for the minimal TDS problem. However, the running time of the algorithm is exponential. Belhoul et al. [7] presented a polynomial self-stabilizing distributed algorithm for the minimal problem.

1.2 Our Contribution

Our major contributions are summarized as follows:

1. Based on the LP relaxation technique, we propose a distributed algorithm for the MTDS problem. For a positive integer k and the maximum degree Δ, the proposed algorithm outputs a fractional TDS in $O(k^2)$ communication rounds, where the expected size of the output TDS is $O(k\Delta^{\frac{2}{k}})|TDS_{OPT}|$ and the size of the messages used in our algorithm is $O(\log \Delta)$. This algorithm is therefore the first distributed algorithm to obtain an approximation ratio in a constant number of communication rounds.
2. We then present a rounding algorithm to obtain an integer TDS from a feasible solution of the linear programming relaxation of this problem.

1.3 Organisation of the Paper

We first introduce some notations and give the linear programming relaxation for the MTDS problem in Sect. 2. In Sect. 3, based on the LP relaxation technique, we design a distributed algorithm for the MTDS problem, and then propose a rounding algorithm for this problem. In Sect. 4, we conclude the paper.

2 Preliminaries

For a given network graph $G = (V, E)$, we assume that $V = \{v_1, v_2, \ldots, v_n\}$. For a node $v_i \in V$, the open neighborhood and closed neighborhood of v_i are $N(v_i) = \{v_j \in V | v_i v_j \in E\}$ and $N[v_i] = N(v_i) \cup \{v_i\}$, respectively. Let $\delta(v_i)$ and Δ be the degree of a node v_i and the maximum degree of G, respectively. Since we will often use the maximum degree of a node v_i in its neighborhood and 2-neighborhood, we denote $\delta^{(1)}(v_i) := \max_{v_j \in N(v_i)} \delta(v_j)$ and $\delta^{(2)}(v_i) := \max_{v_j \in N(v_i)} \delta^{(1)}(v_j)$ as the maximum degree in $N(v_i)$ and the 2-neighborhood of v_i, respectively.

Throughout, we assume a synchronous communication model [6]. That is, in each communication round, every node send a message to its neighbors in graph G. The time complexity of an algorithm is the number of communication rounds. Since we process messages only in the neighborhood of a node, the size of the messages is at most $O(\log \Delta)$ in our proposed algorithm.

Next, we give the linear programming relaxation for the MTDS problem. In a given graph $G = (V, E)$, a subset $T \subseteq V$ is said to be a TDS if each node $v \in V$

is adjacent to some member of T. In other words, for every $v_i \in V$, at least one node in $N(v_i)$ is in T. We introduce a binary variable x_i to each $v_i \in V$: $x_i = 1$ whenever $v_i \in T$. Hence, T is a TDS of G if and only if every $v_i \in V$ satisfies $\sum_{v_j \in N(v_i)} x_j \geq 1$. Let N be the adjacency matrix of G. The MTDS problem be formulated as an integer linear program (ILP_{MTDS}):

$$\begin{aligned} \textbf{minimize} \quad & \sum_{i=1}^{n} x_i \\ \textbf{s.t.} \quad & N \cdot \boldsymbol{x} \geq \mathbf{1} \\ & \boldsymbol{x} \in \{0,1\}^n. \end{aligned}$$

By relaxing the integer constraints to $\boldsymbol{x} \geq \mathbf{0}$, we get the linear programming relaxation (LP_{MTDS}) of ILP_{MTDS}:

$$\begin{aligned} \textbf{minimize} \quad & \sum_{i=1}^{n} x_i \\ \text{s.t.} \quad & N \cdot \boldsymbol{x} \geq \mathbf{1} \\ & \boldsymbol{x} \geq \mathbf{0}. \end{aligned}$$

The dual of this linear programming relaxation is

$$\begin{aligned} \textbf{maximize} \quad & \sum_{i=1}^{n} y_i \\ \text{s.t.} \quad & N \cdot \boldsymbol{y} \leq \mathbf{1} \\ & \boldsymbol{y} \geq \mathbf{0}. \end{aligned}$$

3 Algorithms for the MTDS Problem

Firstly, we will give the distributed algorithm for the MTDS problem, where all the nodes know the maximum degree Δ of G. Then, we present a rounding algorithm for this problem.

3.1 Approximating LP_{MTDS}

In this subsection, we propose a distributed algorithm for the MTDS problem by integrating the techniques in [12,25]. Assume that all the nodes know Δ. Similarly to the assumption about dominating set in [15,22], a node v_i is colored black when the node satisfies $\sum_{v_j \in N(v_i)} x_j \geq 1$. Initially all nodes are white. Let $\tilde{\delta}(v_i)$ denote the *dynamic degree* of a node v_i that is the number of white nodes $v_j \in N(v_i)$. Therefore, $\delta(v_i) = \tilde{\delta}(v_i)$ when the algorithm starts.

Before analyzing the algorithm, we give an overview of Algorithm 1. All nodes execute the algorithm synchronously. The variable x_i is initialized as 0, each x_i is gradually increasing over time. Algorithm 1 has two for-loops. The outer loop is mainly to reduce $\tilde{\delta}(v_i)$ of G. From Lemma 1, we know that the dynamic degree $\tilde{\delta}(v_i)$ is reduced by a constant factor $\Delta^{\frac{1}{k}}$ in each iteration. From Lines 6–7 of Algorithm 1, only the nodes that satisfy $\tilde{\delta}(v_i) \geq \Delta^{\frac{\hbar}{k}}$ increase the corresponding x_i in the inner loop. We call those nodes *active*. In Line 5, for a white node v_i, let $a(v_i)$ denote the number of active nodes in $N(v_i)$. If the node v_i is black, set $a(v_i) = 0$. The inner loop is mainly to reduce $a(v_i)$ in G.

Algorithm 1. Approximating LP_{MTDS}

Input: $G = (V, E)$, a positive integer k and maximum degree Δ.
Output: A feasible solution $\boldsymbol{x}$ for LP_{MTDS}.

1: initially $x_i := 0$, $\tilde{\delta}(v_i) := \delta(v_i)$;
2: **for** $\hbar := k - 1$ to 0 by -1 **do**
3:    ($\star$ $\tilde{\delta}(v_i)$, set $z_i := 0$ $\star$)
4:    **for** $m := k - 1$ to 0 by -1 **do**
5:       ($\star$ $a(v_i)$ $\star$)
6:       **if** $\tilde{\delta}(v_i) \geq \Delta^{\frac{\hbar}{k}}$ **then**
7:          $x_i := \max\{x_i, \frac{1}{\Delta^{\frac{m}{k}}}\}$
8:       **end if**
9:       send x_i to all nodes in $N(v_i)$;
10:      **if** $\sum_{v_j \in N(v_i)} x_j \geq 1$ **then**
11:         the node v_i is colored 'black'
12:      **end if**
13:      send the color of node v_i to all nodes in $N(v_i)$;
14:      update $\tilde{\delta}(v_i) := |\{v_j \in N(v_i) | \textit{the node } v_j \textit{ is white}\}|$
15:   **end for**
16:   ($\star$ z_i $\star$)
17: **end for**

First, some auxiliary lemmas are given for proving the approximation ratio of our algorithm.

Lemma 1. *For each $v_i \in V$, we have*
(i) $\tilde{\delta}(v_i)$ is at most $\Delta^{\frac{\hbar+1}{k}}$ when each outer loop iteration $\hbar$ starts;
(ii) $a(v_i)$ is at most $\Delta^{\frac{m+1}{k}}$ when each inner loop iteration starts.

Next, we analyze the increased weights in the inner loop iterations. It is a bit difficult to directly analyze the sum of all the increased x_i. Hence, for each v_i, we introduce a new variable z_i. Initially all $z_i = 0$ in Line 3 of Algorithm 1. When the node v_i increases x_i over time, the increased weights are equally distributed among the z_j of the nodes v_j in the open neighborhood $N(v_i)$, where the nodes v_j are white before the increment of x_i. Therefore $\sum z_j$ is equal to the sum of all the increased x_i in each outer loop iteration of Algorithm 1. Hence, we only need to prove that all z_i are bounded.

Lemma 2. *For each $v_i \in V$, we have $z_i \leq \frac{1}{\Delta^{(\hbar-1)/k}}$ at the end of each outer loop iteration of Algorithm 1.*

Based on the above lemmas, we will present the approximation ratio and time complexity of Algorithm 1.

Theorem 1. *For a positive integer k and the maximum degree Δ, Algorithm 1 is a $\left(k\Delta^{\frac{2}{k}}\right)$-approximation for the minimum total dominating set problem in G.*

Proof. Because all nodes v_i with dynamic degrees $\tilde{\delta}(v_i) \geq 1$, x_i will become 1 when the iteration ($\hbar = 0, m = 0$) ends, the output $\boldsymbol{x}$ is a feasible solution for LP_{MTDS}.

Next, we consider the approximation ratio. Firstly, we prove that the increased weights is at most $\Delta^{\frac{2}{k}}$ in every outer loop iteration. According to Lemma 1 (i), when the outer loop iteration starts, $\tilde{\delta}(v_i) \leq \Delta^{\frac{\hbar+1}{k}}$ for each v_i. Therefore there are at most $\Delta^{\frac{\hbar+1}{k}}$ non-zero z_j in $N(v_i)$ in the iteration. According to Lemma 2, all z_j satisfy $z_j \leq \frac{1}{\Delta^{(\hbar-1)/k}}$. Hence the sum of the z_j in $N(v_i)$ of v_i in every outer loop iteration is at most

$$\sum_{v_j \in N(v_i)} z_j \leq \Delta^{\frac{\hbar+1}{k}} / \Delta^{\frac{\hbar-1}{k}} = \Delta^{\frac{2}{k}}.$$

Now, we set $y_j := \frac{z_j}{\Delta^{\frac{2}{k}}}$ to each node v_j, we have $\forall v_i \in V : \sum_{v_j \in N(v_i)} y_j \leq 1$, and so the $\bar{y}$ is a feasible solution of DLP_{MTDS}. Therefore, $\sum y_j$ is a lower bound of an optimal TDS (TDS_{OPT}). For each outer loop iteration, we obtain

$$\sum_{j=1}^{n} z_j \leq \Delta^{\frac{2}{k}} |TDS_{OPT}|.$$

According to the above description of z_j, the sum of all z_j is equal to the sum of all increased x_i, and the outer loop have k iterations. Hence we obtain

$$\sum_{i=1}^{n} x_i \leq k\Delta^{\frac{2}{k}} |TDS_{OPT}|.$$

Therefore, we prove that this conclusion is correct. □

Theorem 2. *The number of the communication rounds for Algorithm 1 is $2k^2$.*

Proof. On the one hand, the algorithm involves sending two messages in every inner loop iteration, hence it takes 2 rounds. On the other hand, the algorithm has 2 for-loops. So the number of such iterations is k^2. Hence the number of the communication rounds is $2k^2$. □

3.2 Approximating *MTDS* with a Rounding Algorithm

In this subsection, by exploiting a feasible solution to the linear relaxation program of the MTDS problem, we present a rounding algorithm to obtain an integer TDS. Firstly, we introduce a crucial lemma and two well-known inequalities [25] without proofs, which will be used in the proof of Theorem 3.

Lemma 3. *For any TDS, we have*

$$\sum_{i=1}^{n} \frac{1}{\delta^{(1)}(v_i)} \leq |TDS|$$

Lemma 4. *Let $\mathcal{R} \subset \mathbb{R}^+$ denote a set of positive real numbers, we have*

$$\prod_{x \in \mathcal{R}} x \leq \left(\frac{\sum_{x \in \mathcal{R}} x}{|\mathcal{R}|} \right)^{|\mathcal{R}|}.$$

Lemma 5. *For $1 \leq x \leq l$, we obtain*

$$\left(1 - \frac{x}{l}\right)^{l} \leq \frac{1}{e^x}.$$

Then, we present the rounding algorithm as follows:

Algorithm 2. From LP_{MTDS} to ILP_{MTDS}

Require:
A feasible solution $\boldsymbol{x}^{(\beta)}$ for LP_{MTDS}
Ensure:
An integer feasible solution $\boldsymbol{x}_{TDS}$ for ILP_{MTDS}
1: calculate $\delta^{(2)}(v_i)$ for each node v_i
2: $p_{v_i} := \min\{1, x_i^{(\beta)} \cdot \ln \delta^{(2)}(v_i)\}$
3: $x_{TDS,v_i} := \begin{cases} 1, & \text{with probability} \quad p_{v_i} \\ 0, & \text{otherwise} \end{cases}$
4: send x_{TDS,v_i} to all nodes in $N(v_i)$
5: **if** $x_{TDS,v_j} := 0$ for all nodes $v_j \in N[v_i]$ **then**
6: $\quad x_{TDS,v_i} := 1$ and $x_{TDS,v_j} := 1$ for some $v_j \in N(v_i)$
7: **end if**
8: **if** $x_{TDS,v_i} := 1$ and $x_{TDS,v_j} := 0$ for all $v_j \in N(v_i)$ **then**
9: $\quad x_{TDS,v_j} := 1$ for some $v_j \in N(v_i)$
10: **end if**

Theorem 3. *Suppose TDS_{OPT}, $\boldsymbol{x}^{(\beta)}$ and $\boldsymbol{x}^*$ are a minimum total dominating set, an β-approximation solution and an optimal solution for LP_{MTDS}, respectively. Let $\boldsymbol{x}_{TDS}$ be a feasible solution for ILP_{MTDS} output by Algorithm 2. We obtain that the expected value of the corresponding TDS is at most*

$$E[|TDS|] \leq (2\beta \ln \Delta + 3) \cdot |TDS_{OPT}|.$$

Proof. From the execution of Algorithm 2, it be seen that a node only be added to a TDS in Lines 3, 6 and 9. We use the variables X, $2Y$ and Z to denote the numbers of all nodes which are added to the TDS in Lines 3, 6 and 9, respectively.

Firstly, we consider the expected value of variable X, we obtain

$$\begin{aligned}
E[X] &= \sum_{i=1}^{n} p_{v_i} \\
&\leq \sum_{i=1}^{n} (x_i^{(\beta)} \cdot \ln \delta^{(2)}(v_i)) \\
&\leq \ln \Delta \cdot \sum_{i=1}^{n} x_i^{(\beta)} \\
&\leq \beta \ln \Delta \cdot \sum_{i=1}^{n} x_i^{*} \\
&\leq \beta \ln \Delta \cdot |TDS_{OPT}|.
\end{aligned}$$

The second inequality holds from $\delta^{(2)}(v_i) \leq \Delta$; and the third inequality follows since $\boldsymbol{x}^{(\beta)}$ is an β-approximation solution for LP_{MTDS}.

Secondly, we compute $E[Y]$. Let q_{v_i} denote the probability that no node in $N[v_i]$ is added to the TDS. For a $v_j \in N[v_i]$, if $x_j^{(\beta)} \cdot \ln \delta^{(2)}(v_j) \geq 1$, then $p_{v_j} = 1$, hence $q_{v_i} = 0$. So we only need to consider all $p_{v_j} < 1$, we get

$$\begin{aligned}
q_i = \prod_{v_j \in N[v_i]} (1 - p_{v_j}) &= (1 - p_{v_i}) \prod_{v_j \in N(v_i)} (1 - p_{v_j}) \\
&\underset{(\delta^{(1)}(v_i) \leq \delta^{(2)}(v_i))}{\leq} \prod_{v_j \in N(v_i)} (1 - x_j^{(\beta)} \cdot \ln \delta^{(1)}(v_i)) \\
&\underset{(Lemma\ 4)}{\leq} \left(1 - \frac{\sum_{v_j \in N(v_i)} x_j^{(\beta)} \cdot \ln \delta^{(1)}(v_i)}{\delta_{v_i}} \right)^{\delta_{v_i}} \\
&\leq (1 - \frac{\ln \delta^{(1)}(v_i)}{\delta_i})^{\delta_i} \underset{(Lemma\ 5)}{\leq} e^{-\ln \delta^{(1)}(v_i)} = \frac{1}{\delta^{(1)}(v_i)}.
\end{aligned}$$

Since $\boldsymbol{x}^{(\beta)}$ is a feasible solution for LP_{MTDS}, $\sum_{v_j \in N(v_i)} x_j^{(\beta)} \geq 1$, so the third inequality is true. Combining this with Lemma 3, we obtain

$$\begin{aligned}
E[2Y] = 2E[Y] &= 2 \sum_{i=1}^{n} q_{v_i} \\
&\leq 2 \sum_{i=1}^{n} \frac{1}{\delta^{(1)}(v_i)} \\
&\leq 2|TDS_{OPT}|.
\end{aligned}$$

Finally, we compute $E[Z]$. Similar to the analysis of the expected value of Y, we also obtain

$$\begin{aligned} E[Z] &= \sum_{i=1}^{r} p_{v_i} + \sum_{i=r+1}^{n} q_{v_i} \\ &\le \ln\Delta \cdot \sum_{i=1}^{r} x_i^{(\beta)} + \sum_{i=r+1}^{n} \frac{1}{\delta^{(1)}(v_i)} \\ &\le (\beta \ln\Delta + 1)|TDS_{OPT}|. \end{aligned}$$

By adding $E[X]$, $E[2Y]$ and $E[Z]$, we obtain

$$E[|TDS|] \le (2\beta \ln\Delta + 3) \cdot |TDS_{OPT}|.$$

□

4 Conclusion

In this paper, we initially devise a distributed algorithm for solving the fractional total dominating set problem. Subsequently, we introduce a rounding algorithm aimed at deriving an integer total dominating set by leveraging a feasible solution to the linear relaxation program of this problem. It is intriguing to explore this problem even in scenarios where the maximum degree is unknown or the vertices in the graph are weighted.

Acknowledgments. The research is partially supported by National Natural Science Foundation of China (Nos. 12271259, U20A2068, 12061059), and the Qinghai Key Laboratory of Internet of Things Project (2017-ZJ-Y21), and the Natural Sciences and Engineering Research Council of Canada (NSERC) Grant (2024–03840).

Disclosure of Interests. The authors have no competing interests to declare that are relevant to the content of this article.

References

1. Abu-Khzam, F.N.: An improved exact algorithm for minimum dominating set in chordal graphs. Inf. Process. Lett. **174**, 106206 (2022)
2. Alipour S., Futuhi E., Karimi S.: On distributed algorithms for minimum dominating set problem, from theory to application. arXiv preprint arXiv:2012.04883 (2020)
3. Alipour S., Jafari A.: A local constant approximation factor algorithm for minimum dominating set of certain planar graphs. In Proceedings of the 32nd ACM Symposium on Parallelism in Algorithms and Architectures. July, pp. 501–502 (2020)
4. Allesina, S., Bodini, A.: Who dominates whom in the ecosystem? Energy flow bottlenecks and cascading extinctions. J. Theor. Biol. **230**(3), 351–358 (2004)

5. Alzoubi, K.M., Wan, P.J., Frieder, O.: Maximal independent set, weakly-connected dominating set, and induced spanners in wireless ad hoc networks. Int. J. Found. Comput. Sci. **14**(02), 287–303 (2003)
6. Awerbuch, B., Goldberg, A. V., Luby, M., Plotkin, S. A.: Network decomposition and locality in distributed computation. In Proceedings of the 30th Annual IEEE Symposium on Foundations of Computer Science, pp. 364–369 (1989)
7. Belhoul, Y., Yahiaoui, S., Kheddouci, H.: Efficient self-stabilizing algorithms for minimal total k-dominating sets in graphs. Inf. Process. Lett. **114**(7), 339–343 (2014)
8. Chen, J., He, K., Du, R., Zheng, M., Xiang, Y., Yuan, Q.: Dominating set and network coding-based routing in wireless mesh networks. IEEE Trans. Parallel Distrib. Syst. **26**(2), 423–433 (2013)
9. Chlebík, M., Chlebíková, J.: Approximation hardness of dominating set problems in bounded degree graphs. Inf. Comput. **206**(11), 1264–1275 (2008)
10. Cockayne, E.J., Dawes, R.M., Hedetniemi, S.T.: Total domination in graphs. Networks **10**(3), 211–219 (1980)
11. Das, B., Bharghavan, V. : Routing in ad-hoc networks using minimum connected dominating sets. In: Proceedings of ICC'97-International Conference on Communications Vol. 1, pp. 376–380, IEEE, (1997)
12. Elkin, M.: Distributed approximation: a survey. ACM SIGACT News **35**(4), 40–57 (2004)
13. Garey, M.R., Johnson, D.S.: Computers and Intractability: A Guide to the Theory of NP-Completeness. Freeman, San Francisco (1978)
14. Goddard, W., Hedetniemi, S.T., Jacobs, D.P., Srimani, P.K.: A self-stabilizing distributed algorithm for minimal total domination in an arbitrary system graph. In: Proceedings International Parallel and Distributed Processing Symposium, IEEE (2003)
15. Guha, S., Khuller, S.: Approximation algorithms for connected dominating sets. Algorithmica **20**(4), 374–387 (1998)
16. Haynes T.W., Hedetniemi S.T., Slater P.J.: Fundamentals of Domination in Graphs. Monogr. Textb. Pure Appl. Math. **208** (1998)
17. Haynes T.W., Hedetniemi S.T., Slater P.J.: Domination in Graphs: Advanced Topics, Chapman and Hall/CRC Pure and Applied Mathematics Series, vol.209, Marcel Dekker, NewYork (1998)
18. Haynes, T.W., Hedetniemi, S.M., Hedetniemi, S.T., Henning, M.A.: Domination in graphs applied to electric power networks. SIAM J. Discret. Math. **15**(4), 519–529 (2002)
19. Henning, M.A., Yeo, A.: A transition from total domination in graphs to transversals in hypergraphs. Quaest. Math. **30**(4), 417–436 (2007)
20. Henning, M.A.: A survey of selected recent results on total domination in graphs. Discret. Math. **309**(1), 32–63 (2009)
21. Henning, M.A., Yeo, A.: Total domination in graphs. Springer, New York (2013). https://doi.org/10.1007/978-1-4614-6525-6
22. Jia, L., Rajaraman, R., Suel, T.: An efficient distributed algorithm for constructing small dominating sets. Distrib. Comput. **15**(4), 193–205 (2002)
23. Jiang, P., Liu, J., Wu, F., Wang, J., Xue, A.: Node deployment algorithm for underwater sensor networks based on connected dominating set. Sensors **16**(3), 388 (2016)
24. Khanna, S., Motwani, R., Sudan, M., Vazirani, U.: On syntactic versus computational views of approximability. In: Proceeding of the 35th Annual Symposium on the Foundations of Computer Science, pp. 819–830 (1994)

25. Kuhn, F., Wattenhofer, R.: Constant-time distributed dominating set approximation. Distrib. Comput. **17**(4), 303–310 (2005)
26. Laskar, R., Pfaff, J., Hedetniemi, S.M., Hedetniemi, S.T.: On the algorithmic complexity of total domination. SIAM Journal on Algebraic Discrete Methods **5**(3), 420–425 (1984)
27. Papademitriou, C., Yannakakis M.: Optimization, approximation and complexity classes. In: Proceeding of the 20th Annual ACM Symposium on Theory of Computing, pp.229–234 (1988)
28. Crescenzi, P., Kann, V.: A compendium of NP optimization problems (1995)
29. Schaudt, O., Schrader, R.: The complexity of connected dominating sets and total dominating sets with specified induced subgraphs. Inf. Process. Lett. **112**(24), 953–957 (2012)
30. Zhu J.: Approximation for minimum total dominating set. In: Proceedings of the 2nd International Conference on Interaction Sciences: Information Technology, Culture and Human (ICIS '09). Association for Computing Machinery, New York, NY, USA, pp. 119–124. DOI:https://doi.org/10.1145/1655925.1655948

Revisit the Online Facility Location Problem with Uniform Facility Cost

Mengzhen Li[1], Runjie Miao[2], Chenchen Wu[3(✉)], and Dachuan Xu[1]

[1] Institute of Operations Research and Information Engineering, Beijing University of Technology, Beijing 100124, People's Republic of China
lmz9710@emails.bjut.edu.cn, xudc@bjut.edu.cn

[2] School of Mathematics and Statistics, Zhengzhou University, Zhengzhou 450001, People's Republic of China
runjiemiao@hotmail.com

[3] College of Science, Tianjin University of Technology, Tianjin 300384, People's Republic of China
wu_chenchen_tjut@163.com

Abstract. In the classical facility location problem, all information about potential facilities and clients is provided initially. However, obtaining complete information about all clients is challenging. When client information is provided incrementally, this gives rise to the **online facility location problem**. Both the online facility location problem with general facility costs and the one with uniform facility cost have attracted the attention of researchers. In the existing literature, the online facility location problem with uniform facility cost typically assumes that facilities can be located anywhere within the metric space. However, in practical scenarios where communication networks constructed by the facilities are graph-based, facility locations are limited to specific points on the network, this means that potential facility locations are discrete. In this work, we revisit the online facility location problem with uniform facility cost, focusing on scenarios where facilities can only be placed at discrete locations, known as the potential facility set. Considering clients arriving in a random order, we propose a 9.9-competitive online algorithm for the uniform facility cost case.

Keywords: Approximation algorithms · Online facility location problem · Uniform facility cost · Competitive ratio

1 Introduction

The facility location problem is one of the most fundamental topics in theoretical computer science, combinatorial optimization, and operations research. It originates from the challenge of determining the optimal locations for building facilities such as factories, hospitals, fire stations, and signal towers. This problem has wide-ranging applications, including urban planning, image stitching,

S. Ghosh and Z. Zhang (Eds.): AAIM 2024, LNCS 15179, pp. 133–144, 2024.
https://doi.org/10.1007/978-981-97-7798-3_12

supply chain management, and communication networks. In the classical facility location problem, we are given a set $\mathcal{F}$ of potential facilities and a set $\mathcal{C}$ of clients within a metric space. The objective is to minimize the total cost, which includes the cost of opening facilities and the assignment cost of serving clients. Specifically, the cost of opening facility i is f_i, and the assignment cost for serving client j from facility i is represented by the distance $d(i, j)$, where $i \in \mathcal{F}$ and $j \in \mathcal{C}$. The facility location problem is a well-known NP-hard problem in combinatorial optimization. As such, considerable research has been dedicated to developing approximation algorithms to address it effectively.

The first constant factor approximation algorithm for the facility location problem was introduced by Shmoys *et al.* [16] in 1997. Since then, the approximation ratio has been progressively improved. Chudak and Shmoys [4] advanced it to $(1 + \frac{2}{e})$, followed by Jain *et al.* [10] to 1.61. Mahdian *et al.* [14] further improved this to 1.52, and Byrka and Aardal refined it to 1.5 [3]. Currently, the best known approximation ratio is 1.488, proposed by Li [13] a decade ago. On the hardness side, Guha and Khuller [9] established a lower bound for the approximation ratio at 1.463, unless NP $\subseteq$ DTIME $\left(n^{O(\log\log n)}\right)$ [9], and Sviridenko [17] enhanced that the lower bound is 1.463 unless P = NP in private communication.

In 2001, Meyerson [15] introduced the online variant of the classical facility location problem, i.e., the online facility location problem. The most significant feature of online facility location problem is that the clients arrive one by one and must be irrevocably assigned to an open facility upon arrival. The goal is to establish a set of facilities and an assignment of the clients to them that minimize the total facility opening cost plus the assignment cost for all clients. Following Meyerson's work, the online facility location problem has received considerable attention, see [1,2,5–7,11] and the reference therein. In the online setting, the order in which clients arrive plays an important role. Meyerson [15] presented the first randomized online algorithm for facility location and showed that it was $O(1)$-competitive when clients arrive in random order. If clients are ordered adversarially, Meyerson also showed that no algorithm can be constant-competitive and provided an $O(\log n)$-competitive algorithm[1].Fotakis [6,7] later extended this result, developing an asymptotically optimal $O(\log n/ \log\log n)$-competitive algorithm.

Several online models are mentioned in this paper. In the scenario where the clients arrive in random order, Meyerson studied both the online facility location problem with general facility costs (referred to as the M_1-model) and the online facility location problem with uniform facility cost (referred to as the M_2-model). In the M_1-model, the information of the potential facility set is given initially and the facility opening costs may vary. Meyerson provided a 33-competitive online algorithm for this scenario. In the M_2-model, Meyerson assumed that one can open a new facility anywhere in the entire metric space with a uniform facility

[1] An online algorithm $\mathcal{A}$ is α-competitive if $\text{cost}(\mathcal{A}(I)) \leq \alpha \cdot \text{cost}(\text{OPT}(I))$ for all instance I. Here, $\text{cost}(\text{OPT}(I))$ is the cost of the optimal solution to the corresponding offline instance (i.e., $t_u = 0$ for every client u).

cost. Meyerson presented an 8-competitive online algorithm for the M_2-model, which was recently enhanced to 3 by Kaplan et al. [12].

The assumption that one can open new facilities anywhere in the entire metric space in the M_2-model is somewhat inconsistent with the actual application scenarios of the facility location problem. For example, in urban planning, there are numerous constraints on potential hospital locations, preventing the opening of a new hospital at every client's location. The sites for hospitals are selected in the initial phases of urban planning, considering factors like land properties, geographical location, and environment. As the city grows, only a select few hospitals will be opened from the predetermined potential locations to cater to the increasing demand. The same goes for potential locations for fire stations, signal towers, bus stations, etc. Another example occurs in image stitching, where the size of the client set (i.e., the number of pixels) is often much larger than the size of the potential facility set (i.e., the number of feature points). The third example occurs in constructing communication networks, where the available facilities are limited to a graph, and we are unable to establish new facilities everywhere. These scenarios motivate us to reconsider the assumption that one can open facilities throughout the entire metric space and revisit the online facility location problem with uniform facility cost. To better capture these scenarios, we restrict the selection of facilities to those in the initially provided facility set, and we denote our model as M_3-model.

More precisely, in the M_3-model, the information of potential facility set is initially given and we can only open new facilities from the potential facility set with a uniform facility cost. The clients arrive one by one in random order and must be irrevocably assigned to an open facility upon arrival. At any given time, we know nothing about the future clients. The goal is to open some facilities and assign all clients properly such that the total cost is minimized. Therefore, a main difference between the M_2-model and the M_3-model is that we can open new facilities anywhere in the M_2-model whereas we must open facilities from the initially given potential facility set in the M_3-model. When the initially given potential facility set in the M_3-model is exactly the entire metric space, the M_3-model degenerates to the M_2-model. Besides, it is obvious that the M_3-model studied in this paper is a special case of the M_1-model. The relationship among the three models is shown in Fig. 1.

Our contributions. We revisit the online facility location problem with uniform facility cost and provide a 9.9-competitive online algorithm for the M_3-model. Although our model is a special case of the M_1-model, noting that the best competition ratio of the M_1-model is 33. Our work represents a significant improvement over this special case. When applied to the M_2-model, our algorithm implies a 7.465-competitive online algorithm.

The organization of this paper is as follows. In Sect. 2, we introduce the 9.9-competitive online algorithm for the M_3-model. The analysis of our online algorithm are placed in Sect. 3. We conclude this paper in Sect. 4 with some open problems that we will address in the future.

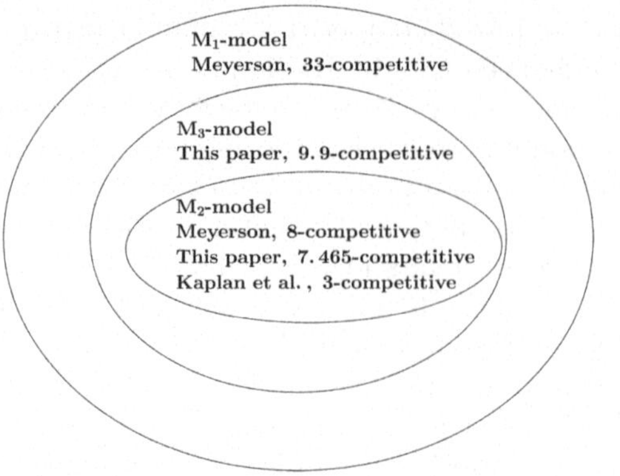

Fig. 1. The relationship among the three models.

2 Online Algorithm

Through scaling, we can assume without loss of generality that the uniform facility cost in the online model is exactly 1. Let $\mathcal{F}$ denote the potential facility set and $d(\cdot,\cdot)$ denote the distance metric. For any client u and facility subset $X \subseteq \mathcal{F}$, we use $d(u, X)$ to denote the shortest distance from client u to the set X. Thus, $d(u, \mathcal{F})$ is the distance between client u and its closest facility (denoted by f_u) in $\mathcal{F}$.

High-level description of our algorithm. When a new client u arrives, we calculate the cost of assigning u to the existing facilities. Based on this cost, we open the new facility f_u with probability (which may be beneficial to the future clients). Finally, we assign client u to the closest open facility. The formal description of our online algorithm for the M$_3$-model is presented below.

Algorithm 1. Online algorithm for the M$_3$-model

Input: Facility set $\mathcal{F}$

Initialize $Y \leftarrow \emptyset$, $\gamma \leftarrow 2.75$
for arriving client u **do**
  Calculate $d(u, Y)$ and $d(u, \mathcal{F})$
  $f_u \leftarrow \arg\min_{f \in \mathcal{F}} d(u, f)$
  $p_u \leftarrow \min\{(d(u, Y) - d(u, \mathcal{F}))/\gamma, 1\}$ /* If $Y = \emptyset$, set $d(u, Y) \leftarrow +\infty$ */
  Flip a coin with probability p_u of Heads
  if Heads **then**
    Set $Y \leftarrow Y \cup \{f_u\}$
    Assign u to f_u and set $\text{cost}(u) \leftarrow 1 + d(u, f_u)$
  else
    Assign u to the closest facility in Y and set $\text{cost}(u) \leftarrow d(u, Y)$
  end if
end for

For the sake of convenience in proofs, we introduce the set of facilities opened before the arrival of each client u as $\bar{F}_u$, corresponding to Y before Line 7 in Algorithm 1, and the set of facilities opened after the arrival of each client as F_u, corresponding to Y after Line 8 or Line 10 in Algorithm 1. Set $\hat{F}_u = F_u \setminus \bar{F}_u$. We claim that the competitive ratio of Algorithm 1 is no more than 9.9.

Theorem 1. *The competitive ratio of Algorithm 1 is no more than 9.9.*

3 Analysis

To prove Theorem 1, we need the following lemmas.

Lemma 1. *For each client u, the expected value of* $\text{cost}(u)$ *satisfies*

$$\mathbb{E}[\text{cost}(u)] \leq (1+\gamma)p_u + d(u, \mathcal{F}).$$

Proof. Note that

$$\mathbb{E}[\text{cost}(u)] = p_u(1 + d(u, \mathcal{F})) + (1 - p_u)d(u, \bar{F}_u)$$

and p_u is either 1 or $(d(u, \bar{F}_u) - d(u, \mathcal{F}))/\gamma$. We distinguish two cases and prove the claimed bound.

i) If $p_u = 1$, we have

$$\mathbb{E}[\text{cost}(u)] = 1 + d(u, \mathcal{F}) \leq (1+\gamma)p_u + d(u, \mathcal{F}),$$

where the final inequality follows from $\gamma > 0$.

ii) If $p_u = [d(u, \bar{F}_u) - d(u, \mathcal{F})]/\gamma \leq 1$, we have

$$\begin{aligned}
&\mathbb{E}[\text{cost}(u)] \\
=&p_u(1 + d(u, \mathcal{F})) + (1 - p_u)d(u, \bar{F}_u) \\
=&p_u(1 + d(u, \mathcal{F})) + (1 - p_u)(d(u, \bar{F}_u) - d(u, \mathcal{F})) + (1 - p_u)d(u, \mathcal{F}) \\
=&p_u + p_u d(u, \mathcal{F}) + (1 - p_u)\gamma p_u + d(u, \mathcal{F}) - p_u d(u, \mathcal{F}) \\
=&p_u + (1 - p_u)\gamma p_u + d(u, \mathcal{F}) \\
\leq&(1+\gamma)p_u + d(u, \mathcal{F}),
\end{aligned}$$

where the final inequality follows from that $p_u \leq 1$ and $\gamma > 0$. This completes the proof. □

Denote by OPT the optimal solution to the corresponding offline instance. For each $c \in \text{OPT}$, denote by D_c the clients served by c in OPT. To prove Theorem 1, it suffices to show that

$$\mathbb{E}\left[\sum_{u \in D_c} \text{cost}(u)\right] \leq 9.9 \cdot \left(1 + \sum_{u \in D_c} d(u, c)\right)$$

for every facility $c \in \mathrm{OPT}$.

In the following discussion, we focus on an arbitrary but fixed facility $c \in \mathrm{OPT}$. For simplicity, for the client $u \in D_c$, we will replace $d(u, c)$ by d_u^*. Set

$$a = \frac{\sum_{u \in D_c} d_u^*}{|D_c|},$$

and choose

$$\alpha = 1.82 > 1, \beta = \frac{\alpha}{\alpha - 1}.$$

We first divide D_c into "good" part G and "bad" part B as follows: G consists of the first $\lceil \frac{|D_c|}{\alpha} \rceil$ closest clients to c in D_c and B consists of the rest clients. Clearly,

$$|G| \geq 1,$$

and

$$1 + \frac{|D_c|}{\alpha} \geq |G| \geq \frac{|D_c|}{\alpha}.$$

Below, Lemma 2 provides a property of clients $u \in G$. For ease of proof, the set D_c is divided according to the definitions of the good part and bad part, as shown in the following figure (Fig. 2).

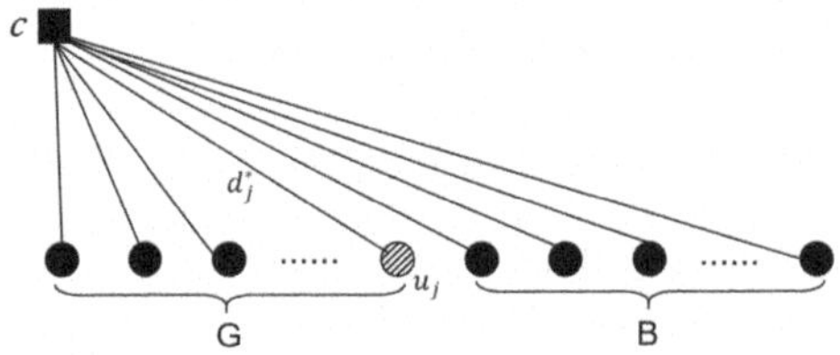

Fig. 2. The partition of the set D_c.

The client u_j^* represents the farthest client in the good part G and d_j^* denotes the distance between facility c and client u_j^*.

Lemma 2. *For each client $g \in G$, $d_g^* \leq \frac{\alpha}{\alpha-1} a = \beta a$.*

Proof. Note that

$$|B| = |D_c| - |G| \geq |D_c| - \frac{|D_c|}{\alpha} - 1 = |D_c|(1 - \frac{1}{\alpha}) - 1.$$

We conclude that

$$d_g^* \leq \frac{\sum_{u \in B} d_u^* + d_j^*}{|D_c|(1 - \frac{1}{\alpha})} \leq \frac{\sum_{u \in D_c} d_u^*}{|D_c|(1 - \frac{1}{\alpha})} = \frac{\alpha}{\alpha - 1} a.$$

This completes the proof. □

Set $g_0 = \arg\min_{t_g}\{g : g \in G, \hat{F}_g \neq \emptyset\}$. In our analysis, we further divide G into two subsets G_1 and G_2 as follows: G_2 consists of the clients in G that arrive later than g_0, and G_1 is the rest of clients in G. We have the following lemmas.

Lemma 3. *For the clients in G_1, we have $\mathbb{E}[\sum_{g \in G_1} p_g] \leq 1$.*

Proof. From the choice of G_1, the first client (say x_1) in G_1 contributes p_{x_1} to $\mathbb{E}[\sum_{g \in G_1} p_g]$. Following from the fact that $x_i (i > 1)$ contributes to $\mathbb{E}[\sum_{g \in G_1} p_g]$ only if we do not open facilities for $x_1, x_2, \cdots, x_{i-1}$, the second client (if it exists, say x_2) contributes $(1 - p_{x_1})p_{x_2}$ to $\mathbb{E}[\sum_{g \in G_1} p_g]$, the third client (if it exists, say x_3) contributes $\prod_{i=1}^{2}(1 - p_{x_i})p_{x_3}$ to $\mathbb{E}[\sum_{g \in G_1} p_g]$, and so on. Thus, the value of $\mathbb{E}[\sum_{g \in G_1} p_g]$ depends on the number of clients in G_1.

Suppose that $|G_1| = k$ and denote by $x_1, x_2, \cdots, x_k$ the clients in G_1 with x_i arrives earlier than x_{i+1} for all $i = 1, 2, \cdots, k-1$. Then we have

$$
\begin{aligned}
&\mathbb{E}[\sum_{g \in G_1} p_g] \\
=&p_{x_1} + (1 - p_{x_1})p_{x_2} + \prod_{i=1}^{2}(1 - p_{x_i})p_{x_3} + \cdots + \prod_{i=1}^{k-1}(1 - p_{x_i})p_{x_k} \\
\leq&p_{x_1} + (1 - p_{x_1}) \\
=&1.
\end{aligned}
$$

Note that the above result hold for any integer k. We conclude that $\mathbb{E}[\sum_{g \in G_1} p_g] \leq 1$. This completes the proof. □

Lemma 4. *For the clients in G_2, we have*

$$\sum_{g \in G_2} p_g \leq \frac{2}{\gamma}\beta a|G_2| + \frac{1}{\gamma}\sum_{g \in G_2}(d_g^* - d(g, \mathcal{F})).$$

Proof. From the choice of G_2 and g_0, for each client $g \in G_2$, we have

$$d(g, \bar{F}_g) \leq d(g, f_{g_0}) \leq d_g^* + d_{g_0}^* + d(g_0, f_{g_0}) \leq d_g^* + 2d_{g_0}^* \leq d_g^* + 2\beta a,$$

where the first inequality follows from the choice of g_0, the second inequality follows from the triangle inequality and the fact that $g_0, g \in D_c$ (as shown in Fig. 3). Thus, $p_g \leq \frac{1}{\gamma}(2\beta a + d_g^* - d(g, \mathcal{F}))$. Consequently,

$$\sum_{g \in G_2} p_g \leq \frac{2}{\gamma}\beta a|G_2| + \frac{1}{\gamma}\sum_{g \in G_2}(d_g^* - d(g, \mathcal{F})).$$

This completes the proof. □

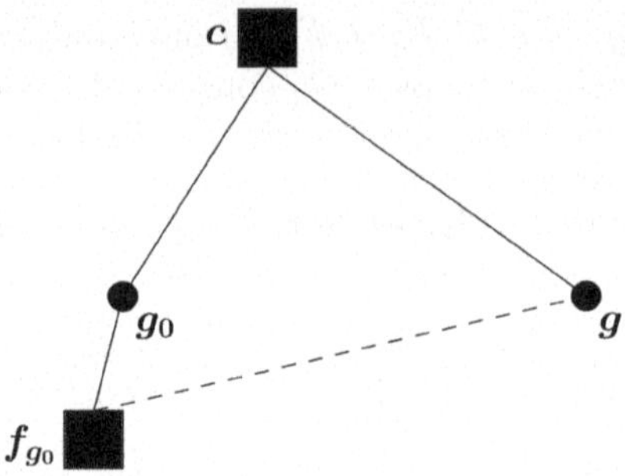

Fig. 3. An illustration of the second inequality in Lemma 4.

The following corollary follows from Lemma 3 and Lemma 4 directly.

Corollary 1. *For the clients in G, we have*

$$\mathbb{E}[\sum_{g \in G} p_g] \leq 1 + \frac{2}{\gamma}\beta a|G_2| + \frac{1}{\gamma}\sum_{g \in G_2}(d_g^* - d(g, \mathcal{F})).$$

Next, we turn to bound the client in B. We need the following lemma.

Lemma 5. *For $u, v \in D_c$, if u arrives later than v, then*

$$p_u \leq p_v + \frac{1}{\gamma}(d(v, \mathcal{F}) + d_u^* + d_v^* - d(u, \mathcal{F})).$$

Proof. Note that p_v is either 1 or $\frac{1}{\gamma}(d(v, \bar{F}_v) - d(v, \mathcal{F}))$. We distinguish two cases.

If $p_v = 1$, the result follows from the fact that $p_u \leq 1$. If $p_v = \frac{1}{\gamma}(d(v, \bar{F}_v) - d(v, \mathcal{F}))$, then we have

$$\begin{aligned}
p_u \leq & \frac{1}{\gamma}[d(u, \bar{F}_u) - d(u, \mathcal{F})] \\
\leq & \frac{1}{\gamma}[d(u, \bar{F}_v) - d(u, \mathcal{F})] \\
\leq & \frac{1}{\gamma}[d(v, \bar{F}_v) + d_u^* + d_v^* - d(u, \mathcal{F})] \\
= & \frac{1}{\gamma}[d(v, \bar{F}_v) - d(v, \mathcal{F}) + d(v, \mathcal{F}) + d_u^* + d_v^* - d(u, \mathcal{F})] \\
= & p_v + \frac{1}{\gamma}[d(v, \mathcal{F}) + d_u^* + d_v^* - d(u, \mathcal{F})].
\end{aligned}$$

This completes the proof. □

Lemma 6. *For every client $b \in B$, we have*

$$\mathbb{E}[p_b] \leq \frac{1}{|G|+1}\left[1 + \sum_{g \in G} p_g + \frac{1}{\gamma}\sum_{g \in G}(d(g, \mathcal{F}) + d_g^*)\right] + \frac{1}{\gamma}(d_b^* - d(b, \mathcal{F})).$$

Proof. Suppose that g is the last good client that arrives before b. If g does not exist, then $p_b \leq 1$. If g exists, from Lemma 5, we have $p_b \leq p_g + \frac{1}{\gamma}(d(g,\mathcal{F}) + d_g^* + d_b^* - d(b,\mathcal{F}))$. Thus,

$$\begin{aligned}\mathbb{E}[p_b] \leq& \frac{1}{|G|+1}\left[1 + \sum_{g\in G}\left(p_g + \frac{1}{\gamma}(d(g,\mathcal{F}) + d_g^* + d_b^* - d(b,\mathcal{F}))\right)\right] \\ \leq& \frac{1}{|G|+1}\left[1 + \sum_{g\in G} p_g + \frac{1}{\gamma}\sum_{g\in G}(d(g,\mathcal{F}) + d_g^*)\right] + \frac{1}{\gamma}(d_b^* - d(b,\mathcal{F})).\end{aligned}$$

This completes the proof. □

Corollary 2. *For the clients in B, we have*

$$\mathbb{E}\left[\sum_{b\in B} p_b\right] \leq (\alpha-1)\left[1 + \sum_{g\in G} p_g + \frac{1}{\gamma}\sum_{g\in G}(d(g,\mathcal{F}) + d_g^*)\right] + \frac{1}{\gamma}\sum_{b\in B}(d_b^* - d(b,\mathcal{F})).$$

Proof. From Lemma 6, we have

$$\begin{aligned}\mathbb{E}\left[\sum_{b\in B} p_b\right] \leq& \frac{|B|}{|G|+1}\left[1 + \sum_{g\in G} p_g + \frac{1}{\gamma}\sum_{g\in G}(d(g,\mathcal{F}) + d_g^*)\right] + \frac{1}{\gamma}\sum_{b\in B}(d_b^* - d(b,\mathcal{F})) \\ \leq& \frac{|B|}{|G|}\left[1 + \sum_{g\in G} p_g + \frac{1}{\gamma}\sum_{g\in G}(d(g,\mathcal{F}) + d_g^*)\right] + \frac{1}{\gamma}\sum_{b\in B}(d_b^* - d(b,\mathcal{F})) \\ \leq& (\alpha-1)\left[1 + \sum_{g\in G} p_g + \frac{1}{\gamma}\sum_{g\in G}(d(g,\mathcal{F}) + d_g^*)\right] + \frac{1}{\gamma}\sum_{b\in B}(d_b^* - d(b,\mathcal{F})),\end{aligned}$$

where the final inequality follows from the fact that $|B| \leq \left(1 - \frac{1}{\alpha}\right)|C|$ and $|G| \geq \frac{1}{\alpha}|C|$. □

From Corollary 2, we obtain the following result.

Corollary 3. *For the clients in D_c, we have*

$$\mathbb{E}\left[\sum_{u\in D_c} p_u\right] \leq \alpha\sum_{g\in G} p_g + (\alpha-1)\left[1 + \frac{1}{\gamma}\sum_{g\in G}(d(g,\mathcal{F}) + d_g^*)\right] + \frac{1}{\gamma}\sum_{b\in B}(d_b^* - d(b,\mathcal{F})).$$

Proof. From Corollary 2, we have

$$\begin{aligned}\mathbb{E}\left[\sum_{u\in D_c} p_u\right] =& \mathbb{E}\left[\sum_{g\in G} p_g\right] + \mathbb{E}\left[\sum_{b\in B} p_b\right] \\ \leq& \alpha\sum_{g\in G} p_g + (\alpha-1)\left[1 + \frac{1}{\gamma}\sum_{g\in G}(d(g,\mathcal{F}) + d_g^*)\right] + \frac{1}{\gamma}\sum_{b\in B}(d_b^* - d(b,\mathcal{F})).\end{aligned}$$

The result follows. □

Now we are ready to prove Theorem 1.

Proof. From Lemma 1 and Corollary 3, we have

$$
\begin{aligned}
&\mathbb{E}\left[\sum_{u\in D_c}\mathrm{cost}(u)\right]\\
&\le (1+\gamma)\mathbb{E}\left[\sum_{u\in D_c}p_u\right]+\sum_{u\in D_c}d(u,\mathcal{F})\\
&\le \alpha(1+\gamma)\sum_{g\in G}p_g+(\alpha-1)(1+\gamma)\left[1+\frac{1}{\gamma}\sum_{g\in G}(d(g,\mathcal{F})+d_g^*)\right]\\
&\quad+\left(1+\frac{1}{\gamma}\right)\sum_{b\in B}(d_b^*-d(b,\mathcal{F}))+\sum_{g\in G}d(g,\mathcal{F})+\sum_{b\in B}d(b,\mathcal{F})\\
&\le \alpha(1+\gamma)\sum_{g\in G}p_g+(\alpha-1)(1+\gamma)\left[1+\frac{1}{\gamma}\sum_{g\in G}(d(g,\mathcal{F})+d_g^*)\right]+\left(1+\frac{1}{\gamma}\right)\sum_{b\in B}d_b^*+\sum_{g\in G}d(g,\mathcal{F})\\
&\le \alpha(1+\gamma)\sum_{g\in G}p_g+(\alpha-1)(1+\gamma)\left[1+\frac{1}{\gamma}\sum_{g\in G}d(g,\mathcal{F})\right]+\left(1+\frac{1}{\gamma}\right)\sum_{u\in D_c}d_u^*+\sum_{g\in G}d(g,\mathcal{F})\\
&\le (2\alpha-1)(1+\gamma)+2\alpha\left(1+\frac{1}{\gamma}\right)|G_2|\beta a+\alpha\left(1+\frac{1}{\gamma}\right)\sum_{g\in G_2}(d_g^*-d(g,\mathcal{F}))\\
&\quad+(\alpha-1)\left(1+\frac{1}{\gamma}\right)\sum_{g\in G}d(g,\mathcal{F})+\left(1+\frac{1}{\gamma}\right)\sum_{u\in D_c}d_u^*+\sum_{g\in G}d(g,\mathcal{F})\\
&\le (2\alpha-1)(1+\gamma)+(2\beta+1)\left(1+\frac{1}{\gamma}\right)\sum_{u\in D_c}d_u^*+\alpha\left(1+\frac{1}{\gamma}\right)\sum_{g\in G_2}d_g^*+\alpha\left(1+\frac{1}{\gamma}\right)\sum_{g\in G_1}d(g,\mathcal{F})\\
&\le (2\alpha-1)(1+\gamma)+(2\beta+1)\left(1+\frac{1}{\gamma}\right)\sum_{u\in D_c}d_u^*+\alpha\left(1+\frac{1}{\gamma}\right)\sum_{g\in G}d_g^*\\
&\le (2\alpha-1)(1+\gamma)+(\alpha+2\beta+1)\left(1+\frac{1}{\gamma}\right)\sum_{u\in D_c}d_u^*\\
&\le \max\left\{(2\alpha-1)(1+\gamma),(\alpha+2\beta+1)\left(1+\frac{1}{\gamma}\right)\right\}\cdot\left(1+\sum_{u\in D_c}d_u^*\right)\\
&\le 9.9\cdot\left(1+\sum_{u\in D_c}d_u^*\right),
\end{aligned}
$$

where the final inequality follows from $\alpha = 1.82$, $\beta = \frac{\alpha}{\alpha-1}$ and $\gamma = 2.75$. □

Remark. In the case that we can open a new facility anywhere in the entire metric space or we can open a new facility at each client's location, we have $d(u, \mathcal{F}) = 0$ for every client u. After an analysis similar to Sect. 3, we can prove that the competitive ratio does not exceed

$$\max\left\{(2\alpha-1)(1+\gamma),(\alpha+\beta+1)\left(1+\frac{1}{\gamma}\right)\right\}.$$

Chooseing $\alpha = 1.566$, $\beta = \frac{\alpha}{\alpha-1}$ and $\gamma = 2.501$, we conclude that the competitive ratio does not exceed 7.465.

4 Conclusion and Open Problems

In this paper, we revisit the online facility location problem with uniform facility cost and provide a 9.9-competitive online algorithm for the M_3-model. For future research, we are interested in improving the competitive ratio of the online facility location problem with general facility costs. We are also interested in designing online algorithms for the online facility location problem with penalties.

Acknowledgments. We sincerely thank the Editor and two anonymous Reviewers for their constructive comments and helpful suggestions on an early version of our paper. This research was supported in part by the National Natural Science Foundation of China under Grant number 12371320.

References

1. Almanza, M., Chierichetti, F., Lattanzi, S., Panconesi, A., Re, G.: Online facility location with multiple advice. Adv. Neural Inf. Proc. Syst. (2021)
2. Anagnostopoulos, A., Bent, R., Upfal, E., Van Hentenryck, P.: A simple and deterministic competitive algorithm for online facility location. Inf. Comput. **194**, 175–202 (2004)
3. Byrka, J., Aardal, K.: An optimal bifactor approximation algorithm for the metric uncapacitated facility location problem. SIAM J. Comput. **39**(6), 2212–2231 (2010)
4. Chudak, F.A., Shmoys, D.B.: Improved approximation algorithms for the uncapacitated facility location problem. SIAM J. Comput. **33**(1), 1–25 (2003)
5. Diveki, G., Imreh, C.: Online facility location with facility movements. Cent. Eur. J. Oper. Res. **19**, 191–200 (2011)
6. Fotakis, D.: A primal-dual algorithm for online non-uniform facility location. J. Discrete Algorithms **5**(1), 141–148 (2007)
7. Fotakis, D.: On the competitive ratio for online facility location. Algorithmica **50**(1), 1–57 (2008)
8. Fotakis, D.: Online and incremental algorithms for facility location. ACM SIGACT News **42**(1), 97–131 (2011)
9. Guha, S., Khuller, S.: Greedy strikes back: improved facility location algorithms. J. Algorithms **31**(1), 228–248 (1999)
10. Jain, K., Mahdian, M., Saberi, A.: A new greedy approach for facility location problems. In: Proceedings of the 34th annual ACM Symposium on Theory of Computing, pp. 731–740 (2002)
11. Jiang, S.H.C., Liu, E., Lyu, Y., Tang Z.G., Zhang, Y.: Online facility location with predictions. In: Proceedings of the tenth International Conference on Learning Representations (2022)
12. Kaplan, H., Naori, D., Raz, D.: Almost tight bounds for online facility location in the random-order model. In: Proceedings of the 2023 Annual ACM-SIAM Symposium on Discrete Algorithms, pp. 1523–1544 (2023)
13. Li, S.: A 1.488 approximation algorithm for the uncapacitated facility location problem. Information and Computation, 222, 45–58 (2013)
14. Mahdian, M., Ye, Y., Zhang, J.: Approximation algorithms for metric facility location problems. SIAM J. Comput. **36**(2), 411–432 (2006)

15. Meyerson, A.: Online facility location. In: Proceedings of the 34th annual ACM Symposium on Theory of Computing, pp. 426–431 (2001)
16. Shmoys, D.B., Tardos, É., Aardal, K.: Approximation algorithms for facility location problems. In: Proceedings of the 29th Annual ACM Symposium on Theory of Computing, pp. 265-274 (1997)
17. Sviridenko, M.: cited as personal communication in [4] (1998)

or-Submodular Maximization Under a Matroid Constraint and a Knapsack Constraint

Haifeng Huang, Qian Liu, Yang Zhou, and Min Li(✉)

School of Mathematics and Statistics, Shandong Normal University, Jinan 250358, People's Republic of China
liminEmily@sdnu.edu.cn

Abstract. As a generalization of k-submodular function, or-submodular functions satisfy r-wise monotone, instead of pair-wise monotone. The problem of maximizing an or-submodular function under different constraints arises in many applications, and it is NP-hard. In this paper, we utilize a greedy algorithm to get a $\frac{1}{r+1}$-approximation solution for maximizing an or-submodular function under a matroid constraint. Moreover, for maximizing an or-submodular function with a knapsack constraint, we also design a $\frac{1}{r+1}(1-e^{-(r+1)})$-approximation algorithm.

Keywords: or-Submodularity · Knapsack · Matroid · Approximation algorithm

1 Introduction

As set functions with special properties, submodular functions and k-submodular functions have a wide range of applications in machine learning, artificial intelligence and other fields. In our paper, we mainly study the functions generalizing the above two classes of functions.

Given a positive integer k, we use $[k]$ to represent the set $\{1, 2, \ldots, k\}$. In this paper, for a finite nonempty set $G = \{e_1, e_2, \ldots, e_n\}$, we define $(k+1)^G :=\{(X_1, \ldots, X_k) \mid X_i \subseteq G, \forall i \in [k]; X_i \cap X_j = \emptyset, \forall i, j \in [k], i \neq j\}$ as a family of k disjoint subsets in G. We regard $\mathbf{X}$ and $\mathbf{Y}$ as two elements of $(k+1)^G$ with the following forms: $\mathbf{X} = (X_1, \ldots, X_k), \mathbf{Y} = (Y_1, \ldots, Y_k)$. For all $i \in [k]$, if all elements in X_i are involved in Y_i, then we say there is a *partial order* relationship between $\mathbf{X}$ and $\mathbf{Y}$, and this relationship is denoted as $\mathbf{X} \preceq \mathbf{Y}$. For any $\mathbf{X}, \mathbf{Y} \in (k+1)^G$ with $\mathbf{X} \preceq \mathbf{Y}$, if there is at least one $i \in [k]$ that satisfies $\mathbf{X}_i \subset \mathbf{Y}_i$, then $\mathbf{X} \prec \mathbf{Y}$. Let $supp(\mathbf{X})$ denote the support set of $\mathbf{X}$, which is the union of all subsets of $\mathbf{X}$, i.e., $supp(\mathbf{X}) = \bigcup_{i \in [k]} X_i$. For any element $e \in G$ and $\mathbf{X} \in (k+1)^G$, the symbol $\mathbf{X}(e)$ can indicate which subset of $\mathbf{X}$ contains this

Supported by Natural Science Foundation of Shandong Province of China (No. ZR2021MA100) and National Science Foundation of China (No. 12371099)

S. Ghosh and Z. Zhang (Eds.): AAIM 2024, LNCS 15179, pp. 145–156, 2024.
https://doi.org/10.1007/978-981-97-7798-3_13

element e. More specifically, for any $e \in G$, if $e \in X_i$, then $\mathbf{X}(e) = i$ and if e is not included in any subset of $\mathbf{X}$, $\mathbf{X}(e) = 0$. For $\forall e \in G$, $\forall i \in [k]$, we introduce a special k-tuple $\mathbf{1}^{(e,i)} = (\emptyset, \ldots, \{e\}, \ldots, \emptyset)$, which satisfies $\mathbf{1}^{(e,i)}(e) = i$ and $\forall a \in G \backslash \{e\}$, $\mathbf{1}^{(e,i)}(a) = 0$. Then for any $\mathbf{X} \in (k+1)^G$, an element $e \in G$, $i \in [k]$, we define $\mathbf{X} + \mathbf{1}^{(e,i)}$ as follows:

$$(\mathbf{X} + \mathbf{1}^{(e,i)})(e) = \begin{cases} 0, & \text{if } \mathbf{X}(e) \neq 0, \mathbf{X}(e) \neq i, \\ \max\{\mathbf{X}(e), i\}, & \text{otherwise.} \end{cases}$$

For any element $a \in G \backslash \{e\}$, $(\mathbf{X} + \mathbf{1}^{(e,i)})(a) = \mathbf{X}(a)$.

Next, we study a class of functions $f\colon (k+1)^G \to \mathbb{R}_+$ defined on $(k+1)^G$. For any $\mathbf{X}, \mathbf{Y} \in (k+1)^G$ with $\mathbf{X} \preceq \mathbf{Y}$, $i \in [k]$ and an element $e \in G \backslash supp(\mathbf{Y})$, we call the function f is *orthant submodular* if

$$f(\mathbf{X} + \mathbf{1}^{(e,i)}) - f(\mathbf{X}) \geq f(\mathbf{Y} + \mathbf{1}^{(e,i)}) - f(\mathbf{Y}).$$

For simplicity, for any $\mathbf{X} \in (k+1)^G$, $i \in [k]$ and $e \in G \backslash supp(\mathbf{X})$, let us define the marginal benefit when an element e is added to the i-th subset of $\mathbf{X}$ as follows,

$$\Delta_{e,i} f(\mathbf{X}) := f(\mathbf{X} + \mathbf{1}^{(e,i)}) - f(\mathbf{X}).$$

Then the orthant submodularity can be equivalently represented by

$$\Delta_{e,i} f(\mathbf{X}) \geq \Delta_{e,i} f(\mathbf{Y}).$$

For any $\mathbf{X} \in (k+1)^G$, $e \in G \backslash supp(\mathbf{X})$, and different i, j in $[k]$, if

$$\Delta_{e,i} f(\mathbf{X}) + \Delta_{e,j} f(\mathbf{X}) \geq 0,$$

we say the function f is *pair-wise monotone*.

If a function f satisfies both orthant submodular and pair-wise monotone, we call it to be a *k-submodular function*. It is not hard to see that a 1-submodular function is reduced to a submodular function. For a general k-submodular function f, and any $\mathbf{X}, \mathbf{Y} \in (k+1)^G$ satisfying $\mathbf{X} \preceq \mathbf{Y}$, if the following inequality $f(\mathbf{X}) \leq f(\mathbf{Y})$ holds, then we say the function f is *monotone*.

Related Work. In 1978, for monotone submodular maximization under a total size constraint, Nemhauser et al. [8] raised a $(1 - 1/e)$-approximation algorithm. Many scholars attempt to change constraints and design different approximation algorithms in [1–5,14]. In recent years, as a promotion of submodular functions, k-submodular functions have been widely studied and applied to a lot of various problems, such as influence maximization, information coverage maximization and so on. For the problem of maximizing a k-submodular function without constraint, Ward and Živný [17] proposed a solution with an approximation ratio of $\frac{1}{3}$ by a deterministic greedy algorithm and gave a randomized greedy algorithm to get an approximate solution with $\frac{1}{1+a}$, where $a = \max\left\{1, \sqrt{\frac{k-1}{4}}\right\}$. Afterwards, the approximation ratio was promoted to $\frac{1}{2}$ by Iwata et al. [6].

There are also lots of results about the maximization of k-submodular under different constraints. In 2015, for a total size constraint in monotone case, Ohsaka and Yoshida [9] presented a $\frac{1}{2}$-approximation algorithm. For a matroid constraint, a $\frac{1}{2}$-approximation algorithm was proposed by Shinsaku Sakaue [12]. Xiao et al. [18] designed a $\frac{1}{2}(1-e^{-2})$-approximation algorithm for a knapsack constraint. Under a knapsack and a matroid constraint, Yu et al. [20] contributed an algorithm to get a solution with $\frac{1}{3}(1-e^{-3})$ approximation ratio. For a general k-submodular function, Sun et al. [13] got a solution with $\frac{1}{3}$ approximation ratio under a matroid constraint. For a knapsack constraint, an approximation guarantee of $\frac{1}{3}(1-e^{-3})$ was also obtained by Xiao et al. [18]. With a matroid and a knapsack, Yu et al. [20] contributed an algorithm to achieve an approximation ratio of $\frac{1}{6}(1-e^{-2})$. More results on maximizing k-submodular functions can be found in [7,10,11,19].

For a positive integer r less than or equal to k and for a set $\mathcal{I}$ that contains r different elements in $\{1, 2, \ldots, k\}$, the function f is *r-wise monotone* if

$$\sum_{i \in \mathcal{I}} \Delta_{e,i} f(\mathbf{X}) \geq 0, \text{where } \mathbf{X} \in (k+1)^G, e \in G \backslash supp(\mathbf{X}).$$

Then, we refer to the function f that satisfies both orthant submodular and r-wise monotone as or-*submodular function*. In fact, this special function has been introduced in [17]. From this definition, we know that or-submodular functions generalize both submodular and k-submodular functions. In fact, if $r = 1$, an or-submodular function is a monotone k-submodular function; if $r = 2$, it is a general k-submodular function; if $r \geq 3$, it is no longer a k-submodular function.

Our Contributions. In this paper, we mainly study three kinds of problems. The first problem is maximizing an or-submodular function f under a matroid constraint (represented as **MorM**). The second problem is maximizing an or-submodular function h subject to a knapsack constraint (denoted by **MorK**). The third problem is maximizing an or-submodular function T under a total size constraint (denoted as **MorTS**), where **MorTS** is a special case of **MorM** and **MorK**.

For various practical problems, there are different constraints. Next, we introduce what is a matroid constraint. Given $\mathcal{A} \subseteq 2^G$, $(G, \mathcal{A})$ is considered as a matroid when it satisfies following three conditions :

(M1) : $\emptyset$ is an element of $\mathcal{A}$,

(M2) : For any element D_1 in $\mathcal{A}$ and any subset D_2 in D_1, we have $D_2 \in \mathcal{A}$,

(M3) : If $D_1, D_2 \in \mathcal{A}$ with $|D_2| < |D_1|$, there exists at least an element $a \in D_1 \backslash D_2$ satisfying $D_2 \cup \{a\} \in \mathcal{A}$.

MorM: For a matroid $(G, \mathcal{A})$, the problem of maximizing an or-submodular function f subject to a matroid constraint is:

$$\max_{\mathbf{X} \in (k+1)^G} f(\mathbf{X}) \qquad \text{s.t.} \bigcup_{i \in [k]} X_i \in \mathcal{A}.$$

For this problem, we prove that a $\frac{1}{r+1}$-approximation solution can be achieved by a greedy algorithm.

MorK: Given an integer $B \in \mathbb{R}_+$, B is interpreted as a budget. Moreover, we introduce a cost function $c\colon G \to \mathbb{R}_+$. For any element $a \in G$, let $c(a)$ be represented as the cost of a. Then the problem of maximizing an or-submodular function h under a knapsack constraint is:

$$\max_{\mathbf{X}\in(k+1)^G} h(\mathbf{X}) \qquad \text{s.t.} \sum_{a\in supp(\mathbf{X})} c(a) \leq B.$$

For this problem, we design a $\frac{1}{r+1}(1-e^{-(r+1)})$-approximation algorithm.

MorTS: For a positive integer I, the problem of maximizing an or-submodular function T with a total size constraint is:

$$\max_{\mathbf{X}\in(k+1)^G} T(\mathbf{X}) \qquad \text{s.t. } |supp(\mathbf{X})| \leq I.$$

For this problem, we also have a $\frac{1}{r+1}$-approximation solution as a corollary.

Organization. The rest of the paper is structured as follows. In Sect. 2, we introduce various relevant lemmas. In Sect. 3, for **MorM** and **MorTS**, we describe an approximation algorithm and show our outcomes. In Sect. 4, for **MorK**, we give an algorithm and prove our outcome. In Sect. 5, we summarize the conclusion of three kinds of problems in the paper.

2 Preliminaries

For a matroid $(G, \mathcal{A})$, each element in $\mathcal{A}$ is independent. For $B \in \mathcal{A}$, if there no exists $C \in \mathcal{A}$ such that $B \subset C$, then we call B is maximal. Moreover, for a matroid $(G, \mathcal{A})$, if $B \in \mathcal{A}$ is maximal, then it is called a basis of $\mathcal{A}$. Moreover, all bases of a matroid have the same cardinality, which is known as the rank in this matroid, denoted by M. And $\mathcal{M}$ is used to denote the set of all bases of $\mathcal{A}$. For **MorM**, the optimal solution with maximal cardinality is called its maximal optimal solution.

In the following lemma, we will show that the support set of the maximal optimal solution of **MorM** should be a basis.

Lemma 1. *For* **MorM**, *M is the size of its maximal optimal solution.*

The following lemma has been given in [12] about the relationship among the bases.

Lemma 2. *([12]) Given $B \in \mathcal{M}$ and $C \in \mathcal{A}$ satisfying $C \subset B$. For any $e \in G\backslash C$ such that $C\cup\{e\} \in \mathcal{A}$, there is an element $\hat{e} \in B\backslash C$ satisfying $(B\backslash\{\hat{e}\})\cup\{e\} \in \mathcal{M}$.*

Because both or-submodular functions and k-submodular functions have the property of orthant submodularity, the following lemma holds, which has already been proved in [15].

Lemma 3. *([15]) For an or-submodular function $f\colon (k+1)^G \to \mathbb{R}_+$, $\forall \mathbf{X}, \mathbf{Y} \in (k+1)^G$ such that for any $i \in [k]$, $X_i \subseteq Y_i$, we have*

$$f(\mathbf{Y}) - f(\mathbf{X}) \leq \sum_{e\in supp(\mathbf{Y})\backslash supp(\mathbf{X})} \Delta_{e,\mathbf{Y}(e)} f(\mathbf{X}).$$

The following lemma can help us solve **MorK**.

Lemma 4. *([16]) For any two* $V, W \in \mathbb{Z}_+$, $l_1 > 0$, *and* $\forall i \in \{2, 3, \ldots, V\}$, $l_i \geq 0$, *then*

$$\frac{\sum_{i\in[V]} l_i}{\min_{z=1,2,\ldots,V}\left(\sum_{i=1}^{z-1} l_i + W l_z\right)} \geq 1 - \left(1 - \frac{1}{W}\right)^V \geq 1 - e^{-V/W}.$$

3 Maximizing an or-Submodular Function with a Matroid Constraint

In this section, we mainly introduce Algorithm 1 to solve **MorM**, whose basic idea is to select an element and its position with maximum marginal gain of f in each iteration. Through analysis, we show that a $\frac{1}{r+1}$-approximation solution can be obtained by Algorithm 1.

Algorithm 1. Greedy algorithm for **MorM**.

Input: An or-submodular function $f\colon (k+1)^G \to \mathbb{R}_+$ and a matroid $(G, \mathcal{A})$.
Output: $\mathbf{s} \in (k+1)^G$ such that $supp(\mathbf{s}) \in \mathcal{A}$.

1: $\mathbf{s} \leftarrow \mathbf{0}$
2: $G(\mathbf{s}) := \{e \in G \backslash supp(\mathbf{s}) | supp(\mathbf{s}) \cup \{e\} \in \mathcal{A}\}$
3: **while** $G(\mathbf{s}) \neq \emptyset$ **do**
4: $\quad (e, i) \leftarrow \arg\max_{e \in G(\mathbf{s}), i \in [k]} \Delta_{e,i} f(\mathbf{s})$
5: $\quad \mathbf{s} := \mathbf{s} + \mathbf{1}^{(e,i)}$
6: $\quad G(\mathbf{s}) := \{e \in G \backslash supp(\mathbf{s}) | supp(\mathbf{s}) \cup \{e\} \in \mathcal{A}\}$
7: **end while**
8: **return** $\mathbf{s}$

3.1 Algorithm and Its Analysis

For any $\mathbf{s} \in (k+1)^G$, we can always define a set $G(\mathbf{s}) \subseteq G$, which includes the elements that can form independent sets with the support set of $\mathbf{s}$.

In Algorithm 1, for any $\mathbf{s}$, as long as $G(\mathbf{s}) \neq \emptyset$, then there must be an element $e \in G(\mathbf{s})$ chosen and added to $supp(\mathbf{s})$ because of Lemma 1. In step 4, in the event of a tie, we select an arbitrary element or position that achieves the maximal increment. Then we continuously select elements until $G(\mathbf{s}) = \emptyset$. For the given matroid $(G, \mathcal{A})$ with rank of M, we know Algorithm 1 has to loop M times. If we suppose that the time of oracle to determine whether a set is independent for this matroid is expressed as OI, and the time of oracle to compute the or-submodular function value is denoted as OF, then we can give the following result:

Theorem 1. *For* **MorM**, *Algorithm 1 produces a* $\frac{1}{r+1}$*-approximation solution in* $O(|G|\, M(OI + kOF))$ *time.*

3.2 Main Analysis for MorM

For $l \in [M]$, let e^l, i^l be the element and location selected in the l-th iteration. To describe how the solution changes during the algorithm loop, we use $\mathbf{s}^l$ to denote the solution after the l-th iteration, i.e., $\mathbf{s}^l = \mathbf{s}^{l-1} + \mathbf{1}^{(e^l, i^l)}$ if we denote $\mathbf{s}^0 = \mathbf{0}$. It is obvious that $\mathbf{s}^M = \mathbf{s}$. For **MorM**, we take $\mathbf{o}$ as a maximal optimal solution.

Then, we will construct a series of vectors $\mathbf{o}^0, \mathbf{o}^1, \ldots, \mathbf{o}^M$, which satisfy

$$\begin{aligned} \mathbf{s}^l \prec \mathbf{o}^l, &\text{ if } l \in \{0, 1, \ldots, M-1\}, \\ \mathbf{s}^l = \mathbf{o}^l = \mathbf{s}, &\text{ if } l = M, \\ supp(\mathbf{o}^l) \in \mathcal{M}, &\text{ if } l \in \{0, 1, \ldots, M\}. \end{aligned}$$

Before the first iteration, $\mathbf{s}^0 = \mathbf{0}$ and $\mathbf{o}^0 = \mathbf{o}$ such that $\mathbf{s}^0 \prec \mathbf{o}^0$ and $supp(\mathbf{o^0}) \in \mathcal{M}$. Suppose that after the $(l-1)$-th iteration, we obtain $\mathbf{o}^{l-1}$ satisfying $\mathbf{s}^{l-1} \prec \mathbf{o}^{l-1}$ and $supp(\mathbf{o}^{l-1}) \in \mathcal{M}$. Then in the l-th iteration, because of Lemma 2, there is an element $\hat{e} \in supp(\mathbf{o}^{l-1}) \backslash supp(\mathbf{s}^{l-1})$ satisfying $(supp(\mathbf{o}^{l-1}) \backslash \{\hat{e}\}) \cup \{e^l\} \in \mathcal{M}$. We assume $o^l = \hat{e}$ and define $\mathbf{o}^{l-1/2}$ as a result that is obtained through assigning 0 to the o^l-th element in $\mathbf{o}^{l-1}$. Similarly, we define $\mathbf{o}^l$ by changing the e^l-th element in $\mathbf{o}^{l-1/2}$ to i^l. By definition of $\mathbf{o}^{l-1/2}$, we know $\mathbf{s}^{l-1} \preceq \mathbf{o}^{l-1/2}$. According to the definition of $\mathbf{o}^{l-1}$, we have $\mathbf{s}^{l-1} \prec \mathbf{o}^{l-1}$ and $supp(\mathbf{o}^l) \in \mathcal{M}$. Through induction, for $l \in \{0, 1, ..., M\}$, we can build a sequence of $\mathbf{o}^l$ that meet the above requirements.

Lemma 5. *In Algorithm 1, for each iteration $l \in [M]$, we obtain*

$$f(\mathbf{o}^{l-1}) - f(\mathbf{o}^l) \leq r[f(\mathbf{s}^l) - f(\mathbf{s}^{l-1})], \tag{1}$$

Proof. Since we fail to determine whether the position of e^l in $\mathbf{o}^{l-1}$ is equal to i^l, the process of proof is divided into two cases.

Case 1 : If $\mathbf{o}^{l-1}(e^l) = i^l$.

There is no doubt that we can obtain $\mathbf{o}^{l-1} = \mathbf{o}^l$, then

$$f(\mathbf{o}^{l-1}) - f(\mathbf{o}^l) = 0.$$

Because $\mathbf{s}^l$ and $\mathbf{s}^{l-1}$ are produced through Algorithm 1, then the value of $f(\mathbf{s}^l) - f(\mathbf{s}^{l-1})$ is non-negative. So, we conclude that

$$f(\mathbf{o}^{l-1}) - f(\mathbf{o}^l) = 0 \leq f(\mathbf{s}^l) - f(\mathbf{s}^{l-1}).$$

Obviously that means

$$f(\mathbf{o}^{l-1}) - f(\mathbf{o}^l) \leq r\left[f(\mathbf{s}^l) - f(\mathbf{s}^{l-1})\right]. \tag{2}$$

Case 2 : If $\mathbf{o}^{l-1}(e^l) \neq i^l$.

$$\begin{aligned} f(\mathbf{o}^{l-1}) - f(\mathbf{o}^l) &= \Delta_{o^l,\mathbf{o}^{l-1}(o^l)} f(\mathbf{o}^{l-1/2}) + \sum_{m=1}^{r-1} \Delta_{e^l,i^m} f(\mathbf{o}^{l-1/2}) \\ &\quad - [\sum_{m=1}^{r-1} \Delta_{e^l,i^m} f(\mathbf{o}^{l-1/2}) + \Delta_{e^l,i^l} f(\mathbf{o}^{l-1/2})] \\ &\leq \Delta_{o^l,\mathbf{o}^{l-1}(o^l)} f(\mathbf{o}^{l-1/2}) + \sum_{m=1}^{r-1} \Delta_{e^l,i^m} f(\mathbf{o}^{l-1/2}) \\ &\leq \Delta_{o^l,\mathbf{o}^{l-1}(o^l)} f(\mathbf{s}^{l-1}) + \sum_{m=1}^{r-1} \Delta_{e^l,i^m} f(\mathbf{s}^{l-1}) \\ &\leq \Delta_{e^l,i^l} f(\mathbf{s}^{l-1}) + (r-1)\Delta_{e^l,i^l} f(\mathbf{s}^{l-1}) \\ &= r[f(\mathbf{s}^l) - f(\mathbf{s}^{l-1})]. \end{aligned}$$

First, we assume that $i^1, \ldots, i^{r-1}$ and i^l are distinct elements in $[k]$. Apparently, it is not hard to get the first equation. Due to the property of r-wise monotonicity, we can obtain the first inequality. The last but one inequality is right because of $\mathbf{s}^{l-1} \preceq \mathbf{o}^{l-\frac{1}{2}}$ and orthant submodularity. The last inequality is due to the fact that the (e^l, i^l) is selected in the l-th iteration of Algorithm 1. Finally, taking into account two different situations, we have completed the proof.

Next we can present the main proof of Theorem 1.

Proof of Theorem 1: For the time complexity of Algorithm 1, it is simple. Then, we mainly study the approximation ratio of the solution output by Algorithm 1. By using conclusion of Lemma 5, we obtain

$$f(\mathbf{o}) - f(\mathbf{s}) = \sum_{l=1}^{M} [f(\mathbf{o}^{l-1}) - f(\mathbf{o}^l)] \leq \sum_{l=1}^{M} (r[f(\mathbf{s}^l) - f(\mathbf{s}^{l-1})]) = rf(\mathbf{s}) - rf(\emptyset).$$

Since $f(\emptyset) \geq 0$, we have

$$f(\mathbf{s}) \geq \frac{1}{r+1} f(\mathbf{o}),$$

then, we complete the proof.

According to Theorem 1, we can easily give the following conclusions about the k-submodular maximization problem:

Corollary 1. *([12]) When $r = 1$, Algorithm 1 returns a $\frac{1}{2}$-approximation solution for monotone k-submodular maximization under a matroid constraint.*

Corollary 2. *([13]) When $r = 2$, Algorithm 1 returns a $\frac{1}{3}$-approximation solution for non-monotone or general k-submodular maximization subject to a matroid constraint.*

Note that for the problem of maximizing a monotone k-submodular function without any constraints, the approximation ratio of $\frac{1}{2}$ is shown to be asymptotically tight [6]. Thus, in the constrained case, one could not obtain a better approximation ratio than $\frac{1}{2}$. In addition, the approximation ratio of $\frac{1}{3}$ is the best result of deterministic algorithms for the non-monotone case, but the fly in the ointment is that there has not been any discussion of the tightness about the $\frac{1}{3}$ approximation ratio.

Since matroids have many forms, we introduce a uniform matroid $(G, \mathcal{A})$, which is a special circumstance of matroids.

$$\mathcal{A}: = \{M \subseteq N \mid \mid M \mid \leq K, K \in \mathbb{N}\}.$$

It is obvious to see that the **MorTS** is a special situation of **MorM**.

Corollary 3. *For* **MorTS**, *we suppose that the solution returned by Algorithm 1 is* $\mathbf{s}$. *Then, we get*

$$T(\mathbf{s}) \geq \frac{1}{r+1} T(\mathbf{o}).$$

4 Maximizing an or-Submodular Function with a Knapsack Constraint

In this section, we mainly solve **MorK** by designing Algorithm 2. Algorithm 2 provides us with an effective approach, whose main idea is to enumerate all feasible solutions of p size and then selects the most cost-effective element to add them until the budget is used up.

For **MorK**, we assume $\mathbf{o}$ is an optimal solution. Let $\mid supp(\mathbf{o}) \mid = u$. If $u < p$, then an optimal solution will be found in Step 1 of Algorithm 2. So we pay attention to the case that $u \geq p$. For all elements in $supp(\mathbf{o})$, we will choose them again by a greedy strategy. Specifically, for all $l \in [u]$, we define $\mathbf{c}^l = \mathbf{c}^{l-1} + \mathbf{1}^{(o^l, i^l)}$ and $\mathbf{c}^0 = \mathbf{0}$, where o^l, i^l such that $\Delta_{(o^l, i^l)} h(\mathbf{c}^{l-1}) \geq \Delta_{(e,j)} h(\mathbf{c}^{l-1})$, $\forall e \in supp(\mathbf{o}) \backslash supp(\mathbf{c}^{l-1})$ and $\forall j \in [k]$. According to the definition of $\mathbf{c}^l$, when $l = u$, we have $supp(\mathbf{c}^l) = supp(\mathbf{o})$.

In addition, for any $l \in [u]$, we set up a series of $\widetilde{\mathbf{o}}^0, \widetilde{\mathbf{o}}^1, \ldots, \widetilde{\mathbf{o}}^u$, with $\widetilde{\mathbf{o}}^l = (\widetilde{\mathbf{o}}^{l-1} + \mathbf{1}^{(o^l, i^l)}) + \mathbf{1}^{(o^l, i^l)}$ and $\widetilde{\mathbf{o}}^0 = \mathbf{o}$. Considering the analysis in [17], for any $l \in [u]$ and an or-submodular function h, we can have

$$h(\widetilde{\mathbf{o}}^{l-1}) - h(\widetilde{\mathbf{o}}^l) \leq r[h(\mathbf{c}^l) - h(\mathbf{c}^{l-1})]. \tag{3}$$

Further more, by adding the first p inequalities, we get

$$h(\mathbf{o}) \leq rh(\mathbf{c}^p) + h(\widetilde{\mathbf{o}}^p), \tag{4}$$

Let $\mathbf{c}^p$ denote an initial feasible solution in Algorithm 2, that is $\mathbf{c}^p = \mathbf{s}^0$. In Algorithm 2, although elements in G will be considered, we are only interested in elements in $supp\,(\mathbf{o}) \cup supp\,(\mathbf{s})$, where $\mathbf{s}$ is a feasible solution in Algorithm 2. In

Algorithm 2. Greedy algorithm for **MorK**.

Input: A positive integer p, a budget B, a finite set G, a cost function c and an or-submodular function $h\colon (k+1)^G \to \mathbb{R}_+$
Output: $\mathbf{s} \in (k+1)^G$ such that $\sum_{a \in supp(\mathbf{s})} c(a) \leq B$

```
1: s ← arg max_{|supp(s)|=p−1, c(supp(s))≤B} h(s) ;
2: for every s^0(|supp(s^0)| = p, c(supp(s^0)) ≤ B) do
3:     G^0 := G\supp(s^0)
4:     l: = 1
5:     while | G^{l−1} |≠ 0 do
6:         (e^l, i^l) ← arg max_{e∈G^{l−1}, i∈[k]} Δ_{e,i}h(s^{j−1}) / c(e)
7:         if c(supp(s^{l−1})) + c(e^l) ≤ B then
8:             s^l := s^{l−1} + 1^{(e^l, i^l)}
9:         else
10:            s^l = s^{l−1}
11:        end if
12:        G^l := G^{l−1}\{e^l}
13:        l := l + 1
14:    end while
15:    s ← arg max {h(s), h(s^l)}
16: end for
17: return s
```

other words, we can ignore elements that belong to neither $supp\,(\mathbf{o})$ nor $supp\,(\mathbf{s})$, since they fail to influence the quality of solution produced by Algorithm 2. During the iteration process, suppose that the first time we encounter an element $e^{d+1} \in supp\,(\mathbf{o})$ that satisfies Step 6 of Algorithm 2 but does not satisfy Step 7 in the $(d+1)$-th iteration. In addition, for all $l \in [d]$, let (e^l, i^l) be selection at the l-th iteration and $\mathbf{s}^l = \mathbf{s}^{l-1} + \mathbf{1}^{(e^l, i^l)}$ be interpreted as result in the end of l-th iteration. Then, we define $\mathbf{o}^0$ by removing the element e^{d+1} from $\widetilde{\mathbf{o}}^p$ denoted by $\mathbf{o}^0 = \widetilde{\mathbf{o}}^p \backslash \mathbf{1}^{(e^{d+1}, \widetilde{\mathbf{o}}^p(e^{d+1}))}$. Based on the above analysis of (e^{d+1}, i^{d+1}), for any $l \in [d]$, we get $\mathbf{o}^l = (\mathbf{o}^{l-1} + \mathbf{1}^{(e^l, i^l)}) + \mathbf{1}^{(e^l, i^l)}$ as the result in the end of the l-th iteration.

Let $\mathbf{o}^{l-\frac{1}{2}}$ denote a new sequence of k-tuples. For any $l \in [d]$, we define $\mathbf{o}^{l-\frac{1}{2}} = \mathbf{o}^{l-1} + \mathbf{1}^{(e^l, i^l)}$. According to the definition of $\mathbf{o}^{l-1/2}$, $\mathbf{s}^l$ and $\mathbf{o}^l$, for any $l \in [d]$, we obtain $\mathbf{s}^{l-1} \preceq \mathbf{o}^{l-1/2}$ and $\mathbf{s}^{l-1} \preceq \mathbf{o}^{l-1}$. Clearly, for $l \in \{0, 1, \ldots, d\}$, we have $supp(\mathbf{o}^0) \backslash supp(\mathbf{s}^l) = supp(\mathbf{o}^l) \backslash supp(\mathbf{s}^l)$.

For the solution produced by each iteration in Algorithm 2, we consider the analysis process in Lemma 5. Then for the first d iteration, we obtain two important lemmas.

Lemma 6. *For each iteration $l \in [d]$ in Algorithm 2, we obtain*

$$h(\mathbf{o}^{l-1}) - h(\mathbf{o}^l) \leq r[h(\mathbf{s}^l) - h(\mathbf{s}^{l-1})]. \tag{5}$$

Lemma 7. *For an or-submodular function h, for any $l \in \{0, 1, \dots, d\}$, we have*

$$h(\mathbf{o}^0) \le (r+1)\,h(\mathbf{s}^l) - rh(\mathbf{s}^0) + \sum_{e \in supp(\mathbf{o}^0) \setminus supp(\mathbf{s}^l)} \Delta_{e,\mathbf{o}^0(e)} h(\mathbf{s}^l).$$

Because or-submodular function h also satisfies orthant submodularity, we present an important lemma shown in [18].

Lemma 8. *([18]) For any $l \in \{1, 2, \dots, k\}$ and the element $e^{d+1} \in supp(\mathbf{o}) \setminus supp(\mathbf{s}^d)$, we can have*

$$\Delta_{e^{d+1},l} h(\mathbf{o}^0) \le \frac{h(\mathbf{s}^0)}{p}.$$

Then we can give the main result about **MorK**.

Theorem 2. *For* **MorK**, *Algorithm 2 generates a $\frac{1}{r+1}(1 - e^{-(r+1)})$ approximation solution with query complexity $O\left(n^{p+2}k^{p+1}\right)$, where $p = \left\lceil \frac{e^{r+1}}{r+1} \right\rceil$.*

Then, we can get the following results about the k-submodular functions based on the relationship between k-submodular functions and or-submodular functions, which have been discussed in [18].

Corollary 4. *([18]) For* **MorK**, *if $r = 1$, we view $\mathbf{s}$ as a solution that is output by Algorithm 2. Then we get*

$$h(\mathbf{s}) \ge \frac{1}{2}(1 - \frac{1}{e^2})h(\mathbf{o}).$$

Corollary 5. *([18]) For* **MorK**, *if $r = 2$, we assume that $\mathbf{s}$ is a solution obtained by Algorithm 2. Then we have*

$$h(\mathbf{s}) \ge \frac{1}{3}(1 - \frac{1}{e^3})h(\mathbf{o}).$$

5 Conclusions

Throughout the paper, we study three kinds of constraints for the or-submodular maximization problems: The first kind of problem is maximizing an or-submodular function f subject to a matroid constraint (**MorM**); The second kind of problem is maximizing an or-submodular function h under a knapsack constraint (**MorK**); The third kind of problem is maximizing an or-submodular function T under a total size constraint (**MorTS**). For **MorM**, we show that a $\frac{1}{r+1}$-approximation solution is achieved by Algorithm 1. For **MorTS**, we also have a $\frac{1}{r+1}$-approximation solution. For **MorK**, through Algorithm 2, we obtain a $\frac{1}{r+1}(1 - \frac{1}{e^{r+1}})$-approximation solution.

Note that for these problems, the deterministic approximation ratio is no better than $\frac{1}{2}$; But when $r = 2$, there is no discussion about whether $\frac{1}{3}$ is the best or not. In the future, we will try to discuss whether $\frac{1}{3}$ is the best for k-submodular functions, so $\frac{1}{r+1}$ for or-submodular functions. Moreover, we will design better randomized approximation algorithms and plan to reduce the complexity by decreasing threshold method.

References

1. Bian, A., Buhmann, J., Krause, A., Tschiatschek, S.: Guarantees for greedy maximization of non-submodular functions with applications. In: the 34th International Conference on Machine Learning Proceedings, pp. 498–507. JMLR, Sidney, Australia (2017)
2. Calinescu, G., Chekuri, C., Pal, M., Vondrák, J.: Maximizing a monotone submodular function subject to a matroid constraint. SIAM J. Comput. **40**(6), 1740–1766 (2011)
3. Ene, A., Nguyễn, H.: A nearly-linear time algorithm for submodular maximization with a knapsack constraint. In: the 46th International Colloquium on Automata, Languages and Programming Proceedings, pp. 53:1–53:12. Schloss DagstuhlCLeibniz-Zentrum fr Informatik, Patras, Greece (2019)
4. Filmus, Y., Ward, J.: Monotone submodular maximization over a matroid via non-oblivious local search. SIAM J. Comput. **43**(2), 514–542 (2014)
5. Huang, C., Kakimura, N., Mauras, S., Yoshida, Y.: Approximability of monotone submodular function maximization under cardinality and matroid constraints in the streaming model. SIAM J. Discret. Math. **36**(1), 355–382 (2022)
6. Iwata, S., Tanigawa, S., Yoshida, Y.: Improved approximation algorithms for k-submodular function maximization. In: The 27th Annual ACM-SIAM Symposium on Discrete Algorithms Proceedings, pp. 404–413. Society for Industrial and Applied Mathematics, Arlington, VA, USA (2016)
7. Liu, Q., Yu, K., Li, M., Zhou, Y.: k-submodular maximization with a knapsack constraint and p matroid constraints. Tsinghua Sci. Technol. **28**(5), 896–905 (2023)
8. Nemhauser, G., Wolsey, L., Fisher, M.: An analysis of approximations for maximizing submodular set functions-I. Math. Program. **14**(1), 265–294 (1978)
9. Ohsaka, N., Yoshida, Y.: Monotone k-submodular function maximization with size constraints. In: the 28th International Conference on Neural Information Proceeding Systems Proceedings, pp. 694–702. MIT Press, Montreal Quebec, Canada (2015)
10. Pham, C., Vu, Q., Ha, D., Nguyen, T., Le, N.: Maximizing k-submodular functions under budget constraint: applications and streaming algorithms. J. Comb. Optim. **44**(1), 723–751 (2022)
11. Rafiey, A., Yoshida, Y.: Fast and private submodular and k-submodular functions maximization with matroid constraints. In: the 37th International Conference on Machine Learning Proceedings, pp. 7887–7897. JMLR, Virtual Conference (2020)
12. Sakaue, S.: On maximizing a monotone k-submodular function subject to a matroid constraint. Discret. Optim. **23**, 105–113 (2017)
13. Sun, Y., Liu, Y., Li, M.: Approximation algorithms for maximization of k-submodular function under a matroid constraint. Tsinghua Sci. Technol. **29**(6), 1633–1641 (2024)
14. Sviridenko, M.: A note on maximizing a submodular set function subject to a knapsack constraint. Oper. Res. Lett. **32**(1), 41–43 (2004)
15. Tang, Z., Wang, C., Chan, H.: On maximizing a monotone k-submodular function under a knapsack constraint. Oper. Res. Lett. **51**(1), 28–31 (2022)
16. Wolsey, L.: Maximizing real-valued submodular functions: primal and dual heuristics for location problems. Math. Oper. Res. **7**(3), 410–425 (1982)
17. Ward, J., Živný, S.: Maximizing k-submodular functions and beyond. ACM Trans. Algorithms **12**(4), 1–26 (2016)

18. Xiao, H., Liu, Q., Zhou, Y., Li, M.: Approximation algorithms for k-submodular maximization subject to a knapsack constraint. Accepted by J. Oper. Res. Soc. China (2024). https://doi.org/10.1007/s40305-024-00539-y
19. Yu, K., Li, M., Zhou, Y., Liu, Q.: On maximizing monotone or non-monotone k-submodular functions with the intersection of knapsack and matroid constraints. J. Comb. Optim. **45**(3), 1–21 (2023)
20. Yu, K., Li, M., Zhou, Y., Liu, Q.: Guarantees for maximization of k-submodular functions with a knapsack and a matroid constraint. In: Ni, Q., Wu, W. (eds.) AAIM 2022. LNCS, vol. 13513, pp. 156–167. Springer Cham (2022). https://doi.org/10.1007/978-3-031-16081-3_14

A Distributed Method for Negative Content Spread Minimization on Social Networks

Ruidong Yan[1(✉)], Zhenhua Guo[1], Weili Wu[2], and Baoyu Fan[1]

[1] IEIT SYSTEMS Co., Ltd., Beijing 100875, China
{yanruidong,guozhenhua,fanbaoyu}@ieisystem.com
[2] Department of Computer Science, The University of Texas at Dallas, Richardson, TX 75080, USA
weiliwu@utdallas.edu

Abstract. Currently, social networks have emerged as significant platforms for individuals to share personal information and social content. However, it is important to recognize that social networks have both positive and negative aspects. To effectively address the dissemination of negative social content such as rumors and misinformation, it is crucial to implement strategies that involve immediate blocking of associated links. This paper introduces a *Negative Content Spread Minimization* (NCSM) problem, which aims to minimize the spread of negative content by removing a set of edges from the network. We begin by demonstrating the NP-hardness of the NCSM problem through reduction from the *Knapsack Problem.* Furthermore, we establish that the objective function is not submodular under the *Independent Cascade* model. To address, we employ a distributed method which includes community partition and influential edges selection. The advantage of this approach is to reduce computational overhead by selecting key edges in parallel in each community. To evaluate proposed algorithm, we conduct experiments using real-world datasets and compare them against existing methods. The experimental results demonstrate that our method outperforms state-of-the-art algorithms.

Keywords: Social network · negative content spread · non-submodularity

1 Introduction

Online *Social Networks* (SNs), such as WeChat and Twitter, have become an integral part of users' daily lives. However, like any other phenomenon, SNs have both positive and negative aspects. While they facilitate the generation and dissemination of positive content, such as new ideas and products, they also

This work was supported in part by the National Key R&D Program of China under Grants 2022YFB4501600 and 2022YFB4501603.

S. Ghosh and Z. Zhang (Eds.): AAIM 2024, LNCS 15179, pp. 157–169, 2024.
https://doi.org/10.1007/978-981-97-7798-3_14

provide a platform for the spread of negative rumors. These negative contents have significant impacts on network users. Firstly, they contribute to the proliferation of unreliable web content. Secondly, they can induce panic and anxiety among users. For instance, during the aftermath of "Hurricane Harvey", social media platforms were inundated with photographs depicting sharks swimming in flooded towns and airports in Houston. These images had the potential to evoke anxiety and social panic. Nevertheless, the complexity and cost associated with tracing and debunking misleading, inaccurate, and inflammatory content across multiple large online social networks pose significant challenges for authorities and governments. Consequently, it has become imperative to explore effective measures to control the spread of negative content, making it a pressing concern.

Extensive research has been conducted on mitigating the dissemination of negative social content from various perspectives. Previous studies can be broadly categorized into two groups. The first category aims to identify key nodes or edges involved in the propagation of information and remove them to impede the spread of negative content [10,12]. The second category focuses on identifying seed sets of positive content to counteract the negative content within the same network [2,11]. However, these methods encounter two challenges: 1) limited scalability due to centralized strategies, and 2) suboptimal performance resulting from the selection of nodes/edges solely based on a global perspective, disregarding local characteristics.

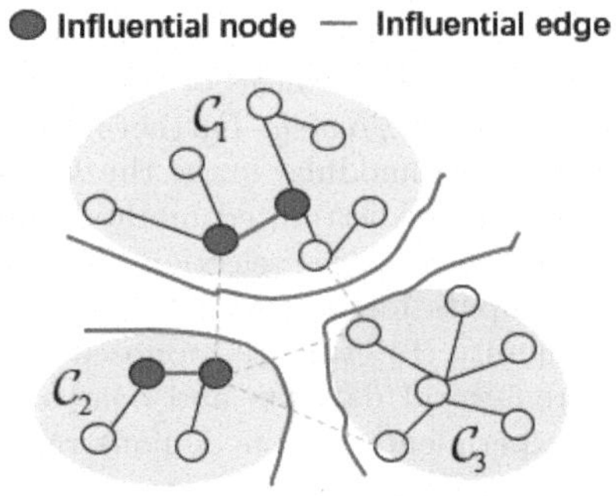

Fig. 1. An instance of removing influential edges in each community.

To address the above challenges, our focus is on a distributed algorithm that utilizes the community structure of social networks to reduce computational complexity and enhance performance. To illustrate our motivation, we present an example in Fig. 1. In this figure, the original network is divided into three distinct and independent communities through community detection. According to [8], users within the same community have frequent interactions, making them more likely to influence one another. Conversely, users across different communities have limited contact and are less likely to influence each other. This observation suggests that selecting influential nodes/edges in parallel within communities could serve as a viable approximation. In our approach, we propose a method to select key edges in parallel based on the assessment of node influence within each community. The red edges in communities C_1 and C_2 represent the identified key edges, which are subsequently removed to restrict the spread of negative social content.

This study investigates the problem of *Negative Content Spread Minimization* (NCSM). More specifically, given a directed network and a seed set of negative content, our objective is to identify and remove certain edges in order to minimize the number of nodes activated by the negative content. In contrast to existing approaches, we propose a novel method that carefully selects influential edges based on community structures. The contributions are outlined below:

- We formalize the NCSM problem and prove that is NP-hard.
- We prove objective function is not submodular. And we propose a community-based distributed algorithm to select influential edges in parallel thereby significantly reducing computational overheads.
- We compare proposed algorithm with existing methods on real-world datasets. Experimental results show the effectiveness and efficiency of our method.

The structure of this paper is organized as follows: In Sect. 2, we provide a comprehensive review of existing related work. Section 3 presents the problem description and the information propagation model. The properties of the objective function are discussed in Sect. 4. The proposed algorithms are detailed in Sect. 5. Section 6 presents the experimental results, and finally, in Sect. 7, we draw our conclusions.

2 Related Work

The problem of minimizing the negative content spread is a significant research topic in the field of social networks. There are mainly two types of strategies include blocking influential nodes/edges [12,14] or clarifying the negative contents by spreading the positive truths such as [13].

On one hand, the strategy of removing nodes or edges alters the network structure in a way that minimises the spread of negative content. For example, [10] effectively controls the propagation process in the network by proposing a NIIP algorithm that dynamically selects immune nodes in response to multiple possible propagation outbreaks over a sustained period of time. [14] designs a sub-modular upper and lower bounds to approximate the non-submodular objective function and proposes an efficient heuristic algorithm to filter the set of key edges for rumour propagation. On the other hand, [11] investigates the rumor blocking problem, which seeks to identify k seed users to rigger the spread of a positive cascade, such that the number of the users who are not influenced by rumor can be maximized. They propose an approximate algorithm with a $(1 - 1/e)$ factor based on the *Random Inverse Tuple.*

Recently, [15] proposes a hybrid control strategy for rumor propagation incorporating protective and blocking measures. And an optimal control model is developed to maximize the control of rumor propagation while minimizing control costs. [1] designs an AHP-TOPSIS method to select the important nodes and host agents. It proposes a counter-rumor diffusion model *Susceptible-Infected-Recovered-Prevented-Agent* (SIRPA). Furthermore, researchers have applied deep learning methods to the problem of negative content control such as [5]. It proposes a *Confidence-Based Opinion Adoption* (CBOA) model that considers opinions and confidence levels according to the traditional linear threshold (LT) model. Based on this model, it proposes a *Directed Graph Convolutional Network* (DGCN) method for selecting the most influential positive cascade nodes to suppress rumour propagation.

Note that the majority of these approaches employ a centralized strategy, which is not directly applicable to the context of this paper.

3 Diffusion Model and Problem Definition

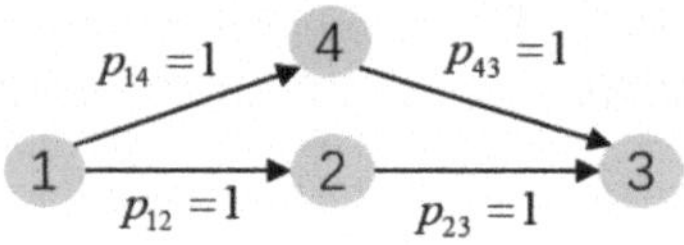

Fig. 2. An instance for non-submodularity.

Influence Diffusion Model: We provide a brief introduction to the information diffusion model known as the *Independent Cascade* (IC) model. The model is based on a directed social network $G = (V, E, p)$, where V represents the set of nodes (users) and $E \subseteq V \times V$ represents the set of edges (relationships between users). The probability p_{vu} associated with an edge (v, u) or e_{vu} indicates the likelihood of node u being successfully activated by node v. Nodes are classified as either *active* if they adopt or share content from other nodes, or *inactive* otherwise. The information propagation process occurs in discrete time steps. Initially, the seed set is defined as $S_0 = S$. At each time step T, S_T represents the set of *active* nodes. In this step, each node v in S_T has a single chance to activate each *inactive* neighbor u through its out-edge, with a probability p_{vu}, at time step $T+1$. This process is repeated until no new nodes can be activated. It is important to note that nodes can transition only from an *inactive* state to an *active* state, but not vice versa.

Problem Statement: The goal of this paper is to identify a set of K edges, denoted as $\mathcal{E}$, from the edge set E, and remove these edges from the original graph G. The purpose of this removal is to minimize the number of nodes that are activated by negative content.

Definition 1. *(Negative Content Spread Minimization (NCSM)). Given a directed social network $G = (V, E, p)$, a diffusion model $\mathcal{M}$, a predetermined seed set S for the negative content, the NCSM problem is to find a set $\mathcal{E} \subseteq E$ with K edges such that $L(\mathcal{E}) = \sum_{v \in V} \sigma_{E \backslash \mathcal{E}}(v, S)$, i.e., the number of nodes activated by negative content is minimized. It can be rewritten as follows:*

$$\mathcal{E}^* = \arg \min_{\mathcal{E} \subseteq E, |\mathcal{E}|=K} \sum_{v \in V} \sigma_{E \backslash \mathcal{E}}(v, S, \mathcal{M}), \tag{1}$$

where $\sigma_{E \backslash \mathcal{E}}(v, S, \mathcal{M})$ (we omit $\mathcal{M}$ if context is clear, e.g., $\sigma_{E \backslash \mathcal{E}}(v, S)$) denotes the probability that the seed set S activates node v successfully on topology $E \backslash \mathcal{E}$.

4 Properties of NCSM Problem

We first show the hardness of NCSM problem, and then we discuss the properties of objective function (1).

Theorem 1. *The NCSM problem is NP-hard under the IC model.*

Proof is in the Appendix

Theorem 2. *The object function (1) is not submodular.*

Proof. If E is a finite set, a submodular function is a set function $F : 2^E \rightarrow \Re$, where 2^E denotes the power set of E, which satisfies $F(A \cup \{e\}) - F(A) \geq F(B \cup \{e\}) - F(B)$ for every $A \subseteq B \subseteq E$ and $e \in E \backslash B$. We prove the function (1) is not submodular by the counterexample in Fig. 2.

Suppose that node 1 is the only seed node. That is, node 1 has a piece of negative content initially. Each node shares the content with its neighbors with the probability of 1. Let $A = \{e_{14}\}$, $B = \{e_{14}, e_{34}\}$, and $e = \{e_{23}\}$. Note that $A \subseteq B \subseteq E$ and $e \in E \backslash B$. The function $L(X) = \sum_{v \in V} \sigma_{E \backslash X}(v, S)$ represents the total probabilities of nodes being activated by the seed set when we remove the edge set X. We have $L(A) = \sum_{v \in V} \sigma_{E \backslash A}(v, S) = 4$ because there are four nodes must be activated by the node 1 even though we remove the edge e_{14}. Similarly, we have $L(A \cup \{e\}) = 2$, $L(B) = 3$ and $L(B \cup \{e\}) = 2$. Therefore $L(A \cup \{e\}) - L(A) < L(B \cup \{e\}) - L(B)$. The above analysis indicates function $L(\cdot)$ is not submodular.

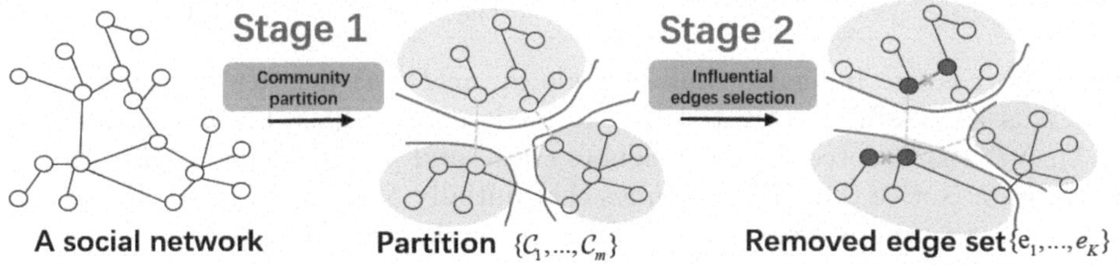

Fig. 3. The overview of our distributed strategy for NCSM problem.

5 Algorithm

We present a distributed strategy including two components. Firstly, we employ a **community partition** method dividing the network into m non-overlapping communities. By doing so, we limit the influence within each community. Secondly, we introduce an **influential edge selection** method selecting K influential edges from these communities to minimize $L(\mathcal{E}) = \sum_{v \in V} \sigma_{E \backslash \mathcal{E}}(v, S)$ (see Fig. 3).

5.1 Community Partition

Here, we employ [9] to identify m non-overlapping communities due to low time complexity and suitability for a given social network $G = (V, E, p)$.

Definition 2 *(Non-overlapping Community). Given a network* $G = (V, E, p)$, *the non-overlapping community partition* $\mathcal{C} = \{\mathcal{C}_1, \mathcal{C}_2, ..., \mathcal{C}_m\}$ *satisfies the conditions:* $\bigcup_{1 \le j \le m} \mathcal{C}_j = V$ *and* $\forall\ i \neq j \in [m]$, $\mathcal{C}_i \cap \mathcal{C}_j = \emptyset$, *where* m *is the number of communities and* $[m] = \{1, 2, ..., m\}$.

5.2 Influential Edges Selection

In this part, we first propose an improved node influence rank method based on [6] (Algorithm 1), and then we provide a method (formula (5)) for defining key edges and a distributed K edges selection approach (Algorithm 2).

A Influence Rank Algorithm: IRIE-2. Calculating the exact influence spread of a node u on a network based on the IC model is not feasible [4]. As a result, researchers have proposed various approximation strategies, including the *Maximum Influence Path* (MIP) [3] and the *Influence Rank and Influence Estimation* (IRIE) [6]. In this context, we provide a brief overview of IRIE, which utilizes formula (2) to compute the influence rank (i.e., influence spread) of node u on a network.

$$\mathbf{r}(u) = (1 - ActP_S(u)) \cdot \left\{ 1 + \eta \cdot \left[\sum_{v \in N^{out}(u)} p_{uv} \cdot \mathbf{r}(v) \right] \right\}, \tag{2}$$

where $\mathbf{r}(u)$ is the influence rank of node u, $ActP_S(u)$ denotes the probability that u is successfully activated by the seed set S, adjustable parameter $\eta \in (0, 1)$, $N^{out}(u)$ denotes the out-neighbors of u, and p_{uv} is the probability that u activates v. Notice that IRIE assumes that initially S is empty, so $ActP_s(u) = 0$. Furthermore, we initialize $\mathbf{r}(u) = 1$ for every node u in the network.

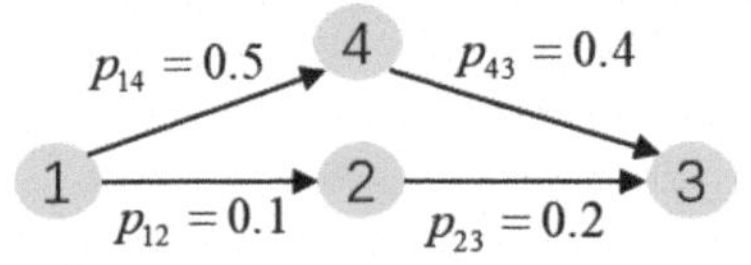

Fig. 4. A network for comparison.

Method	node 1	node 2	node 3	node 4
IC	2.584	1.2	1	1.4
IRIE	1.3	1.1	1	1.2
IRIE-2	**1.6**	1.1	1	1.2

Fig. 5. The influence rank of nodes ($\eta = 0.5$).

Here, we present a comparative analysis of the influence spread of nodes using both the IRIE and IC models, as illustrated in Fig. 4. The corresponding results are depicted in Fig. 5. It is worth noting that, in most cases (with $\eta = 0.5$),

Algorithm 1: IRIE-2 with $S = \emptyset$

Input: $G = (V, E, p)$, θ, η.
Output: $\mathbf{r}(u)$ and $u \in V$.
1 $S \leftarrow \emptyset$;
2 $\mathbf{r}(u) \leftarrow 1$, $ActP_S(u) \leftarrow 0$ for any $u \in V$;
3 **for** *each* $u \in V$ **do**
4   update $\mathbf{r}(u)$ with formula (3) and (4) ;
5 **return** $\mathbf{r}(u)$ *and* $u \in V$

the influence spread obtained through the IRIE method tends to be smaller than that obtained through the IC method. For instance, the influence spread of node 1 is 1.3 according to IRIE, whereas it is 2.584 according to the IC model. This example highlights a limitation of the IRIE method, which focuses solely on the one-hop neighbors of a node. To address this limitation, we propose an enhancement strategy for IRIE, referred to as IRIE-2, which incorporates not only the one-hop neighbors but also the two-hop neighbors. Consequently, the influence rank of nodes under our IRIE-2 method is as follows.

$$\mathbf{r}(u) = (1 - ActP_S(u)) \cdot \left\{ 1 + \eta \cdot \left[\sum_{v \in N^{out}(u)} p_{uv} \cdot \mathbf{r}(v) + \sum_{w \in N^{out}(v)} p_{vw} \cdot \mathbf{r}(w) \right] \right\}, \tag{3}$$

where v and w are the one-hop and two-hop neighbours of u, respectively. Furthermore, $ActP_S(u)$ can be estimated by

$$ActP_S(u) = \sum_{s \in S} ActP_s(u), \tag{4}$$

where $ActP_S(u)$ denotes the probability that the seed nodes $s \in S$ activates u. To simplify the calculations, we used the MIP [3] to estimate $ActP_S(u)$.

Based on the Eqs. (3) and (4), we propose that IRIE-2 algorithm, outlined in Algorithm 1, with the initialization $S \leftarrow \emptyset$. The algorithm takes the network G as input, along with the influence threshold θ used to filter paths with values smaller than θ in MIP, and the parameter η. The output of the algorithm is the influence rank $\mathbf{r}(u)$ for each node $u \in V$. In the algorithm, each node u is initially assigned $\mathbf{r}(u) \leftarrow 1$ and $ActP_S(u) \leftarrow 0$. Subsequently, each node u updates its $\mathbf{r}(u)$ using the formulas (3) and (4).

Let us revisit the example in Fig. 4 and recalculate the influence rank of the nodes using the enhanced IRIE-2 algorithm with $S = \emptyset$. The updated results are depicted in Fig. 5. It is evident that, in comparison to IRIE, our node influence rank obtained through IRIE-2 aligns more closely with the IC model. This observation serves as validation for the effectiveness of our proposed method.

Influential Edges. We define the influence edges based on the influence rank of a node. Specifically, the influence of an edge e_{uv} (i.e., $Inf\_edge(u, v)$) can be determined by the influence rank of node u and node v as follows.

$$Inf\_edge(u, v) = \mathbf{r}(u) + \mathbf{r}(v). \tag{5}$$

Algorithm 2: Distributed Influential Edges Selection in Communities

---

Input: $\mathcal{C} = \{\mathcal{C}_1, ..., \mathcal{C}_m\}$, θ, η, S, $\{K_1, ..., K_m\}$.
Output: $\mathcal{E}$.

initialize influential edges set $\mathcal{E}_i \leftarrow \emptyset$, ..., $\mathcal{E}_m \leftarrow \emptyset$;
for *each* $\mathcal{C}_i \subseteq \mathcal{C}$ *and* $i \in [m]$ ***parallel*** **do**
    initialize index $t \leftarrow 1$;
    if $S \cap \mathcal{C}_i = M_i$ *and* $M_i \neq \emptyset$ **then**
        for *each* $s \in M_i$ **do**
            use MIP to compute $ActP_s(u)$ with θ and $u \in \mathcal{C}_i$;
        use formula (4) to update $ActP_S(u) \leftarrow \sum_{s \in M_i} ActP_s(u)$;
        call Algorithm 1 to compute initial $\mathbf{r}(u)$ with η and $u \in \mathcal{C}_i$;
        for $t \leq K_i$ **do**
            use formula (5) to compute $Inf\_edge(u, v) \leftarrow \mathbf{r}(u) + \mathbf{r}(v)$ in $\mathcal{C}_i$;
            $e \leftarrow \arg_{(u,v) \in \mathcal{C}_i} \max Inf\_edge(u, v)$;
            remove e from $\mathcal{C}_i$, i.e., $\mathcal{C}_i \leftarrow \mathcal{C}_i \backslash e$;
            $\mathcal{E}_i \leftarrow \mathcal{E}_i \cup \{e\}$;
            call formula (3) and (4) to recompute $\mathbf{r}(u)$ and $ActP_S(u)$;
            $t \leftarrow t + 1$;

return $\mathcal{E} = \mathcal{E}_1 \cup \mathcal{E}_2 ... \cup \mathcal{E}_m$.

The formula (5) is the key condition to select influential edges and remove them from the network.

Remove Influential Edges in Communities. In the distributed scenario, we also address the allocation of a budget K across m communities. Specifically, we distribute a budget K_i to each community $\mathcal{C}_i$ based on the distribution of the seed set S within the communities. This allocation is determined by the equation $K \approx \sum_{i=1}^{m} K_i$ where $K_i = \lfloor (|M_i|/|\mathcal{C}_i|) \cdot K \rfloor$. Here, $M_i = S \cap \mathcal{C}_i$ and $|M_i|$ is the size of M_i. We propose an influential edges selection algorithm (see Algorithm 2).

The algorithm takes several inputs: a community partition $\mathcal{C} = \{\mathcal{C}_1, ..., \mathcal{C}_m\}$, budget allocations $\{K_1, ..., K_m\}$, a negative content seed set S, an influence threshold parameter θ, and an adjustable parameter η. Its outputs is the set of edges $\mathcal{E}$ moved from network. The algorithm consists of the following key steps:

- **Step 1**: It initialises the set of influential edges $\mathcal{E}_i \leftarrow \emptyset$, ..., $\mathcal{E}_m \leftarrow \emptyset$;
- **Step 2**: It selects K_i key edges($\mathcal{E}_i$) in parallel for each community $\mathcal{C}_i$;
- **Step 3**: It aggregates local results and outputs the global $\mathcal{E} = \mathcal{E}_1 \cup \mathcal{E}_2 ... \cup \mathcal{E}_m$.

We discuss **Step 2** in detail. In this step, we first determine whether the negative content seed node is in $\mathcal{C}_i$. If $S \cap \mathcal{C}_i = \emptyset$, it means that no seed node is in $\mathcal{C}_i$, and we do not remove any edge in the current community. If $S \cap \mathcal{C}_i = M_i$ and $M_i \neq \emptyset$, we compute $ActP_S(u)$ for each seed using MIP with formula (4) and θ, and compute influence rank $\mathbf{r}(u)$ for node u in $\mathcal{C}_i$ using formula (3). In each $\mathcal{C}_i$, our goal is to pick the most influential K_i edges. Therefore, in the loop structure, the algorithm evaluates the influence rank of each edge in $\mathcal{C}_i$ using formula (5) and finds the most critical edge e in the current iteration t. After that, it removes e from the community $\mathcal{C}_i$. Furthermore, we need to recalculate $\mathbf{r}(u)$ and $ActP_S(u)$ for the nodes in $\mathcal{C}_i$ due to the change of topology in this community.

6 Numerical Experiments

In this section, we conduct experiments on three real-world datasets to evaluate the effectiveness and efficiency of the proposed algorithm. We first introduce the datasets and the experiment setup, and then analyse and discuss the results.

6.1 Experiment Setup

Datasets. The following three datasets are collected from the Stanford Large Datasets http://snap.stanford.edu/data/index.html.

Congress-Twitter: This network represents the Twitter interaction network for the 117th United States Congress, both House of Representatives and Senate. It comprises 475 nodes and 13,289 edges.

Wiki-Vote: The network contains all the Wikipedia voting data from the inception of Wikipedia till January 2008. Nodes in the network represent wikipedia users and a directed edge from node u to node v represents that user u voted on user v. It comprises 7,115 nodes and 103,689 edges.

HEP-PH: This citation graph is from the e-print arXiv and covers all the citations within a dataset of 34,546 papers with 421,578 edges. If a paper u cites paper v, the graph contains a directed edge from u to v.

Parameter Setting. In all experiments, we randomly select 1% nodes on networks as the seed nodes of the negative content, i.e., $|S| = 1\% * |V|$ where $|V|$ denotes the total number of nodes. We randomly select a p_{uv} from (0,1) for each edge (u, v) in each network. Furthermore, we set $\theta = 0.45$ to filter out paths in MIP whose influence is less than θ. In our IRIE-2, we let parameter $\eta \in [0.1, 0.7]$. We aim to remove K edges from the network and we set K between $1\% * |E|$ and $5\% * |E|$, where $|E|$ denotes the total number of edges in a network.

Comparison Methods. The following methods will be compared in terms of their performance:

- **IRIE** [6]: The method evaluates the influence of a node considering only 1-hop neighbours.
- **BP-Greedy** [7]: This method removes the top K edges from the network based on *Bond Percolation.*

Evaluation Metric: We measure the effectiveness of the various methods by defining a metric called *Protect Ratio* (PR) as follows.

$$PR(\mathcal{E}) = \frac{L(\emptyset) - L(\mathcal{E})}{L(\emptyset)}. \tag{6}$$

Here, $L(\emptyset)$ represents the case where no edges are removed, while $L(\mathcal{E})$ denotes the removal of the edge set $\mathcal{E}$ consisting of K edges. It is worth noting that the PR falls within the range of $(0, 1)$. Furthermore, a higher PR score signifies greater effectiveness of the algorithm.

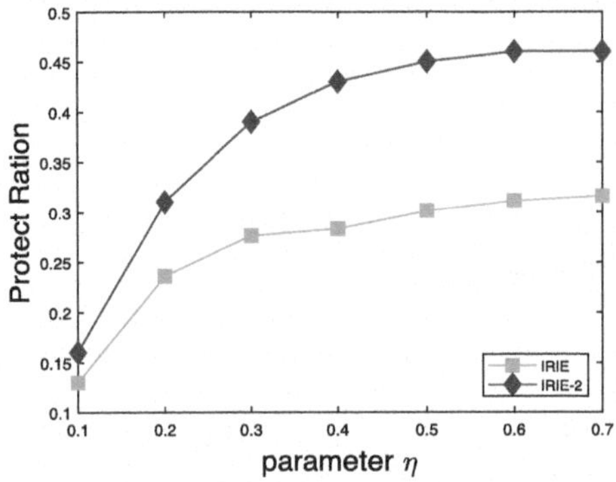

(a) Wiki-Vote dataset

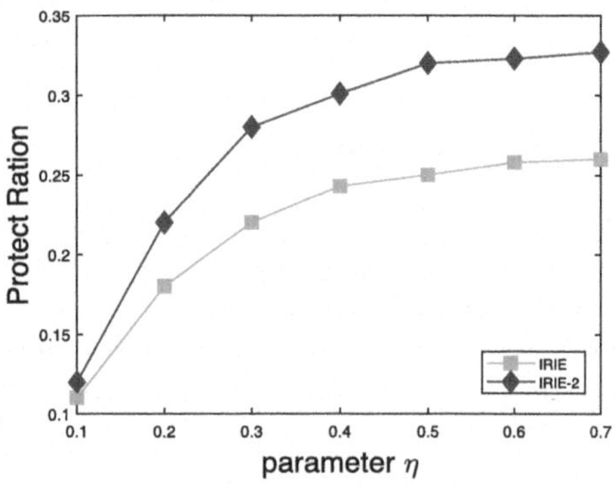

(b) HEP-PH dataset

Fig. 6. The relationship of parameter η to IRIE-2/IRIE.

6.2 Relationship of Parameter η to IRIE-2

In this section, we investigate the relationship between the parameter η and IRIE-2. In this experiment, we let η gradually increase from 0.1 to 0.7, and we observe the trend of *Protect Ration* ($PR(\mathcal{E})$) calculated using the IRIE-2 algorithm. Here, we fix the total number of edges removed to be $3\% * |E|$. Furthermore, we compare the performance of the IRIE algorithm. The experimental results are shown in Fig. 6. In the figure, the horizontal axis represents $\eta \in [0.1, 0.7]$, and the vertical axis represents PR score.

We have the following observations: 1) $PR(\mathcal{E})$ increases with η on both Wiki-Vote and HEP-PH social networks. The reason for the above phenomenon is that the update rules of IRIE and IRIE-2 are closely related to η, and increasing η

amplifies the influence rank of nodes as well as edges. This promotes IRIE and IRIE-2 to screen key edges, thus proving their effectiveness. 2) Keeping η the same, the $PR(\mathcal{E})$ obtained by our IRIE-2 is much larger than that of IRIE, especially when $\eta \in [0.3, 0.6]$. This is due to the fact that our IRIE-2 takes into account not only 1-hop neighbours, but also 2-hop neighbours when calculating the influence rank of a node. This makes the $PR(\mathcal{E})$ larger, in other words, our algorithm is better at negative content control.

6.3 Comparison with Existing Methods

In this part, we compare the proposed distributed algorithm based on IRIE-2 with existing methods including IRIE and BP-Greedy. For this comparison, we set $\eta = 0.5$ across three datasets. We systematically vary the total number of removed edges K (from $1\% * |E|$ to $5\% * |E|$) and observe the $PR(\mathcal{E})$ performance of different methods. The experimental results are shown in Fig. 7.

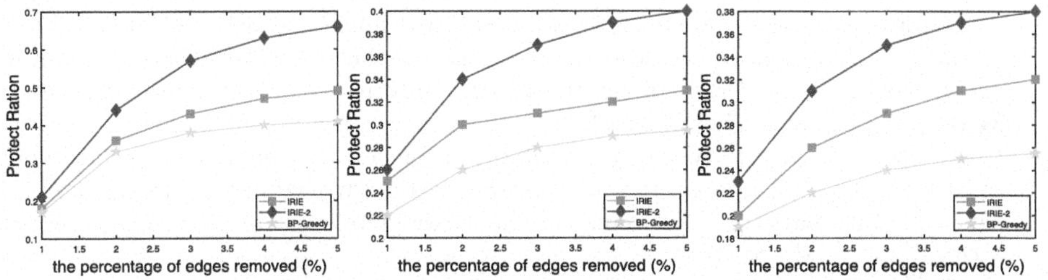

(a) Congress-Twitter dataset (b) Wiki-Vote dataset (C) HEP-PH dataset

Fig. 7. Compare with other existing algorithm on datasets.

In the figure, the horizontal axis indicates the percentage of edges removed (1% to 5% of the total number of edges $|E|$), and the vertical axis indicates PR score. In our experiments, we initially partition the original network into m sub-networks using a community partition algorithm [9]. Subsequently, our distributed approach is employed to independently select K_i key edges within each community in parallel in each community $\mathcal{C}_i$, where $i \in [m]$. The experimental results demonstrate the superior performance of our algorithm compared to the comparison methods, as evidenced by our higher PR scores across all three datasets. Additionally, our distributed strategy achieves significant speed improvements of 2.1×, 7.4×, and 13.6× on the Congress-Twitter, Wiki-Vote, and HEP-PH datasets, respectively, when compared to the traditional centralized approach [7].

7 Conclusion

This paper addresses the problem of *Negative Content Spread Minimization* (NCSM), which aims to minimize the spread of negative content by removing a set of edges from the network. We establish the NP-hardness of the NCSM problem and prove that its objective function is not submodular. Considering factors such as computational complexity and the performance of centralized algorithms, we propose a distributed approach consisting of two phases: community partition and influential edge selection. Drawing inspiration from the IRIE algorithm, we devise a node influence ranking strategy based on two-hop neighbors, upon which we construct a key edge screening rule. To assess the effectiveness of our proposed methods, we conduct extensive experiments on real-world networks. The experimental results demonstrate that our algorithm outperforms state-of-the-art approaches in terms of both effectiveness and efficiency. In future work, we plan to delve into the theoretical analysis of distributed approximation algorithms.

References

1. Bajpai, S., Chaturvedi, A.: A multi-criteria decision making based integrated approach for rumor prevention in social networks. Multimedia Tools Appl., 1–26 (2024)
2. Budak, C., Agrawal, D., El Abbadi, A.: Limiting the spread of misinformation in social networks. In: Proceedings of the 20th International Conference on World Wide Web, pp. 665–674. ACM (2011)
3. Chen, W., Wang, C., Wang, Y.: Scalable influence maximization for prevalent viral marketing in large-scale social networks. In: Proceedings of the 16th ACM SIGKDD International Conference on Knowledge Discovery and Data Mining, pp. 1029–1038. ACM (2010)
4. Chen, W., Wang, Y., Yang, S.: Efficient influence maximization in social networks. In: Proceedings of the 15th ACM SIGKDD International Conference on Knowledge Discovery and Data Mining, pp. 199–208. ACM (2009)
5. He, Q., et al.: Graph convolutional network-based rumor blocking on social networks. IEEE Trans. Comput. Soc. Syst. (2022)
6. Jung, K., Heo, W., Chen, W.: IRIE: scalable and robust influence maximization in social networks. In: 2012 IEEE 12th International Conference on Data Mining, pp. 918–923. IEEE (2012)
7. Kimura, M., Saito, K., Motoda, H.: Minimizing the spread of contamination by blocking links in a network. In: Proceedings of the 23nd AAAI Conference of Artificial Intelligence, pp. 1175–1180. AAAI (2008)
8. Ni, Q., Guo, J., Huang, C., Wu, W.: Community-based rumor blocking maximization in social networks: algorithms and analysis. Theoret. Comput. Sci. **840**, 257–269 (2020)
9. Raghavan, U.N., Albert, R., Kumara, S.: Near linear time algorithm to detect community structures in large-scale networks. Phys. Rev. E **76**(3), 036106 (2007)
10. Song, C., Hsu, W., Lee, M.L.: Node immunization over infectious period. In: Proceedings of the 24th ACM International on Conference on Information and Knowledge Management, pp. 831–840. ACM (2015)
11. Tong, G., et al.: An efficient randomized algorithm for rumor blocking in online social networks. IEEE Trans. Network Sci. Eng., **7**, 845–854 (2017)
12. Wang, B., Chen, G., Fu, L., Song, L., Wang, X.: DRIMUX: dynamic rumor influence minimization with user experience in social networks. IEEE Trans. Knowl. Data Eng. **29**(10), 2168–2181 (2017)

13. Yan, R., Li, Y., Li, D., Wang, Y., Zhu, Y., Wu, W.: A stochastic algorithm based on reverse sampling technique to fight against the cyberbullying. ACM Trans. Knowl. Discovery Data (TKDD) **15**(4), 1–22 (2021)
14. Yan, R., Li, Y., Wu, W., Li, D., Wang, Y.: Rumor blocking through online link deletion on social networks. ACM Trans. Knowl. Discovery Data (TKDD) **13**(2), 1–26 (2019)
15. Zi, H., Wu, S., Wu, K.: Hybrid control strategy for rumor spread in social networks. In: Jin, H., Pan, Y., Lu, J. (eds.) IAIC 2023. CCIS, vol. 2060, pp. 198–210. Springer, Singapore (2023). https://doi.org/10.1007/978-981-97-1332-5_16

Approximation Algorithms for the Minimum Weight Cycle/Path Partition Problem

Yaqi Li, Wei Yu(✉), and Zhaohui Liu

School of Mathematics, East China University of Science and Technology, Shanghai 200237, China
y30221658@mail.ecust.edu.cn, {yuwei,zhliu}@ecust.edu.cn

Abstract. Let $G = (V, E)$ be an undirected complete graph on kn vertices. Each edge is associated with a non-negative weight. The edge weights satisfies the triangle inequality. A k-cycle partitioning is a set of n vertex-disjoint k-cycles, i.e. cycles containing exactly k vertices (and thus $k-1$ edges). The minimum weight k-cycle partition problem (MinWkCP) aims to compute a k-cycle partition with minimum total edge weight. The minimum weight k-path partition problem (MinWkPP) is obtained by replacing cycles with paths in the MinWkCP.

In this paper, we first devise a tight $\frac{3}{2}$-approximation algorithm for the MinW4CP, improving on the best-known 3-approximation algorithm by Goemans and Williamson. Then we deal with a special case of the MinWkCP and MinWkPP, where the edge weights are either 1 or 2, and devise approximation algorithms with ratios $\frac{8k^2+14k-8}{7k^2}$ and $\frac{8(k+1)}{7k}$, respectively. For the MinW3PP and MinW3CP on $\{1, 2\}$-edge-weighted graphs, we propose two matching based algorithms with tight approximation ratios $\frac{3}{2}$ and $\frac{5}{3}$, respectively. Finally, for the $\{1, 2\}$-edge-weighted MinW3CP, we design a local search algorithm to further improve the ratio to $\frac{8}{5}$.

Keywords: Approximation Algorithm · Cycle Partition · Path Partition · Matching · Local Search

1 Introduction

Let $G = (V, E)$ be an undirected complete graph on kn vertices, where k and n are positive integers and E indicates the edge set. Each edge $e \in E$ is associated with a nonnegative weight $w(e)$ satisfying the triangle inequality. A cycle (path) containing precisely q vertices is called a q-cycle (q-path). The minimum weight k-cycle partition problem (MinWkCP) is to determine a set of n vertex-disjoint k-cycles covering all the vertices such that the total edge weight of these cycles is minimized. By replacing k-cycles with k-paths in the MinWkCP, we obtain the minimum weight k-path partition problem (MinWkPP). We use $\text{MinWkCP}_{1,2}$ ($\text{MinWkPP}_{1,2}$) to denote the special case of the MinWkCP (MinWkPP) where

S. Ghosh and Z. Zhang (Eds.): AAIM 2024, LNCS 15179, pp. 170–181, 2024.
https://doi.org/10.1007/978-981-97-7798-3_15

$w(e) \in \{1, 2\}$ for each $e \in E$. These problems arise frequently in the design of communication networks, vehicle routing, and cluster analysis [5].

Unfortunately, all the above-mentioned problems are NP-hard for any fixed $k \geq 3$ [4,9,10]. As a result, we focus on the related approximability results in this paper. Goemans and Williamson [5] gave a $4(1-\frac{1}{k})$-approximation algorithm for the MinWkCP via primal-dual methods. This implies a 3-approximation algorithm and a $\frac{8}{3}$-approximation algorithm for MinW4CP and MinW3CP, respectively. Monnot and Toulouse [9] presented a $\frac{3}{2}$-approximation algorithm for the MinW4PP based on the result of Hassin and Rubinstein [6]. They also proposed a specialized algorithm for the MinW4PP$_{1,2}$ that achieves an approximation ratio of $\frac{7}{6}$.

In the literature, there exist plenty of approximability results [1,3,7–9,11,12] on the maximization version of the MinWkCP (MinWkPP), whose objective is to find a k-cycle partition (k-path partition) such that the total edge weight is maximized. Closely related to this paper, Hassin and Schneider [7] considered the $\{0, 1\}$-edge-weighted maximization version of the MinW3PP and showed that a local search algorithm with depth 9 that achieves an approximation ratio of 0.55. Here depth informally represents the size of the neighborhood. Another related problem is the (1,2)-TSP, which is actually a special case of the MinWkCP$_{1,2}$ with $k = |V|$ (or equivalently $n = 1$). The best-known approximation ratio $\frac{8}{7}$ for the (1,2)-TSP is due to Berman and Karpinski [2].

The main contribution of this paper is that we obtain better approximation algorithms for the MinW4CP, MinWkCP$_{1,2}$ and MinWkPP$_{1,2}$. For the MinW4CP, we develop a matching-based $\frac{3}{2}$-approximation algorithm, improving on the above 3-approximation by Goemans and Williamson [5]. We also show the ratio $\frac{3}{2}$ is tight. For the MinWkCP$_{1,2}$ and MinWkPP$_{1,2}$, we propose approximation algorithms with ratios $\frac{8k^2+14k-8}{7k^2}$ and $\frac{8(k+1)}{7k}$, respectively, by invoking the state-of-the-art algorithm for the (1,2)-TSP as a subroutine. For the MinW3PP$_{1,2}$ and MinW3CP$_{1,2}$, we give two simple matching-based approximation algorithms, achieving tight ratios of $\frac{3}{2}$ and $\frac{5}{3}$, respectively. In addition, we devise a local search algorithm to further improve the approximation ratio for the MinW3CP$_{1,2}$ to $\frac{8}{5}$.

The remainder of the paper is organized as follows. Section 2 introduces some notations used throughout the paper. In Sect. 3, we propose a matching-based algorithm for the MinW4CP and analyze its approximation ratio. In Sect. 4.1, we present algorithms for the MinWkCP$_{1,2}$ and MinWkPP$_{1,2}$. In Sect. 4.2, we give two simple matching-based algorithm for the MinW3PP$_{1,2}$ and MinW3CP$_{1,2}$, which is followed by a local search algorithm for the MinW3CP$_{1,2}$ in Sect. 4.3. Finally, we conclude the paper in Sect. 5.

2 Preliminaries

Given an edge-weighted undirected graph $G = (V, E)$ with $|V| = kn$, a set $\mathcal{C}$ of k-cycles covering all the vertices of V is called a k-cycle partition. Similarly, we may define a k-path partition. For a subset $E' \subseteq E$ of edges, $w(E') = \sum_{e \in E'} w(e)$

indicates the total weight of edges in E'. In particular, when $E' = \{e_1, \ldots, e_h\}$ we also represent $w(E')$ by $w(e_1, \ldots, e_h)$. For a subgraph H of G, we use $V(H)$ and $E(H)$ to indicate the set of vertices and the set of edges of H, respectively. $w(H)$ represents the total weight of the edges in $E(H)$, i.e. $w(H) = w(E(H))$. Throughout this paper, we use APX to denote the approximate solution, i.e. a k-cycle (k-path) partition, generated by an algorithm in discussion for the MinWkCP/MinWkPP. OPT indicates the corresponding optimal solution. We denote by $apx = w(APX)$ and $opt = w(OPT)$ the total edge weight of the cycles/paths in APX and OPT, respectively.

3 Minimum Weight 4-Cycle Partition Problem

In this section, we present a matching-based algorithm for the MinW4CP. We show that this algorithm has a tight approximation ratio of $\frac{3}{2}$.

Given an instance of the MinW4CP whose input consists of an undirected complete graph $G = (V, E)$, where $|V| = 4n$, and a nonnegative weight function $w : E \rightarrow \mathbb{R}^+$. We suggest the following algorithm:

Algorithm A

1. Compute a minimum weight perfect matching M_1 in G.
2. Contract the edges of M_1 and form a complete graph $G' = (V', E')$. Each edge $e \in M_1$ corresponds to a vertex $v_e \in V'$, thus $|V'| = 2n$. For each edge $(u, v) \in E'$ with u and v corresponding to the edge $(a_u, b_u) \in M_1$ and $(a_v, b_v) \in M_1$, respectively, we define a new weight function $w' : E' \rightarrow \mathbb{R}^+$ as

$$w'(u, v) = \min\{w(a_u, a_v) + w(b_u, b_v), w(a_u, b_v) + w(b_u, a_v)\}.$$

3. Determine a minimum weight perfect matching M' in G'. By definition each edge $e' = (u, v) \in M'$ corresponds to two edges e_1, e_2 in G, which are either $(a_u, a_v), (b_u, b_v)$ or $(a_u, b_v), (b_u, a_v)$, such that $w'(e') = w(e_1) + w(e_2)$. Let M_2 be the set of all edges in G corresponding to all edges of M'.
4. Output APX consisting of the n 4-cycles induced by the edges in $M_1 \cup M_2$.

Theorem 1. *Algorithm A is a $\frac{3}{2}$-approximation algorithm for the MinW4CP.*

Proof. To prove the theorem, it is sufficient to show that $apx \leq \frac{3}{2} opt$.

Since OPT consists of n vertex-disjoint 4-cycles, we can obtain two disjoint perfect matchings A and B of G by putting two opposite edges in each 4-cycle of OPT to A and adding the remaining two opposite edges to B. By definition, the edges of each 4-cycle in APX alternate between M_1 and M_2, which are both perfect matchings.

Consider now the graph $H = (V, M_1 \cup A)$. Since M_1 and A are both perfect matchings, each connected component of H is either single edge from $M_1 \cap A$ or a simple cycle whose edges alternate between M_1 and A (Fig. 1).

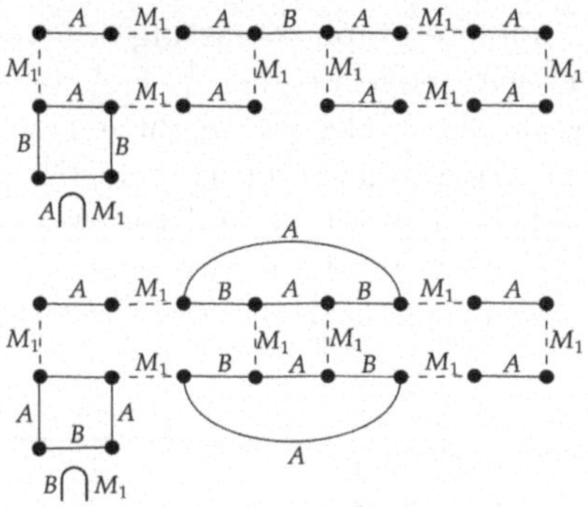

Fig. 1. Exchange the edges from A and the edges from B in $C_{OPT}(e)$.

For any edge $e \in E(OPT)$, we denote by $C_{OPT}(e)$ the 4-cycle in OPT that contains e. If there exists an edge $e \in B$ that links two distinct connected components of H, we exchange the edges from A and the edges from B in $C_{OPT}(e)$. Note that after this operation, A and B are still two disjoint perfect matchings of G using only edges in $E(OPT)$. By repeatedly performing this operation, we will end with a new graph $H' = (V, M \cup A)$ such that each connected component of H is a cycle whose edges alternate between M_1 and A and the four vertices of each 4-cycle of OPT lie on the same cycle of H'. As a result, the number of vertices of each cycle in H' is a multiple of 4.

One can see that A can be partitioned into two n-sized matchings A_1 and A_2 such that $M_1 \cup A_i$ $(i = 1, 2)$ is a 4-path partition. This can be done by choosing edges in A_1 and A_2 alternately along each connected component of H', which is a cycle. Connecting the two endpoints of each 4-path, we can obtain a 4-cycle partition. Let $\bar{A}_i$ be the set of edges used to connect the endpoints of the 4-paths. Due to the triangle inequality, the weight of each edge in $\bar{A}_i$ is no more than the weight of the corresponding 4-path whose endpoints are connected by it. As a consequence, $w(\bar{A}_i) \leq w(A) + w(M_1)$. Since M_2 corresponds to a minimum weight perfect matching M' in $G' = (V', E')$, we obtain that

$$w(M_2) \leq w(A_i) + w(\bar{A}_i) \leq 2w(A_i) + w(M_1)$$

for $i = 1, 2$. Summation over i gives that $2w(M_2) \leq 2w(A) + 2w(M_1)$, which implies that $w(M_2) \leq w(A) + w(M_1)$.

Moreover, since M_1 is a minimum weight perfect matching of G and B is a perfect matching of G, we deduce that $w(M_1) \leq w(B)$. Hence we have $w(M_2) \leq w(A) + w(M_1) \leq w(A) + w(B) = opt$.

Again, A is a perfect matching of G implies that $w(M_1) \leq w(A)$. Then we know that $2w(M_1) \leq w(A) + w(B) = opt$ and $w(M_1) \leq \frac{1}{2} opt$.

Finally, by definition $apx = w(M_1) + w(M_2) \leq \frac{1}{2} opt + opt = \frac{3}{2} opt$. □

The following example shows that the ratio $\frac{3}{2}$ in the above theorem is tight.

Example 1. Consider a complete graph consisting of 8 vertices. There is a cycle containing all the vertices with edge weights $\frac{1}{2}$, 1, $\frac{1}{2}$, 0, $\frac{1}{2}$, 1, $\frac{1}{2}$, 0 in this cyclic order. For each edge not on the cycle, its weight is equal to the weight of the shortest path between its two endpoints on this cycle. Clearly, $opt = 4$. A possible M_1 consists of the two edges of weight 1 and the two edges of weight 0 from the cycle. In this case, M' consists of the two edges of weight 2 in G'. Thus, $apx = w(M_1) + w(M_2) = w(M_1) + w(M') = 6 = \frac{3}{2}opt$ (Fig. 2).

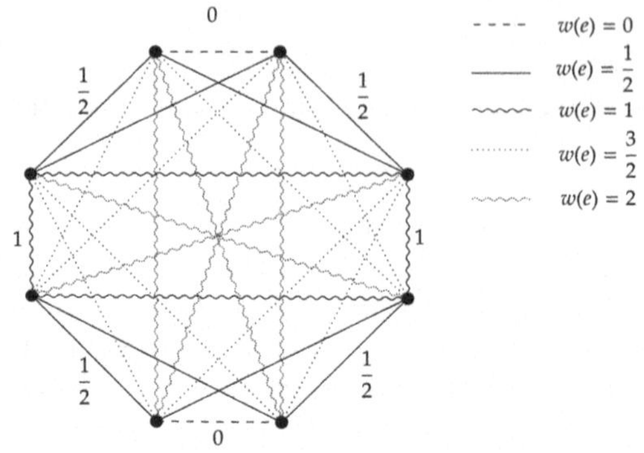

Fig. 2. The tight example for the MinW4CP.

4 Minimum Weight k-Cycle/Path Partition Problem on $\{1, 2\}$-Edge-Weighted Graphs

In this section we first develop approximation algorithms for the $\text{MinWkCP}_{1,2}$ and $\text{MinWkPP}_{1,2}$ with arbitrary k. After that we devise a better approximation algorithm for the $\text{MinW3CP}_{1,2}$ based on matching and further improve approximation ratio for the $\text{MinW3CP}_{1,2}$ using local search method.

4.1 $\text{MinWkCP}_{1,2}$ and $\text{MinWkPP}_{1,2}$ with Arbitrary k

We will devise an algorithm for the $\text{MinWkCP}_{1,2}/\text{MinWkPP}_{1,2}$ by invoking a subroutine for its special case with $n = 1$, known as (1,2)-TSP in the literature [2]. This is inspired by the work of Hassin and Rubinstein [6] and Li and Yu [8] on the maximization version of the MinWkCP and MinWkPP.

Let us focus on the $\text{MinWkCP}_{1,2}$ first. Suppose we are given an instance of the $\text{MinWkCP}_{1,2}$ consisting of an integer $k \geq 3$ and a $\{1, 2\}$-edge-weighted undirected complete graph G on kn vertices. Our algorithm first runs an α-approximation algorithm, denoted as Algorithm $\mathcal{A}_{tsp}$, for the (1,2)-TSP on the graph G to generate a Hamiltonian cycle. Then it removes properly n edges from this Hamiltonian cycle to derive n k-paths. By rotating along the Hamiltonian cycle we can choose these n edges such that their total weight is maximum. Eventually, connecting the endpoints of the obtained k-paths produces a feasible k-cycle partition. Figure 3 gives the full description of the algorithm.

Algorithm B:

Input: An $\{1,2\}$-edge-weighted undirected complete graph $G=(V,E)$ with $|V|=kn$, where $k \geq 3$ and n are positive integers.
Output: A k-cycle partition of G.

1. Run Algorithm $\mathcal{A}_{tsp}$ on an instance of the (1,2)-TSP defined on G to derive an α-approximate solution, which is a Hamiltonian cycle denoted as C;
2. Starting with an arbitrary edge of C, number the kn edges of C clockwise as $e_{1,1}\cdots e_{1,k},\cdots\cdots,e_{n,1}\cdots e_{n,k}$;
3. For each $i=1,\ldots,k$, set $S_i := \{e_{1,i},\ldots,e_{n,i}\}$ and determine i^* such that $w(S_{i^*}) = \max_{1\leq i\leq k} w(S_i)$;
4. Delete the edges in S_{i^*} from C to derive a k-path partition of G composed of n paths $P_1,\ldots,P_n$;
5. Return the k-cycle partition of G, denoted as APX, obtained by connecting the two endpoints of each path in $\{P_1,\ldots,P_n\}$.

Fig. 3. Algorithm for the MinWkCP$_{1,2}$.

Theorem 2. *Algorithm B is an* $\frac{\alpha(k^2-1)+2k}{k^2}$*-approximation algorithm for the MinWkCP*$_{1,2}$.

Proof. One can see that Algorithm B returns a feasible k-cycle partition in polynomial time. It remains to prove that $apx \leq \frac{\alpha(k^2-1)+2k}{k^2}\cdot opt$.

Let opt_{tsp} be the optimal value for the instance of the (1,2)-TSP in step 1. Denote by $C_1^*,\ldots,C_n^*$ all the k-cycles in OPT. By definition $opt=\sum_{i=1}^n w(C_i^*)$. For each $i=1,\ldots,n$, we remove the edge in C_i^* of largest weight to obtain a k-path P_i'. Due to the fact that C_i is a k-cycle and the nonnegativity of the edge weights, we deduce that $w(P_i') \leq \frac{k-1}{k}\cdot w(C_i^*)$. Then we have

$$\sum_{i=1}^{n} w(P_i') \leq \frac{k-1}{k}\sum_{i=1}^{n} w(C_i^*) = \frac{k-1}{k}opt.$$

Now we can complete the n paths $P_1',\ldots,P_n'$ into a Hamiltonian cycle C' by connecting their endpoints with n edges $e_1,\ldots,e_n$, where e_i is an edge connecting the last vertex of P_{i-1}' with the first vertex of P_i' (set $P_0'=P_n'$). Since the edges weights are either 1 or 2, $\sum_{i=1}^n w(e_i)\leq 2n$. As C' is a feasible solution for the instance of the (1,2)-TSP defined on G, we obtain that

$$opt_{tsp} \leq \sum_{i=1}^{n} w(P_i') + \sum_{i=1}^{n} w(e_i) \leq \frac{k-1}{k}\cdot opt + 2n. \tag{1}$$

On the other hand, by construction $w(C) = \sum_{i=1}^{k} w(S_i) = \sum_{i=1}^{k}\sum_{h=1}^{n} w(e_{h,i})$ and APX consists of the n paths $P_1, \ldots, P_n$ constructed in step 4 and n edges (of weight no more than 2) connecting their endpoints. Therefore,

$$\begin{aligned} apx &\leq \sum_{i=1}^{n} w(P_i) + 2n \\ &\leq \frac{k-1}{k} \cdot w(C) + 2n \\ &\leq \alpha \cdot \frac{k-1}{k} \cdot opt_{tsp} + 2n \\ &\leq \alpha \cdot \frac{k-1}{k} \left(\frac{k-1}{k} \cdot opt + 2n \right) + 2n \\ &\leq \frac{\alpha(k^2-1) + 2k}{k^2} \cdot opt, \end{aligned}$$

where the second inequality follows the definition of i^* in step 3, the third inequality holds by the fact that C is an α-approximate solution for the instance of the (1,2)-TSP, the fourth inequality follows from Eq. (1) and the last inequality follows from trivial lower bound nk of opt (each edge in OPT has a weight of at least one). □

Due to the result of Berman and Karpinski [2], we can take $\alpha = \frac{8}{7}$. As a result, Theorem 2 yields the following conclusion.

Corollary 1. *For the MinWkCP$_{1,2}$, there is an approximation algorithm with ratio* $\frac{8k^2+14k-8}{7k^2}$.

To solve the MinWkPP$_{1,2}$, we can modify Algorithm B by cancelling step 5 and returning APX as the k-path partition consisting of the n k-paths $P_1, \ldots, P_n$. Denote by $P_1^*, \ldots, P_n^*$ all the k-paths in OPT. Then $opt = \sum_{i=1}^{n} w(P_i^*)$. The paths $P_1^*, \ldots, P_n^*$ can be completed with n edges into a Hamiltonian cycle, which implies $opt_{tsp} \leq opt + 2n$. Combining this with the trivial lower bound $n(k-1)$ of opt leads to

$$opt_{tsp} \leq opt + 2n \leq \left(1 + \frac{2}{k-1}\right) opt = \frac{k+1}{k-1} \cdot opt. \tag{2}$$

and

$$\begin{aligned} apx &\leq \sum_{i=1}^{n} w(P_i) \\ &\leq \frac{k-1}{k} \cdot w(C) \\ &\leq \alpha \cdot \frac{k-1}{k} \cdot opt_{tsp} \\ &\leq \alpha \cdot \frac{k-1}{k} \cdot \frac{k+1}{k-1} \cdot opt \\ &= \frac{\alpha(k+1)}{k} \cdot opt. \end{aligned}$$

Again, taking $\alpha = \frac{8}{7}$ leads to the following result.

Theorem 3. *For the MinWkPP$_{1,2}$, there is an $\frac{8(k+1)}{7k}$-approximation algorithm.*

4.2 Matching-Based Algorithm for the MinW3PP$_{1,2}$ and MinW3CP$_{1,2}$

In this section, we present two simple matching algorithms for the MinW3PP$_{1,2}$ and MinW3CP$_{1,2}$, which achieve tight approximation ratios $\frac{3}{2}$ and $\frac{5}{3}$, respectively. These algorithms are better than the algorithm in Theorem 1 and Corollary 1.

Given an instance of the MinW3PP$_{1,2}$ whose input consists of an undirected complete graph $G = (V, E)$, where $|V| = 3n$, and a weight function $w : E \to \{1, 2\}$. The algorithm is given as follows:

Algorithm C

1. Compute a minimum weight matching M_1 of size n in G.
2. Contract the edges of M_1 and form a complete bipartite graph $G' = (V', V \setminus V(M_1), E')$. Each edge $e \in M_1$ corresponds to a vertex $v_e \in V'$, which implies that $|V'| = n$. For each edge $(u, v) \in E'$ with $u \in V'$ corresponding to the edge $(a_u, b_u) \in M_1$, we define a new weight function $w'(\cdot)$ on E' as

$$w'(u, v) = \min\{w(a_u, v), w(b_u, v)\}.$$

3. Find a minimum weight perfect matching M' in G'. By definition each edge $e' = (u, v) \in M'$ corresponds to an edge e in G, which are either (a_u, v) or (b_u, v), such that $w'(e') = w(e)$. Let M_2 be the set of all edges in G corresponding to all edges of M'.
4. Output APX consisting of the n 3-paths induced by the edges in $M_1 \cup M_2$

Theorem 4. *Algorithm C is a $\frac{3}{2}$-approximation algorithm for the MinW3PP$_{1,2}$.*

Proof. To prove the theorem, we will show that $apx \leq \frac{3}{2}opt$.

Suppose that OPT consists of n 3-paths $P_1^*, \ldots, P_n^*$. For any 3-path P_i^* ($i = 1, \ldots, n$), let e_i be the edge from $E(P_i^*)$ with the minimum weight. By definition $2w(e_i) \leq w(P_i^*)$, which leads to $2\sum_{i=1}^{n} w(e_i) \leq \sum_{i=1}^{n} w(P_i^*)$ by summing over i.

Note that $e_1, \ldots, e_n$ form an n-sized matching in G. Due to the optimality of M_1, we obtain $2w(M_1) \leq 2\sum_{i=1}^{n} w(e_i) \leq \sum_{i=1}^{n} w(P_i^*) = opt$.

Since all the edge weights of G are either 1 or 2, it holds that $1 \leq w'(u,v) \leq 2$ for any edge $(u,v) \in E'$. Therefore, for any edge $(u,v) \in M_1$ with $u \in V'$, we deduce that $w'(u,v) \leq 2 \leq 2w(a_u, b_u)$. This implies that

$$w(M_2) = \sum_{(u,v)\in M_1, u\in V'} w'(u,v) \leq 2\sum_{u\in V'} w(a_u, b_u) = 2w(M_1).$$

Consequently, we have $apx = w(M_1) + w(M_2) \leq 3w(M_1) \leq \frac{3}{2}opt$. □

For the MinW3CP$_{1,2}$, we can modify Algorithm C by simply redefining the weight function $w'(\cdot)$ of G' in Step 2 as follows: for any $(u,v) \in E'$ with $u \in V'$, $w'(u,v) = w(a_u, v) + w(b_u, v)$.

As a result, each edge $e' = (u,v) \in M'$ in Step 3 corresponds to two edges (a_u, v) and (b_u, v) in G, such that $w'(e') = (a_u, v) + (b_u, v)$. Moreover, the edges in $M_1 \cup M_2$ in Step 4 induce n 3-cycles and hence APX is a feasible solution.

The modified algorithm, which we call Algorithm D, can be analyzed in a similar way as in the proof of Theorem 4.

Theorem 5. *Algorithm D is a $\frac{5}{3}$-approximation algorithm for the MinW3CP$_{1,2}$.*

Proof. Assume that OPT consists of n 3-cycles $C_1^*, \ldots, C_n^*$. For each $i = 1, \ldots, n$, we define e_i as the edge from $E(C_i^*)$ with the minimum weight. Then $3w(e_i) \leq w(C_i^*)$, which implies that $3\sum_{i=1}^{n} w(e_i) \leq \sum_{i=1}^{n} w(C_i^*)$.

Note that $e_1, \ldots, e_n$ form an n-sized matching in G. Due to the optimality of M_1, we get $3w(M_1) \leq 3\sum_{i=1}^{n} w(e_i) \leq \sum_{i=1}^{n} w(C_i^*) = opt$.

Since G consists of edges of weights of 1 or 2, we have that $2 \leq w'(u,v) \leq 4$ for any edge $(u,v) \in E'$. So for any edge $(u,v) \in M_2$ we can deduce that $w'(u,v) \leq 4 \leq 4w(a_u, b_u)$. Thus we have

$$w(M_2) = \sum_{(u,v)\in M_1, u\in V'} w'(u,v) \leq 4\sum_{u\in V'} w(a_u, b_u) = 4w(M_1).$$

To sum up, $apx = w(M_1) + w(M_2) \leq 5w(M_1) \leq \frac{5}{3}opt$. □

Example 2. The following example demonstrates the tightness of Algorithm C and Algorithm D. We consider a complete graph $G = (V, E)$ with a weight function $w : E \to \{1, 2\}$, where $|V| = 9$. The vertex set V is partitioned into two sets V_1 and V_2 such that $|V_1| = 3$ and $|V_2| = 6$. All the edges with its endpoints either both in V_1 or both in V_2 are of weight 1. The remaining edges with one endpoint in V_1 and the other endpoint in V_2 are of weight 2.

For both Algorithm C and Algorithm D, a possible n-sized minimum weight matching M_1 consists of three edges of weight 1 whose both endpoints are from V_2. In this case, M_2 consists of three edges of weight 2 for the MinW3PP$_{1,2}$. Similarly, M_2 is composed of six edges of weight 2 for the MinW3CP$_{1,2}$. Clearly, $opt = 6$ for the MinW3PP$_{1,2}$ and $opt = 9$ for the MinW3CP$_{1,2}$. Therefore we deduce that $apx = w(M_1) + w(M_2) = 3 + 6 = 9 = \frac{3}{2}opt$ for the MinW3PP$_{1,2}$ and $apx = w(M_1) + w(M_2) = 3 + 12 = 15 = \frac{5}{3}opt$ for the MinW3CP$_{1,2}$.

4.3 Local Search Algorithm for the MinW3CP$_{1,2}$

In this section, we propose a local search algorithm for the MinW3CP$_{1,2}$, further improving upon the $\frac{5}{3}$-approximation algorithm based on matching in the previous section.

Given an instance of the MinW3CP$_{1,2}$ consists of an undirected complete graph $G = (V, E)$ with $|V| = 3n$ and a weight function $w : E \to \{1, 2\}$. Let $\mathcal{C}$ be a 3-cycle partition of G, we define its neighborhood of depth t as the set of all the 3-cycle partitions that can be obtained from $\mathcal{C}$ by replacing at most $p \leq t$ cycles of $\mathcal{C}$ with p new cycles covering the same vertices.

Our local search algorithm starts with an arbitrary initial solution, i.e. a 3-cycle partition $\mathcal{C}$. Then the algorithm checks whether its neighborhood of depth t contains a 3-cycle partition of lesser weight. If so, it update $\mathcal{C}$ as this 3-cycle partition and repeat the process; otherwise, the algorithm terminates the search and output the current 3-cycle partition $\mathcal{C}$, which is called a t-optimal solution, as the final approximate solution APX.

In what follows, we analyze the above local search algorithm, called Algorithm LS(d), for the case of $d = 3$.

In order to analyze the approximation ratio of Algorithm LS(d), we require a key lemma proposed by Hassin and Schneider [7]. Before the statement of this lemma, we give the following definition.

Definition 1. *Denote by $o_1, \ldots, o_n$ the 3-cycles in OPT. For each 3-cycle $o \in OPT$, we define*

$$A(o) = \{t \in APX \mid V(t) \cap V(o) \neq \emptyset\}$$

as the set of 3-cycles in APX that are adjacent to o.

Given a 3-cycle $o \in OPT$ and an edge $e \in E(t)$ with $t \in APX$, let

$$\overline{w}_o(e) = (|V(t) \cap V(o)| + |V(e) \cap V(o)|)w(e)$$

be a new weight of e with respect to o. Figure 4 illustrates the possible values of $\overline{w}_o(e)$.

Let

$$\overline{W}(o) = \sum_{t \in A(o)} \overline{w}_o(t) = \sum_{t=\{e_1,e_2,e_3\} \in A(o)} [\overline{w}_o(e_1) + \overline{w}_o(e_2) + \overline{w}_o(e_3)].$$

For cycles $o_{i_1}, \ldots, o_{i_h} \in OPT$, set $\overline{W}(o_{i_1}, \ldots, o_{i_h}) = \sum_{j=1}^{h} \overline{W}(o_{i_j})$ and $w(o_{i_1}, \ldots, o_{i_h}) = \sum_{j=1}^{h} w(o_{i_j})$.

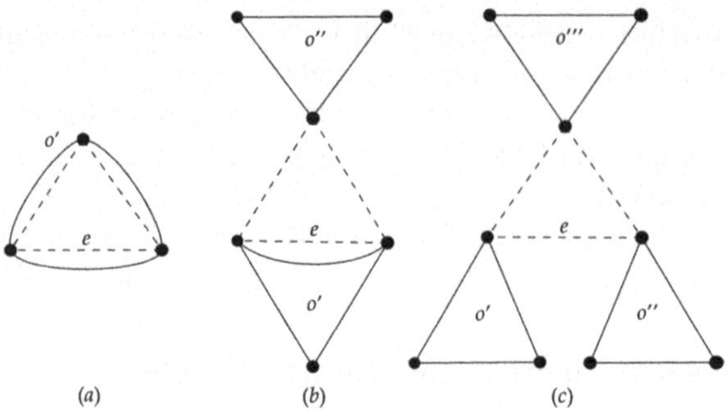

Fig. 4. Possible values of $\overline{w}_o(e)$. In picture (a), $\overline{w}_{o'}(e) = 5w(e)$. In picture (b), $\overline{w}_{o'}(e) = 4w(e)$ and $\overline{w}_{o''}(e) = w(e)$. In picture (c), $\overline{w}_{o'}(e) = \overline{w}_{o''}(e) = 2w(e)$ and $\overline{w}_{o'''}(e) = w(e)$.

Lemma 1. [7] $\sum_{o \in OPT} \overline{W}(o) = 5apx$.

The main proof idea of this lemma is that the weight of each edge in $E(APX)$ is counted 5 times on the left-hand-side: thrice due to the 3-cycle in APX it belongs to and twice due to the edge itself intersects some 3-cycles in OPT (see Fig. 4 for an illustration). Note that the original version of Lemma 1 in [7] is proved for the case of 3-paths. However, the same conclusion also holds for 3-cycles.

For now on, APX is a 3-optimal solution. We have the following result.

Lemma 2. *For any $o \in OPT$, it holds that $\overline{W}(o) \leq 8w(o)$.*

Theorem 6. *Algorithm LS(3) is a $\frac{8}{5}$-approximation algorithm for the MinW3CP$_{1,2}$.*

Proof. By Lemma 2, we know that $\overline{W}(o) \leq 8w(o)$ for all 3-cycles $o \in OPT$. Lemma 1 says that $\sum_{o \in OPT} \overline{W}(o) = 5apx$. Then

$$5apx = \sum_{o \in OPT} \overline{W}(o) \leq 8 \sum_{o \in OPT} w(o) = 8opt.$$

As a result, we obtain $apx \leq \frac{8}{5}opt$. This completes the proof. □

5 Conclusions

We obtain better approximation algorithms for the MinW4CP, MinWkCP$_{1,2}$ and MinWkPP$_{1,2}$. For the MinW4CP, we develop a matching-based approximation algorithm with tight ratio $\frac{3}{2}$. This improves on the previous $4\left(1-\frac{1}{k}\right)$-approximation for the MinWkCP by Goemans and Williamson [5] when $k = 4$. For the MinWkCP$_{1,2}$ and MinWkPP$_{1,2}$, we propose approximation algorithms with ratios $\frac{8k^2+14k-8}{7k^2}$ and $\frac{8(k+1)}{7k}$, respectively. For the MinW3PP$_{1,2}$

and $\text{MinW3CP}_{1,2}$, we give two simple matching-based approximation algorithm, achieving tight ratios of $\frac{3}{2}$ and $\frac{5}{3}$, respectively. Finally, we devise a local search algorithm to further improve the approximation ratio for the $\text{MinW3CP}_{1,2}$ to $\frac{8}{5}$.

For future work, it would be interesting to improve the approximation ratio $4\left(1-\frac{1}{k}\right)$ of Goemans and Williamson for the minimum weight k-path/cycle partition problem for values of $k \neq 4$. Moreover, our local search algorithm is only analyze for depth 3, it would be interesting to show better performance of this local search algorithm with depth greater than 3.

Acknowledgements. This research is supported by the National Natural Science Foundation of China under grant number 11871213.

References

1. Bar-Noy, A., Peleg, D., Rabanca, G., Vigan, I.: Improved approximation algorithms for weighted 2-path partitions. Discret. Appl. Math. **239**, 15–37 (2018)
2. Berman, P., Karpinski, M.: 8/7-approximation algorithm for (1,2)-TSP. In: Proceedings of the Seventeenth Annual ACM-SIAM Symposium on Discrete Algorithms, pp. 641–648 (2005)
3. Chen, Y., Chen, Z.Z., Lin, G.H., Wang, L.S., Zhang, A.: A randomized approximation algorithm for metric triangle packing. J. Comb. Optim. **41**, 12–27 (2021)
4. Garey, M.R., Johnson, D.S.: Computers and Intractability: A Guide to the Theory of NP-Completeness. Freeman, CA (1979)
5. Goemans, M.X., Williamson, D.P.: A general approximation technique for constrained forest problems. SIAM J. Comput. **24**(2), 296–317 (1995)
6. Hassin, R., Rubinstein, S.: An approximation algorithm for maximum packing of 3-edge paths. Inf. Process. Lett. **63**(2), 63–67 (1997)
7. Hassin, R., Schneider, O.: A local search algorithm for binary maximum 2-path partitioning. Discret. Optim. **10**(4), 333–360 (2013)
8. Li, S., Yu, W.: Approximation algorithms for the maximum-weight cycle/path packing problems. Asia-Pacific J. Oper. Res. **40**(4), 2340003 (2023)
9. Monnot, J., Toulouse, S.: Approximation results for the weighted P4 partition problem. J. Discrete Algorithms **6**(2), 299–312 (2008)
10. Steiner, G.: On the k-path partition of graphs. Theoret. Comput. Sci. **290**, 2147–2155 (2003)
11. Zhao, J., Xiao, M.: Improved approximation algorithms for cycle and path packings. In: Proceedings of the International Conference and Workshops on Algorithms and Computation, pp. 179–193 (2024)
12. Zhao, J., Xiao, M.: An improved approximation algorithm for metric triangle packing. In: Chen, X., Li, B. (eds.) TAMC 2024. LNCS, vol. 14637, pp. 50–62 (2024). https://doi.org/10.1007/978-981-97-2340-9_5

Iterative Rounding for Bag of Tasks Scheduling with Rejection in High Performance Computing

Ruiqing Sun(✉)

School of Mathematics and Statistics, Yunnan University, Kunming, China
ruiqing2020@126.com

Abstract. In this paper, we consider a scheduling model, called bag of tasks scheduling problem with rejection, which schedules n jobs on m unrelated parallel machines in high performance computing, where each job consists of a bag of tasks with a processing time and a rejection penalty. Each job is either accepted and all the tasks must be processed and incurs a cost, or rejected which incurs a rejection penalty. Finally, we obtain bi-criteria $(\frac{e}{e-1}, \frac{2e-1}{e-1})$-approximation for bag of tasks scheduling problem with rejection by using iterative rounding and linear programming rounding.

Keywords: Iterative rounding · Bag of tasks scheduling · Approximation algorithm

1 Introduction

High Performance Computing (HPC) is a computing technology and field that aims to enable fast, large-scale, and efficient computing tasks through the use of parallel computing and dedicated hardware. The core idea of HPC is to decompose computing tasks into multiple subtasks and execute them simultaneously on multiple processing units to significantly improve computing speed and efficiency. In modern large HPC systems, each user (or job) submits a large number of independent tasks known as a bag-of-tasks. Tarplee et al. [1] designed a linear programming based algorithm for a heterogeneous computing system with bag-of-tasks, to efficiently compute high quality solutions for simultaneously minimizing energy and makespan. Then, Li et al. [2] designed a polynomial-time algorithm for the energy-aware profit maximizing problem in heterogeneous computing systems with bag-of-tasks to maximize the profit per unit time. Recently, Li [3] considered bag-of-tasks scheduling with rejection problem to minimize the makespan plus the total rejection penalty of all rejected jobs. They presented a polynomial time approximation scheme (PTAS).

S. Ghosh and Z. Zhang (Eds.): AAIM 2024, LNCS 15179, pp. 182–193, 2024.
https://doi.org/10.1007/978-981-97-7798-3_16

Unrelated parallel machine scheduling problem to minimize makespan is a hot topic in the last few decades. Lenstra et al. [4] considered this problem, they designed a 2-approximation algorithm in their seminal work [4]. If each job is processed by machine incurred a cost and the objective is to determine whether there exists a schedule with a makespan of at most a given threshold and a total cost of at most a given threshold, this problem is classic generalized assignment problem (GAP). Shmoys and Tardos [5] presented the classic 2-approximation algorithm to the GAP. Recently, Li and Ou [6] considered parallel machine scheduling with energy consumption which incurred if a job is processed by a machine. This problem is an important special case of GAP on parallel machines such that energy consumption is no more than a given threshold. They designed a $(\frac{4}{3}+\varepsilon)$-approximation algorithm for this problem.

In a cloud and edge collaboration computing environment, due to the limited computing capacity of edge servers, some jobs can be processed by the edge servers (or machines), and other jobs have to be offloaded to the cloud computing center (or rejected) which incurs the penalty costs. Scheduling with rejection, which was first proposed by Bartal et al. [7]. They presented a $(2-1/m)$-approximation algorithm for scheduling with rejection which runs in time $O(n\log n)$ and a PTAS. Li et al. [8] considered penalty cost constrained identical parallel machine scheduling problem such that the makespan is minimized and the total rejection penalty of all rejected jobs is no more than a given upper bound. They presented a 2-approximation algorithm, followed by a PTAS. Recently, Li and Ou [9] considered machine scheduling with restricted rejection such that if a job is rejected, then it incurs a cost. The objective is to minimize the makespan of the accepted jobs and the total rejection penalty of the rejected jobs such that the total cost of rejected jobs no greater than a given upper bound. They designed a PTAS for this problem. For more results on scheduling with rejection, we refer the readers to the surveys provided by Shatbay et al. [10].

Now, iterative rounding is a powerful and flexible tool in the design of approximation algorithms for combinatorial optimization problems (for more details see [11]). This technique was introduced by Jain [12] for approximating survivable network design problems. More papers using iterative rounding method can be found in [13–15].

Motivated by [3,7,14], we propose a model, called bag of tasks scheduling problem with rejection, which schedules n jobs on m unrelated parallel machines in HPC, where each job consists of a bag of tasks with a processing time and a rejection penalty. More precisely, each job J_j is either accepted and all the b_j tasks must be processed and incurs a processing time p_{ij} and a cost c_{ij} on machine M_i, or rejected which incurs a rejection penalty w_j. The objective is to minimize the total cost of jobs of accepted jobs plus total penalty of rejected jobs, conditional that total processing time on each machine no great than the given threshold. In this paper, we obtain the following main results: using iterative rounding method for the problem where the job rejection is not allowed obtain a 2-approximation algorithm. Then, use this result as mention above, we obtain bi-criteria $(\frac{e}{e-1}, \frac{2e-1}{e-1})$-approximation for bag of tasks scheduling problem with rejection. The remainder of this paper is structured as follows. In Sect. 2, we give some preliminaries. In Sect. 3, we design a 2-approximation algorithm for a special case of this problem when job rejection is not allowed by technique of iterative rounding. In Sect. 4, we design a $(\frac{e}{e-1}, \frac{2e-1}{e-1})$-approximation algorithm for our problem by technique of LP-rounding. In the last section, we present conclusions and future work.

2 Preliminaries

2.1 Extreme Point Solutions and Rank Lemma

In this subsection, we discuss the definition and properties of the linear programmings extreme point solution will be used in this paper. More comprehensive details can be found in the book [11].

Definition 1 *[11]. Let $P = \{x : Ax \geq b, x \geq 0\} \subseteq \mathbb{R}^n$. Then $x \in \mathbb{R}^n$ is an* **extreme point solution** *of P if there does not exist a non-zero vector $y \in \mathbb{R}^n$ such that $x + y, x - y \in P$.*

Lemma 1 ***(Rank Lemma).*** *Let $P = \{x : Ax \geq b, x \geq 0\}$ be a polytope and let $x = (x_1, \cdots, x_n)$ be an extreme point solution of P such that $x_i > 0$ for each $i = 1, \cdots, n$. Then any maximal number of linearly independent tight constraints of the form $A_i x = b_i$ for some row i of A equals the number of variables.*

Theorem 1 *[16]. There is an algorithm which returns an optimal extreme point solution to a linear programming. Moreover, the running time of the algorithm is polynomial in the size of the linear programming.*

2.2 The Linear Programmings of the Unrelated Parallel Machines Bag-of-Tasks Scheduling with Rejection

The scheduling model studied in this paper is described below. Given a set of m unrelated parallel machines $\mathcal{M} = \{M_1, M_2, \cdots, M_m\}$, a set of n jobs $\mathcal{J} = \{J_1, J_2, \cdots, J_n\}$. Each job J_j consists of a bag of b_j identical tasks which have two m-vector $\boldsymbol{p}_j = (p_{1j}, \cdots, p_{mj})$ and $\boldsymbol{c}_j = (c_{1j}, \cdots, c_{mj})$, and also has a rejection penalty w_j. Each job J_j is either accepted and all the b_j tasks must be processed with a processing time p_{ij} and incurs a cost c_{ij}, or rejected which incurs a rejection penalty w_j. Each machine can process at most one task at a time, and preemption is not allowed during task processing. The objective is to minimize the total cost of accepted jobs plus the total rejection penalty of rejected jobs, conditional that total processing time on each machine no great than the given threshold T. We introduce a binary variable $z_j \in \{0, 1\}$ for each job $J_j \in J$, where $z_j = 1$ if and only if job J_j is accepted. Let $x_{ij} \in \{0, 1, \cdots, b_j\}$ be the number of tasks of job J_j processed on machine M_i. Here, we get an integer program for our problem:

$$\begin{cases} \min \sum_{i=1}^{m}\sum_{j=1}^{n} c_{ij}x_{ij} + \sum_{j=1}^{n} w_j(1 - z_j) \\ \sum_{i=1}^{m} x_{ij} = b_j z_j, \ \forall j = 1, 2, \cdots, n, \\ \sum_{j=1}^{n} p_{ij}x_{ij} \leq T, \ \forall i = 1, \cdots, m, \\ z_j \in \{0, 1\}, \ \forall j = 1, 2, \cdots, n, \\ x_{ij} \in \{0, 1, \cdots, b_j\}, \ \forall j = 1, 2, \cdots, n, \ \forall i = 1, \cdots, m. \end{cases} \tag{1}$$

Here, the first constraint makes sure that all the tasks of each accepted job must be processed by the machines. The second constraint expresses the fact that the total processing time on each machine is not exceed the given threshold T. In this paper, given an undirected graph $G = (V, E)$ and a vertex $v \in V$, we denote $\delta_G(v)$ be the set of edges which have exactly one endpoint is v. We also denote $d_G(v)$ be degree of v, i.e., $|\delta_G(v)|$.

3 An Iterative Algorithm for a Special Case of Bag of Tasks Scheduling with Rejection Problem

We consider a special case of our problem in this section. That is, the bag of tasks scheduling problem where the job rejection is not allowed. Then, we use the obtained results in this section to design an approximation algorithm for the bag of tasks scheduling with rejection problem in the next section. Let $x_{ij} \in \{0, 1, \cdots, b_j\}$ be the number of tasks of job J_j processed on machine M_i, we first formulate an LP_{bts} relaxation for bag of tasks scheduling problem where the job rejection is not allowed.

$$\begin{cases} \min \sum_{i=1}^{m}\sum_{j=1}^{n} c_{ij}x_{ij} \\ \sum_{i=1}^{m} x_{ij} = b_j, \ \forall j = 1, 2, \cdots, n, \\ \sum_{j=1}^{n} p_{ij}x_{ij} \le T, \ \forall i = 1, \cdots, m, \\ x_{ij} \ge 0, \ \forall j = 1, 2, \cdots, n, \ \forall i = 1, \cdots, m. \end{cases} \tag{2}$$

Here, the first constraint makes sure that all the tasks of each job must be processed by the machines. Although this problem a 2-approximation result can be obtained in polynomial time using Li et al. [2]'s algorithm. Our algorithm does not improve on the approximation factors of the previous algorithm, but we derive a natural iterative rounding algorithm. Lau et al. [11] considered the GAP and gave a 2-approximation algorithm. But their algorithm cannot be directly applied to this problem, otherwise it would result in a pseudo-polynomial time algorithm.

We model this case problem as a bipartite matching problem. We start with a complete weight bipartite graph G with vertex set $\mathcal{J} \cup \mathcal{M}$ and edge set $\mathcal{J} \times \mathcal{M}$ of the bipartite graph. The edge between job $J_j \in \mathcal{J}$ and machine $M_i \in \mathcal{M}$ has cost c_{ij}. The problem can be reduced to finding a subgraph B of G with vertex set $\mathcal{J} \cup \mathcal{M}$ and edge set $E \subseteq \mathcal{J} \times \mathcal{M}$, where an edge in $e_{ij} = (i, j) \in E$ denotes that job J_j is assigned to machine M_i. To make sure that the first constraint, we require $b_j \ge d_B(J_j) \ge 1$. To make sure that the second constraint, we require $\sum_{e \in \delta(M_i) \cap B} p_{ij} \le T$. In addition, we removing all edges e_{ij} from G if $p_{ij} > T$, since no optimal solution will assign the job J_j to machine M_i.

3.1 Iterative Algorithm

In this subsection, we present an algorithm (Algorithm 1) that returns a schedule with the optimal cost, by possibly violating threshold constraint of machines, by a factor of at most 2. Let $e_{ij} = (i, j) \in E$ be the job J_j is assigned to machine M_i. Let $F = \{(v_{ij}, e_{ij}) | i = 1, \cdots, m; j = 1, \cdots, n\}$ be a set where each element is a vector and the first component represents how many tasks are assigned to the machine and the second component represents the machine-job edge e_{ij}.

The condition at line 20,22 and the update at line 21,23 should be understood as a relaxation, since by those two lines we relax the threshold constraint of a machine. However, in the sequel we show that the load of any machine M_i stays within $2T$. Finally, we can use edges in F to return a subgraph B, or the corresponding job-machine matching.

Algorithm 1. Iterative algorithm

$v_{ij} = 0$, for $i = 1, \cdots, m; j = 1, \cdots, n$, $F = \{(v_{ij}, e_{ij}) | i = 1, \cdots, m; j = 1, \cdots, n\}$, $T_i = T$, $i = 1, 2, \cdots, m$, $\mathcal{I} = \{1, 2, \cdots, m\}$
Solve LP_{bts} to obtain an optimal extreme point solution x
for $j = 1, 2, \cdots, n$ **do**
  for $i = 1, 2, \cdots, m$ **do**
    $F \leftarrow F \setminus \{(0, e_{ij})\} \cup \{(\lfloor x_{ij} \rfloor, e_{ij})\}$
    $b_j \leftarrow b_j - \lfloor x_{ij} \rfloor$
    $T_i \leftarrow T_i - \lfloor x_{ij} \rfloor \cdot p_{ij}$
  end for
end for
while $\mathcal{J} \neq \emptyset$ **do**
  Solve LP_{bts} to obtain an optimal extreme point solution x
  if $\exists e_{ij}$ such that $x_{ij} = 0$ **then**
    remove e from G and $F \leftarrow F \setminus \{(0, e_{ij})\}$
  else if $b_j = 0$ **then**
    $\mathcal{J} = \mathcal{J} \setminus J_j$
  else if $\exists e_{ij}$ such that $x_{ij} \geq 1$ **then**
    $F \leftarrow F \setminus \{(v_{ij}, e_{ij})\} \cup \{(v_{ij} + \lfloor x_{ij} \rfloor, e_{ij})\}$
    $b_j \leftarrow b_j - \lfloor x_{ij} \rfloor$
    $T_i \leftarrow T_i - \lfloor x_{ij} \rfloor \cdot p_{ij}$
  else if $\exists i \in \mathcal{I}$ such that $d(M_i) = 1$ **then**
    $\sum_{j=1}^{n} p_{ij} x_{ij} \leq T_i \leftarrow \sum_{j=1}^{n} p_{ij} x_{ij} \leq T_i + p_{ij}$, $\mathcal{I} \leftarrow \mathcal{I} \setminus i$.
  else if $\exists i \in \mathcal{I}$ such that $d(M_i) = 2$ and $\sum_{J_j \in \mathcal{J}} x_{ij} \geq 1$ **then**
    $\sum_{j=1}^{n} p_{ij} x_{ij} \leq T_i \leftarrow \sum_{j=1}^{n} p_{ij} x_{ij} \leq T_i + p_{ij_1} + p_{ij_2}$, where we assume that the jobs adjacent with machine M_i are J_1 and J_2, $\mathcal{I} \leftarrow \mathcal{I} \setminus i$.
  end if
end while
return F

3.2 Characterization of Extreme Point Solutions

The following lemma follows from a direct application of the Rank Lemma.

Lemma 2. *Let x be an extreme point solution to the linear programming LP_{bts} with $x_{ij} > 0$ for each edge e_{ij}. Then there exist $\mathcal{J}' \subseteq \mathcal{J}$ and $\mathcal{M}' \subseteq \mathcal{M}$ such that*

(i) $\sum_{i=1}^{m} x_{ij} = b_j$ *for each* $J_j \in \mathcal{J}'$ *and* $\sum_{j=1}^{n} p_{ij} x_{ij} = T$ *for each* $M_i \in \mathcal{M}'$.
(ii) The constraints corresponding to $\mathcal{J}'$ and $\mathcal{M}'$ are linearly independent.
(iii) $|\mathcal{J}'| + |\mathcal{M}'| = |E(G)|$.

Lemma 3. *Let x be an extreme point solution to LP_{bts}. Then, one of the following must hold.*

(a) There exists an edge $e_{ij} \in E$ such that $x_{ij} = 0$.
(b) There exists an edge $e_{ij} \in E$ such that $x_{ij} \geq 1$.
(c) There exists a machine $i \in \mathcal{I}$ such that $d(M_i) = 1$.
(d) There exists a machine $i \in \mathcal{I}$ such that $d(M_i) = 2$ and $\sum_{J_j \in \mathcal{J}} x_{ij} \geq 1$.

Proof. Suppose for sake of contradiction that all conditions do not hold. By line 12 and line 16, we have $0 < x_{ij} < 1$ for each edge e_{ij}. Each job J_j has degree at least two since $\sum_{i=1}^{m} x_{ij} = b_j \geq 1$ and there is no edge with $x_{ij} = 1$ by line 16. Next, we show that machine M_i has degree at least two, for $M_i \in \mathcal{M}'$.

Claim. Each machine M_i has degree at least two, for $M_i \in \mathcal{M}'$.

Proof. Suppose for sake of contradiction that there exists a machine M_i, where $d(M_i) = 1$ but $i \notin \mathcal{I}$. That is, this machine with degree equal to 1 have been relaxation for some job J_j, since this case is $d(M_i) = 1$ but $i \notin \mathcal{I}$. We also have $M_i \in \mathcal{M}'$ for machine M_i, i.e. the constraint is tight. Then, we have $x_{ij} \geq 1$. The condition (b) holds, a contradiction. The claim holds.

Thus, each machine M_i has degree at least two for $i = 1, \cdots, m$. From Lemma 2 we have that $|\mathcal{J}'| + |\mathcal{M}'| = |E(G)|$. This implies that:

$$
\begin{aligned}
|\mathcal{J}'| + |\mathcal{M}'| = |E| = \frac{\sum_{J_j \in \mathcal{J}} d(j) + \sum_{M_i \in \mathcal{M}} d(i)}{2} &\geq \frac{\sum_{J_j \in \mathcal{J}} d(j) + \sum_{M_i \in \mathcal{M}'} d(i)}{2} \\
&\geq |\mathcal{J}| + |\mathcal{M}'| \geq |\mathcal{J}'| + |\mathcal{M}'|.
\end{aligned}
$$

Thus, all the inequalities can be replaced by equalities. It implies that $\mathcal{J}' = \mathcal{J}$; $\mathcal{M}' = \mathcal{M}$; $d(J_j) = 2$ for each $J_j \in \mathcal{J}$; $d(M_i) = 2$ for each $M_i \in \mathcal{M}'$.

The last sequence of observations implies that E is a cycle cover of the vertices in $\mathcal{J}' \cup \mathcal{M}'$ (tight constraints). Consider any cycle Y in the cover. The total number of jobs in Y is equal to the total number of machines in Y. Since $\sum_{M_i \in \mathcal{M}'} \sum_{J_j \in \mathcal{J}'} x_{ij} = \sum_{J_j \in \mathcal{J}'} \sum_{M_i \in \mathcal{M}'} x_{ij} = b_j|\mathcal{J}'| = b_j|\mathcal{M}'|$, and $\sum_{M_i \in \mathcal{M}'} x_{ij} = b_j$ for each $J_j \in \mathcal{J}'$, there must exist a machine $M_i \in \mathcal{M}'$ with $\sum_{J_j \in \mathcal{J}'} x_{ij} \geq 1$, since $b_j \geq 1$. Otherwise, $\sum_{J_j \in \mathcal{J}'} x_{ij} < 1$. We have $\sum_{M_i \in \mathcal{M}'} \sum_{J_j \in \mathcal{J}'} x_{ij} < b_j|\mathcal{M}'| = b_j|\mathcal{J}'| = \sum_{J_j \in \mathcal{J}'} \sum_{M_i \in \mathcal{M}'} x_{ij}$. This is clearly a contradiction. Then, it imply that condition (d) hold. This completes the analysis.

3.3 Analysis of Iterative Algorithm

Analysis on our problem from the previous subsections is summarized in the following theorem.

Theorem 2. *Iterative algorithm runs in polynomial time, and returns an assignment of optimal cost. The total processing time of each machine M_i for $i = 1, \cdots, m$ is at most $T + \max_j p_{ij}$ for machine M_i and $\max_j p_{ij} \leq T$.*

Proof. We split the proof into three lemmas: analysis of the running time; showing returns an optimal cost and proving that the makespan threshold is at most $T + \max_j p_{ij}$ for machine M_i.

Lemma 4. *The algorithm running time is polynomial.*

Proof. Since at each step of the loop either $O(nm)$ or E or b_j (at most remaining $O(nm)$ at while loop and this will cause $\mathcal{J}$ to decrease) or $\mathcal{I}$ will be decreasing.

Lemma 5. *The returned schedule has the optimal cost.*

We claim that at any iteration of the algorithm the cost of assignment given by F plus the cost of the current linear programming solution to LP_{bts} is at most the cost of the initial linear programming solution. We show that by an inductive argument on the number of iterations. Let $c(LP^*)$ and $c(LP')$ be the optimal cost of initial and executes algorithm line 2 linear programming. Let F' be the set of F at the end of for loop. Let $e_{ij} = (i, j) \in E$ be the job J_j is assigned to machine M_i.

Algorithm first executes line 3 and 4, at the end of for loop, we have

$$F' = F \setminus \{(0, e_{ij})\} \cup \{(\lfloor x_{ij} \rfloor, e_{ij})\} \setminus \cdots \cup \cdots, \text{ for } i = 1, \cdots, m, j = 1, \cdots, n.$$

$$c(LP') \le c(LP^*) - \sum_{x_{ij} \ge 1} \lfloor x_{ij} \rfloor \cdot c_{ij},$$

since we take away the integer part for every $x_{ij} \ge 1$ and the new solution of x^* such that $x_{ij} \leftarrow x_{ij} - \lfloor x_{ij} \rfloor$ also is a feasible solution of $b_j \leftarrow b_j - \lfloor x_{ij} \rfloor$ and $T_i \leftarrow T_i - \lfloor x_{ij} \rfloor \cdot p_{ij}$. Thus, we have

$$c(F') + c(LP') \le c(F) + \sum_{x_{ij} \ge 1} \lfloor x_{ij} \rfloor \cdot c_{ij} + c(LP^*) - \sum_{x_{ij} \ge 1} \lfloor x_{ij} \rfloor \cdot c_{ij} = c(LP^*).$$

Next, we analysis while loop. This can be shown by a simple inductive argument on the number of iterations. In the zero iteration, it is trivially holds. In any iteration, if algorithm executes line 12, then F' and current linear programming solution to LP_{bts} is not change. Since $x_{ij} = 0$. If algorithm executes line 16, then the cost of F' increases by $\lfloor x_{ij} \rfloor c_{ij}$, and the current linear programming solution decreases at least by $c_{ij} \lfloor x_{ij} \rfloor$. Since the residual solution restricted to residual graph also is the feasible solution of new update LP_{bts}. If algorithm executes line 20 or 22, then the cost of F' remains the same, while the cost of the current linear program can only decrease. Since we have relaxed the corresponding constraint. Therefore, the lemma holds.

Lemma 6. *Load of each machine $M_i \in \mathcal{M}$ is at most $T + \max_j p_{ij}$.*

Proof. Fix any machine $M_i \in \mathcal{M}$ and $i \in \mathcal{I}$.

Let T_i' be the remaining time at the end of for loop, and $T_i(F')$ be the total processing time used by jobs assigned to machine M_i in F'. Algorithm first executes line 3 and 4, we must have $T_i(F') + T_i' \le T_i$. The claim can be proved by a simple induction as the lemma 5 for the costs in two for loop.

Then, we argue that if algorithm executes line 12 or 16, at any iteration we must have $T_i^{(k)} + T_i(F^{(k)}) \le T_i(F') + T_i' = T_i$, where $T_i^{(k)}$ is the remaining time at the end of k-th iteration of algorithm left on this machine, and $T_i(F^{(k)})$ is the total processing time used by jobs assigned to machine M_i in $F^{(k)}$. The claim can also be proved by a induction as the lemma 5 for the costs in while loop.

Now, consider algorithm has been executed line 20 or 22, we have $i \notin \mathcal{I}$ for fixed machine M_i. Without loss of generality, we can assume that algorithm executes line 20 or 22 at the k-th iteration for fixed machine M_i. Then, we distinguish the two possible cases that machine M_i constraint is relaxation, and proved that the claim also holds until the algorithm terminates. Note that $0 < x_{ij} < 1$ in this time, otherwise, this algorithm will execute line 12 or line 16.

Case 1. $d(M_i) = 1$. The total processing time on machine M_i, until the end of the algorithm, is at most $T_i(F^{(k)}) + T_i^{(k)} + p_{ij} \le T_i + p_{ij} \le T_i + \max_j p_{ij}$, where $J_j \in \mathcal{J}$ is the only job M_i is connected to.

Case 2. $d(M_i) = 2$ and $\sum_{J_j \in \mathcal{J}} x_{ij} \ge 1$. Let x denote the solution to LP_{bts} when M_i constraint has been relaxation. Let j_1 and j_2 be the two jobs adjacent to machine M_i. We can assume that algorithm executes line 20 or 22 at the k-th iteration. Then, we have the following inequalities

$$\begin{aligned} load(M_i) &\le T_i(F^{(k)}) + p_{ij_1} + p_{ij_2} \\ &\le (T_i - x_{ij_1}p_{ij_1} - x_{ij_2}p_{ij_2}) + p_{ij_1} + p_{ij_2} \\ &\le T_i + (1 - x_{ij_1})p_{ij_1} + (1 - x_{ij_2})p_{ij_2} \\ &\le T_i + (2 - (x_{ij_1} + x_{ij_2}))\max\{p_{ij_1}, p_{ij_2}\} \\ &\le T_i + \max_j p_{ij}. \end{aligned}$$

This completes the theorem analysis.

4 A $(\frac{e}{e-1}, \frac{2e-1}{e-1})$-Approximation Algorithm for Bag of Tasks Scheduling with Rejection Problem

We will give a $(\frac{e}{e-1}, \frac{2e-1}{e-1})$-approximation for our problem in this section. Here, we formulate an LP_{btsr} relaxation for linear programming (1).

$$\begin{cases} \min \sum_{i=1}^{m}\sum_{j=1}^{n} c_{ij}x_{ij} + \sum_{j=1}^{n} w_j(1 - z_j) \\ \sum_{i=1}^{m} x_{ij} = b_j z_j, \ \forall j = 1, 2, \cdots, n, \\ \sum_{j=1}^{n} p_{ij}x_{ij} \le T, \ \forall i = 1, \cdots, m, \\ 0 \le z_j \le 1, \ \forall j = 1, 2, \cdots, n, \\ 0 \le x_{ij} \le b_j, \ \forall j = 1, 2, \cdots, n, \ \forall i = 1, \cdots, m. \end{cases} \tag{3}$$

Proposition 1. *There exists a polynomial time algorithm which given a fractional vector $\boldsymbol{x}$ that satisfies all constraints of linear programming (2), returns* *a solution of cost at most C that each machine M_i makespan threshold is at most $T + \max_j p_{ij}$ and $\max_j p_{ij} \le T$, where C is the optimal cost of linear programming (2) and each machine M_i makespan threshold is at most $T = T_i$.*

Next, we give $(\frac{e}{e-1}, \frac{2e-1}{e-1})$-approximation algorithms for the job rejection case problem.

Algorithm

1. Solve the linear programming (1) with all $p_{ij} \le T$. Let $\boldsymbol{x}^*$ and $\boldsymbol{z}^*$ be an optimal solution of this program.
2. Defining a new solution $\boldsymbol{x}$, $\boldsymbol{z}$ and $\mathcal{A} \subseteq \mathcal{J}$: for every J_j, for $j = 1, 2, \cdots, n$. We randomly choose a threshold β from the uniform distribution over $[1/e, 1]$.
   (a). If $z_j^* \ge \beta$, let $z_j \leftarrow 1$. Let $x_{ij} = b_j \frac{x_{ij}^*}{\sum_{i=1}^m x_{ij}^*}$, for all $i = 1, 2, \cdots, m$. Add J_j to $\mathcal{A}$.
   (b). Otherwise, $z_j^* < \beta$, let $z_j \leftarrow 0$. Let $x_{ij} = 0$, for all $i = 1, 2, \cdots, m$.
3. Define a solution $\boldsymbol{x}'$, T' to the linear programming (2) with the set of jobs $\mathcal{A}$. Let $x'_{ij} = x_{ij}$ for $J_j \in \mathcal{A}$, $i = 1, 2, \cdots, m$, and $T' = \max_i \sum_{j=1}^n p_{ij} x_{ij}$.
4. Apply the algorithm of Proposition 1 to obtain a schedule of the jobs in $\mathcal{A}$.

Theorem 3. *The algorithm is a $(\frac{e}{e-1}, \frac{2e-1}{e-1})$-approximation algorithm for bag of tasks scheduling with rejection problem.*

Proof. The variables $\boldsymbol{x}'$ are defined for $J_j \in \mathcal{A}$ by $x'_{ij} = x_{ij}$ for all $i = 1, 2, \cdots, m$. The linear constraints $\sum_{i=1}^m x_{ij} = b_j z_j$ hold for every $J_j \in \mathcal{A}$ by the definition of x_{ij} in step 2(a), since $\sum_{i=1}^m x_{ij} = b_j \sum_{i=1}^m \frac{x_{ij}^*}{\sum_{i=1}^m x_{ij}^*} = b_j$ and we also have $x_{ij} \le b_j$ due to $\sum_{i=1}^m x_{ij}^* \ge x_{ij}^*$.

Let C and W are the processing cost and rejection cost induced by the optimal solution x^*, z^*, respectively. We next argue that $E(\sum_{i=1}^m \sum_{j=1}^n c_{ij} x_{ij}) \le \frac{e}{e-1} \sum_{i=1}^m \sum_{j=1}^n c_{ij} x_{ij}^* = \frac{e}{e-1} C$ and $E(T') \le \frac{e}{e-1} T$. To see this claim consider a machine M_i, we have

$$x_{ij} = b_j \frac{x_{ij}^*}{\sum_{i=1}^m x_{ij}^*} = b_j \frac{x_{ij}^*}{b_j z_j^*} \le \frac{1}{\beta} b_j \frac{x_{ij}^*}{b_j} = \frac{1}{\beta} x_{ij}^*,$$

when $z_j^* \ge \beta$. It is clear that $x_{ij} \le \frac{1}{\beta} x_{ij}^*$ when $z_j^* < \beta$, since we let $x_{ij} = 0$. Then $x_{ij} \le \frac{1}{\beta} x_{ij}^*$ for all i, j. Thus, we conclude that

$$\begin{aligned} E(\sum_{i=1}^m \sum_{j=1}^n c_{ij} x_{ij}) &\le \sum_{i=1}^m \sum_{j=1}^n c_{ij} E(\frac{1}{\beta} x_{ij}^*) \\ &= \sum_{i=1}^m \sum_{j=1}^n c_{ij} x_{ij}^* E(\frac{1}{\beta}) \\ &= \sum_{i=1}^m \sum_{j=1}^n c_{ij} x_{ij}^* \int_{1/e}^1 \frac{1}{\beta} \cdot \frac{1}{1 - 1/e} d\beta \\ &= \frac{e}{e-1} \sum_{i=1}^m \sum_{j=1}^n c_{ij} x_{ij}^* = \frac{e}{e-1} C. \end{aligned}$$

In the same way, we have

$$E(\sum_{j=1}^{n} p_{ij}x_{ij}) \leq \sum_{j=1}^{n} p_{ij}x_{ij}^{*}E(\frac{1}{\beta}) \leq \frac{e}{e-1}T,$$

thus $E(T') = \max_i \sum_{j=1}^{n} p_{ij}x_{ij} \leq \frac{e}{e-1}T$.

Furthermore, if $z_j < 1/e$, we have $z_j = 0$ and

$$E(w_j(1-z_j)) = w_j \leq \frac{1}{1-1/e}w_j(1-z_j^*).$$

Clearly, we also have

$$E(w_j(1-z_j)) = w_j \cdot Pr(z_j^* < \beta) + w_j \cdot Pr(z_j^* \geq \beta) = \frac{1}{1-1/e}w_j(1-z_j^*),$$

if $z_j^* \geq 1/e$. Thus, $E(\sum_{j=1}^{n} w_j(1-z_j)) \leq \frac{e}{e-1}\sum_{j=1}^{n} w_j(1-z_j^*) = \frac{e}{e-1}W$.

Thus, the expected objective value of $\boldsymbol{x}$ is

$$\begin{aligned} E(\sum_{i=1}^{m}\sum_{j=1}^{n} c_{ij}x_{ij} + w_j(1-z_j)) &\leq \frac{e}{e-1}\sum_{i=1}^{m}\sum_{j=1}^{n} c_{ij}x_{ij}^{*} + \frac{e}{e-1}\sum_{j=1}^{n} w_j(1-z_j^*) \\ &= \frac{e}{e-1}C + \frac{e}{e-1}W. \end{aligned}$$

The expected makespan threshold of $\boldsymbol{x}$ is

$$E(\sum_{j=1}^{n} p_{ij}x_{ij}) \leq \frac{e}{e-1}T.$$

Now, we show how to obtain a determined polynomial time $(\frac{e}{e-1}, \frac{e}{e-1})$-approximation algorithm. Without loss of generality, assume that $z_1^* \leq z_2^* \leq \cdots \leq z_n^*$. Note that the randomized method will produce the same solution $(0, \cdots, z_i^* = 0, z_{i+1}^* = 1, \cdots, 1)$, when β lies in (z_i^*, z_{i+1}^*) with $z_i^* \geq 1/e$. In other words, there are at most $n+1$ different solutions with expected objective value no more than $\frac{e}{e-1}(C+W)$ and expected makespan value no more than $\frac{e}{e-1}T$. Thus, this is a determined polynomial time $(\frac{e}{e-1}, \frac{e}{e-1})$-approximation algorithm and there is a desired solution in $n+1$ different solutions.

Since $\boldsymbol{x}' = \boldsymbol{x}$ is a feasible solution to linear programming (2), by Proposition 1, the makespan of the resulting solution of this algorithm is at most $T' + \max_j p_{ij} \leq \frac{e}{e-1}T + \max_j p_{ij} \leq \frac{2e-1}{e-1}T$. Meanwhile, the cost is at most optimal cost by Proposition 1. Thus, the objective value of resulting solution of this algorithm is at most $\frac{e}{e-1}(C+W)$.

5 Conclusion

In this paper, we consider bag of tasks scheduling problem with rejection. We design a bi-criteria $(\frac{e}{e-1}, \frac{2e-1}{e-1})$-approximation for bag of tasks scheduling with rejection problem by using iterative rounding. Moreover, it is interesting to consider the other constraint such as the bounded degree constraint and matroid constraint. Finally, we want to use the iterative rounding method to solve more scheduling problems.

References

1. Tarplee, K.M., Friese, R., Maciejewski, A.A., Siegel, H.J., Chong, E.K.: Energy and makespan tradeoffs in heterogeneous computing systems using efficient linear programming techniques. IEEE Trans. Parallel Distrib. Syst. **27**(6), 1633–1646 (2015)
2. Li, W., Liu, X., Cai, X., Zhang, X.: Approximation algorithm for the energy-aware profit maximizing problem in heterogeneous computing systems. J. Parallel Distrib. Comput. **124**, 70–77 (2019)
3. Li, W.: Bag-of-tasks scheduling with rejection in large computing systems. Asia-Pac. J. Oper. Res., 2340011 (2023)
4. Lenstra, J.K., Shmoys, D.B., Tardos, É.: Approximation algorithms for scheduling unrelated parallel machines. Math. Program. **46**, 259–271 (1990)
5. Shmoys, D.B., Tardos, É.: An approximation algorithm for the generalized assignment problem. Math. Program. **62**(3), 461–474 (1993)
6. Li, W., Ou, J.: Approximation algorithms for scheduling parallel machines with an energy constraint in green manufacturing. Eur. J. Oper. Res. **314**(3), 882–893 (2024)
7. Bartal, Y., Leonardi, S., Spaccamela, A.M., Stougie, J.: Multi-processor scheduling with rejection. SIAM J. Discret. Math. **13**, 64–78 (2000)
8. Li, W., Li, J., Zhang, X., Chen, Z.: Penalty cost constrained identical parallel machine scheduling problem. Theoret. Comput. Sci. **607**, 181–192 (2015)
9. Li, W., Ou, J.: Machine scheduling with restricted rejection: an application to task offloading in cloud-edge collaborative computing. Eur. J. Oper. Res. **314**(3), 912–919 (2023)
10. Shabtay, D., Gaspar, N., Kaspi, M.: A survey on off-line scheduling with rejection. J. Sched. **16**, 3–28 (2013)
11. Lau, L.C., Ravi, R., Singh, M.: Iterative Methods in Combinatorial Optimization, vol. 46, Cambridge University Press (2011)
12. Jain, K.: A factor 2 approximation algorithm for the generalized Steiner network problem. Combinatorica **21**, 39–60 (2001)
13. Byrka, J., Grandoni, F., Rothvoss, T., Sanità, L.: Steiner tree approximation via iterative randomized rounding. J. ACM **60**(1), 1–33 (2013)
14. Grandoni, F., Ravi, R., Singh, M., Zenklusen, R.: New approaches to multi-objective optimization. Math. Program. **146**, 525–554 (2014)
15. Bansal, N.: On a generalization of iterated and randomized rounding. In: Proceedings of the 51st Annual ACM SIGACT Symposium on Theory of Computing, pp. 1125–1135 (2019)
16. Nemhauser, G.L., Wolsey L.A.: Integer and Combinatorial Optimization. Wiley-Interscience (1999)

Submodularity, Management and Others

Integrated Airline Aircraft Routing and Crew Pairing by Alternating Lagrangian Decomposition

Cong Li[1], Suixiang Gao[1,2(✉)], Wenguo Yang[1], and Zhipeng Jiang[1]

[1] School of Mathematical Sciences, University of Chinese Academy of Sciences, Beijing 100049, China
licong21@mails.ucas.ac.cn, {yangwg,jiangzhipeng}@ucas.ac.cn
[2] Zhongguancun Laboratory, Beijing 100089, China
sxgao@ucas.ac.cn

Abstract. For the aircraft routing and crew pairing problems, a sequential approach is usually used to solve they. When solving the crew pairing problem, the impact of aircraft routing problem is often neglected so that these two problems are independent Although this approach reduces the complexity of the solution process, it may lead to a suboptimal solution. In this paper, we consider an integrated aircraft routing and crew pairing problem. We propose an integrated model that integrates the aircraft routing and crew pairing problems. We propose a solution algorithm based on alternating Lagrangian decomposition. The solution process iterates between the first Lagrangian subproblem about aircraft routing and the second Lagrangian subproblem about crew pairing. These two Lagrangian subproblems are solved by a branch-and-price algorithm. In the branch-and-price algorithm, we present a heuristic branching strategy. The computational experiments are conducted on several real-world data sets.

Keywords: Aircraft routing · Crew pairing · Alternating Lagrangian decomposition · Branch and price

1 Introduction

The airline planning problem is a very challenging problem for airline companies. Because of its complexity, this problem usually is segmented into four stages: flight scheduling, fleet assignment, aircraft routing, and crew scheduling. The first stage aims to determine specific departure and arrival airports, as well as the departure and arrival times of each flight. The second stage aims to determine each flight to be assigned to an aircraft type for the maximization of profit. The third stage aims at finding minimum-cost aircraft routes such that maintenance requirements are satisfied and each flight is assigned to an aircraft. When the aircraft routing problem is solved, airlines may take into account through values that is viewed as negative costs, which represent the extra revenues derived by

S. Ghosh and Z. Zhang (Eds.): AAIM 2024, LNCS 15179, pp. 197–209, 2024.
https://doi.org/10.1007/978-981-97-7798-3_17

assigning the same aircraft to two successive flights such that passengers can't change aircraft on both flights.

In the last stage, the crew scheduling problem finds crew schedules that satisfy all constraints and each flight is covered by one crew. Because of the large crew cost after fuel cost, the crew scheduling problem is one of the most significant problems in airline planning. This problem usually is segmented into two subproblems: the crew pairing problem and the crew assignment problem. The crew pairing problem is devoted to determining a subset of pairings by given scheduled flights so that each flight is covered and the cost of pairings is minimized. For the crew assignment problem, its aim is to construct schedules assigning to crews from the previously computed pairings so that each pairing is covered exactly once.

Traditionally, the sequential approach is used to solve the aircraft routing and crew pairing problems. Although this approach reduces the complexity of the solution process, it may lead to a suboptimal solution. Furthermore, when solving the crew pairing problem, the impact of aircraft routing problem is often neglected so that these two problems are independent. For the aircraft routing and crew pairing problems, many researches have proposed mathematical models and optimization algorithms. Desaulniers et al. [5] proposed a nonlinear multi-commodity network flow formulation and used a Dantizig-Wolfe decomposition and branch-and-bound algorithm for solving a crew pairing problem. Saddoune et al. [10] proposed a set partition model and an alternative solution approach combining a column generation for solving a crew pairing problem. Haouari et al. [8] proposed a compact polynomial-sized model solved by a solver for a daily aircraft routing problem. Bulbul and Kasimbeyli [2] formulated an aircraft routing problem as an asymmetric travelling salesman problem and proposed an augmented Lagrangian algorithm based on a hybrid subgradient algorithm to solve it.

Up to now, some researches also have focused on the integrated aircraft routing and crew pairing problem. Cordeau et al. [3] proposed an integer programming model and proposed a Benders decomposition combining a branch-and-price algorithm for solving an integrated aircraft routing and crew pairing problem. Ahmed et al. [1] proposed an integrated robust model that is mixed-integer and nonlinear for a robust integrated aircraft routing and crew pairing problem. Díaz-Ramírez et al. [6] proposed an integer programming model for the integrated aircraft routing and crew pairing problem with a single fleet and a single maintenance and crew base.

In this paper, we consider the scheduled flights operated by the same aircraft type in a short planning horizon. In addition, we assume that all crews are qualified and operate this aircraft type. Furthermore, we don't consider deadheads that treat crews as a passenger on certain flights to reposition for the crew pairing. We focus on an integrated aircraft routing and crew pairing problem (IARCPP). Because of our devotion to the short planning horizon in the problem, we will assign crews to pairings. The problem finds minimum-cost aircraft routs and crew pairings such that side constraints are satisfied, and each flight is covered by one aircraft and one crew. We propose an integrated model that integrates the interdependent parts of aircraft routing and crew pairing. The variable sit times are the main source of interdependence between aircraft routing and crew pairing. Because the short sit time between two successive flights covered by the same crew relies on whether these two flights is covered by the same aircraft, the set of feasible pairings relies on previously computed aircraft routs. For solving the integrated model, we propose an algorithm that combines alternating Lagrangian decomposition and branch and price. In the branch-and-price algorithm, we also present a heuristic branching strategy. Finally, the computational experiments are conducted on several real-world data sets. The results of the experiments show that our integrated approach is better than the sequential approach in terms of cost savings.

2 Mathematical Formulation

The IARCPP is to find minimum-cost aircraft routs and crew pairings such that side constraints are satisfied, and each flight is covered by one aircraft and one crew. An aircraft rout is a sequence of flights separated by turn and maintenance periods, and starts and ends at the same location that is aircraft base in this paper. The turn and maintenance times can be subject to minimum turn time, and the minimum maintenance time, respectively. An aircraft rout must contain the maximum flying time, the maximum number of takeoffs (or landings), and the maximum duration between two consecutive maintenance. A pairing is a sequence of duties separated by rest periods, and starts and ends at the same crew base. A duty a sequence of flights separated by sit periods. The rules of pairing construction contain the maximum flying time per duty, the maximum number of landings per duty, the maximum duty duration, the maximum number of duties in a pairing, the maximum pairing duration. The rest and sit times can be limited to the minimum rest time, and the minimum and maximum sit time, respectively. Generally, the minimum turn time between two successive flights is smaller than the minimum sit time. Hence, the sit time is allowed to be smaller than the minimum sit time in this problem.

Given a set $F = \{f_1, f_2, ..., f_n\}$ of flights that are operated by the same aircraft type in a several-day period, a set A of aircraft bases, a set M of maintenance airports, the number n_a of aircraft at aircraft base $a \in A$, a set B of crew bases, and the number n_b of crews at crew base $b \in B$. Based on the description of Cordeau et al. [3] for the flight network, we define an acyclic digraph $G^a = (N^a, E^a)$ for each aircraft base $a \in A$, where $N^a = N_D \cup N_A \cup \{o(a), d(a)\}$ and $E^a = E^a_D \cup E^a_T \cup F \cup E^a_M \cup E^a_E \cup \{(o(a), d(a))\}$. Sets N_D and N_A represent the departure node and the arrival node sets of all flights, respectively. The source node $o(a)$ and the sink node $d(a)$ represent the start and end of aircraft routs at aircraft base a, respectively. Each flight $f \in F$ corresponds to an arc composed of the departure node and the arrival node of the flight f. Sets E^a_D, E^a_T, E^a_M, and E^a_E represent the set of all start arcs of aircraft routes, the set of all turn arcs, the set of all maintenance arcs, and the set of all end arcs of aircraft routes, respectively. In other words, firstly, each start arc $(o(a), j_D) \in E_E^a$ of the aircraft route satisfies that the departure airport of flight f_j corresponding to the departure node j_D is aircraft base $o(a)$, and each end arc $(i_A, d(a)) \in E_M^a$ of aircraft routes is also defined in a similar fashion; secondly, each turn arc $(i_A, j_D) \in E_T^a$ satisfies that the arrival airport of flight f_i corresponding to the arrival node i_A is the departure airport of flight f_j corresponding to the arrival node j_D and the connection time (i.e., turn time) between flights f_i and f_j is greater than or equal to the minimum turn time, and each maintenance arc $(i_A, j_D) \in E_M^a$ is also defined in a similar fashion. The empty arc $(o(a), d(a))$ represents a complete aircraft rout that doesn't contain any flights and is assigned to the unused aircraft. In addition, the costs of a turn arc and a maintenance arc are the negative cost (i.e., through value) and maintenance expenditure, respectively, and the costs of other type arcs in G^a are zero. A feasible path from the source node $o(a)$ to the sink node $d(a)$ in G^a corresponds to a feasible aircraft rout. For each crew base $b \in B$, we also define an acyclic digraph $G^b = (N^b, E^b)$, where $N^b = N_D \cup N_A \cup \{o(b), d(b)\}$ and E^b is the arc set that is composed of the set E_D^b of all start arcs of pairings, the set F of flight arcs, the E_S^b of sit arcs, the set E_R^b of rest arcs, the set E_A^b of all end arcs of pairings, and the empty arc $(o(b), d(b))$ (i.e., $E^b = E_D^b \cup E_S^b \cup F \cup E_R^b \cup E_A^b \cup \{(o(b), d(b))\}$). In addition, these subsets of the arc set E^b are defined in a similar fashion to how sets E_D^a, E_T^a, F, E_M^a, E_E^a, and $\{(o(a), d(a))\}$ are defined. The costs of a sit arc and a rest arc are described in Saddoune et al. [10]. And the costs of other type arcs in G^b are also zero. A feasible path from the source node $o(b)$ to the sink node $d(b)$ in G^b corresponds to a feasible pairing.

A turn arc $(i_A, j_D) \in E_T^a$ or sit arc $(i_A, j_D) \in E_S^a$ is said to be short if the turn time or sit time between flight f_i corresponding to the arrival node i_A and flight f_j corresponding to the departure node j_D is smaller than minimum sit time but greater than or equal to minimum turn time. In this case, flights f_i and f_j must be covered by same crew and not be allowed to be covered by different aircraft; otherwise, because of the short time, the departure time of flight f_j is delayed. For each short arc, it is contained in N^a $(a \in A)$ and N^b $(b \in B)$ by the definitions of the turn arc and sit arc. Let $C \subset N^a$ $(C \subset N^a)$ be the set of all short arcs in G^a $(a \in A)$ $(G^b$ $(b \in B))$.

For each aircraft base $a \in A$, let R_a be the set of all feasible paths (i.e., aircraft routs) from the source $o(a)$ to the sink $d(a)$ in G^a. We denote by r_a the empty aircraft rout corresponding to the empty arc $(o(a), d(a))$ and assume $r_a \notin R_a$. For each aircraft rout $r \in R_a$, we define binary parameters a_{fr} and b_r^{ij} that are equal to 1 if flight f and arc (i, j) are contained in aircraft rout r, and 0 otherwise, respectively. Let c_r^a be the cost of aircraft rout r that is equal to the sum of costs of all arcs contained in the corresponding path r. We define a binary variable x_r^a that take value 1 if aircraft rout r is chose in solution; otherwise 0. Similarly, for each crew base $b \in B$, let P_b be the set of all feasible paths (i.e., pairings) from the source $o(b)$ to the sink $d(b)$ in G^b. We also denote by p_a the empty pairing corresponding to the empty arc $(o(b), d(b))$ and assume $p_b \notin P_b$.

We also define corresponding parameters a_{fp}, b_p^{ij}, and c_p^b, as well as a binary variable y_p^b for each pairing $p \in P_b$.

By these notations, the IARCPP is formulated as follows:

$$\text{Min} \sum_{a \in A} \sum_{r \in R_a} c_r^a x_r^a + \sum_{b \in B} \sum_{p \in P_b} c_p^b y_p^b \tag{1}$$

Aircraft constraints:

$$\sum_{a \in A} \sum_{r \in R_a} a_{fr} x_r^a = 1 \quad f \in F \tag{2}$$

$$\sum_{r \in R_a \cup \{r_a\}} x_r^a = n_a \quad a \in A \tag{3}$$

Coupling constraint:

$$\sum_{b \in B} \sum_{p \in P_b} b_p^{ij} y_p^b - \sum_{a \in A} \sum_{r \in R_a} b_r^{ij} x_r^a \leq 0 \quad (i, j) \in C \tag{4}$$

Crew constraints

$$\sum_{b \in B} \sum_{p \in P_b} a_{fp} y_p^b = 1 \quad f \in F \tag{5}$$

$$\sum_{p \in P_b \cup \{p_b\}} y_p^b = n_b \quad b \in B \tag{6}$$

$$x_r^a \in \{0, 1\} \quad r \in R_a, a \in A \tag{7}$$

$$x_r^a \in \mathbb{Z}^+ \cup \{0\} \quad r \in \cup_{a \in A} \{r_a\} \tag{8}$$

$$y_p^b \in \{0, 1\} \quad p \in P_b, b \in B \tag{9}$$

$$y_p^b \in \mathbb{Z}^+ \cup \{0\} \quad p \in \cup_{b \in B} \{p_b\} \tag{10}$$

The objective function (1) represents that the sum of costs of aircraft routs and pairings is minimized. Constraints (2) and (5) ensure that each flight is covered by one aircraft and one crew. Constraints (3) and (6) ensure that each aircraft and crew are assigned to an aircraft route and a pairing, respectively. Coupling constraint (4) ensure that a crew isn't allowed to change aircraft when the sit time is short. Formulas (7)–(10) indicate the domains of variables x_r^a and y_p^b, respectively.

3 Solution Methodology

Because the integrated model (1)–(10) contains coupling constraint (4), in this section, an algorithm based an alternating Lagrangian decomposition (ALD) is developed for solving the IARCPP. The ALD has been applied by Zeighami et al. [11] for solving integrated airline crew scheduling problem. The algorithm iterates between the first Lagrangian subproblem for the aircraft (called ALD-aircraft) and the second Lagrangian subproblem for the crews (called ALD-crew). When the number of flights is increased, the number of variables in ALD-aircraft and

ALD-crew rises exponentially. To overcome this difficulty, we use a branch-and-price (BP) algorithm, that is, a column generation (CG) algorithm embedded in a branch-and-bound scheme. The initial value of the Lagrangian multiplier is set to 0. At the start of the algorithm, we solve ALD-aircraft by BP. We then solve ALD-crew by BP and considering the dual information from ALD-aircraft. The algorithm alternates between the ALD-aircraft and ALD-crew until the optimal value decreases by less than 0.01 over 2 iterations. Next, we describe the branch-and-price algorithm and then explain how the alternating Lagrangian algorithm is applied to solve this problem.

3.1 Branch and Price

At each node contained the branch-and-bound tree, the integrated aircraft model considering constraints (2)–(3) and the first term of objective function (1) (or the integrated crew model considering constraints (5)–(6) and the second term of objective function) is linearly relaxed and solved by CG. A branching strategy is then applied to the current node. The linear relaxation (LR) of the integrated aircraft model (or the integrated crew model) can usually be decomposed into a restricted master problem (RMP) and pricing problems (also called subproblems) by CG. The RMP is derived by replacing the set R_a (or P_b) with the subsets $R_a^t \subseteq R_a$ (or $P_b^t \subseteq P_b$) at iteration $t = 0, 1, 2, ...$. At the start of CG, we will add artificial variables with penalty costs to the RMP to ensure its feasibility. The RMP, solved by an linear programming solver, obtains optimal primal solutions and dual solutions over a subset of the columns (i.e., variables). Previously obtained dual solutions are used in each subproblem to find new negative-reduced-cost columns. These columns are then added to the RMP. The iterative process of CG stops when no negative-reduced-cost columns are found. Actually, found new columns with negative reduced costs may not have much impact on the RMP optimal value (i.e. tailing-off effect) so that the CPU time is very long. Therefore, to avoid this case, the iterative process of CG must be stopped when the RMP optimal value decreases by less than 0.0001 over 10 iterations. Here, each subproblem of CG corresponds to a shortest-path problem with resource constraints solved by a labeling algorithm [9].

In fact, there is one subproblems for each aircraft base and each crew base. Each feasible aircraft rout (or pairing) corresponds to a path from the source node to the sink node in G^a ($a \in A$) (or G^b ($b \in B$)). To guarantee the feasibility of each aircraft rout, the following three resource constraints are used in G^a ($a \in A$) (or five resource constraints are used in G^b ($b \in B$)): the maximum flying time, maximum number of landings, and the maximum duration between two consecutive maintenance (or the maximum fying time per duty, maximum number of landing per duty, the maximum duty duration, the maximum number of duties in a pairing, and the maximum pairing duration). A resource among a path is a variable quantity and its value at every node must be within a resource window (i.e., a given interval). Every resource is consumed at arcs of G^a (or G^b). The sum of all arc costs among a path from the source node to the sink node represents the cost of the path corresponding to a aircraft rout (or pairing). Based on the RMP dual solutions, the costs of flight arcs can be updated in the solving process. In the labeling algorithm, a label is related with a partial path beginning from the source node. Every label contains three (or five) resource components and one reduced-cost component, and corresponds to a partial path. Every resource value at last node of the partial path is determined by a resource component, and the path reduced cost is computed by the reduced-cost component. At the source node, all components of the label are set to zero. A new path is preserved if a new arc is added to the previously computed partial path and each resource value of the path is within the resource window at new node; otherwise, it is discarded. For a large graph, the number of generated labels is enormous so that it can cause a long CPU time in this way. To avoid this, a dominance rule is used to eliminate labels. Let $\Delta_1 = \{T_1^{\xi} | \xi \in \Xi \cup \{c\}\}$ and $\Delta_2 = \{T_2^{\xi} | \xi \in \Xi \cup \{c\}\}$ are two different labels, where Ξ, c, and T_i^{ξ} $(i = 1, 2)$ represent the set of all resources, reduced cost, and the value of the resource ξ or reduced cost c, respectively. If $T_1^{\xi} \leq T_2^{\xi}$ $(\xi \in \Xi)$ and $T_1^c \geq T_2^c$ is satisfied, then the label Δ_2 dominates the label Δ_1. Because of dominated labels not found optimal paths, they will be discarded. We must keep one of two labels if they dominate each other. Let ω_f and γ_a (or φ_f and θ_b) be the dual variables of constraints (2)–(3) (or (5)–(6)), respectively. The reduced cost of variables x_r^a (or y_p^b) is defined by $\bar{c}_r^a = c_r^a - \sum_{f \in F} a_{fr} \omega_f - \gamma_a$ (or $\bar{c}_p^b = c_p^b - \sum_{f \in F} a_{fp} \varphi_f - \theta_b$).

Finally, to obtain integer solutions of the integrated aircraft model and the integrated crew model, we propose a heuristic branching strategy. The heuristic branching strategy has two stages. In the first stage, if the columns taking a fractional value are greater than a predefined threshold (it is 0.65 for our test), they are fixed to 1. Secondly, if no suitable column exists, we choose a column whose fractional value is maximum among all columns with fractional values, and the column is fixed to 1. We create a single branching node if the obtained solution is fractional at a node contained the branch-and-bound tree. The exploration stops if the obtained solution is integer.

3.2 Alternating Lagrangian Decomposition

We can decompose model (1)–(10) into two subproblems (called ALD-aircraft and ALD-crew) via Lagrangian decomposition [7] to reduce the complexity of the solution process. These two Lagrangian subproblems are solved by BP. To obtain an optimal solution, a subgradient or bundle algorithm is usually used in the classical Lagrangian decomposition, but it may require many iterations [4]. Therefore, we use an alternating Lagrangian decomposition that update the Lagrangian multipliers by a similar way used in Zeighami et al. [11]. The solution process iterates between ALD-aircraft and ALD-crew by estimating the effect of decisions made in one problem on the other. Firstly, we obtain an optimal solution $X^* = (\bar{x}_r^a)_{r \in R_a, a \in A}$ by solving the integrated aircraft model considering constraints (2)–(3) and the first term of objective function (1). Let $C_A^1 = \{(i,j) \in C | \sum_{a \in A} \sum_{r \in R_a} b_r^{ij} \bar{x}_r^a = 1\}$ and $C_A^0 = \{(i,j) \in C | \sum_{a \in A} \sum_{r \in R_a} b_r^{ij} \bar{x}_r^a = 0\}$ the set of short arcs covered and uncovered, respectively. The coupling constraint (4) becomes

$$\sum_{b \in B} \sum_{p \in P_b} b_p^{ij} y_p^b - 1 \leq 0 \quad (i,j) \in C_A^1 \tag{11}$$

$$\sum_{b \in B} \sum_{p \in P_b} b_p^{ij} y_p^b - 0 \leq 0 \quad (i,j) \in C_A^0 \tag{12}$$

We prove the following lemma that constraint (11) is redundant for the integrated crew model.

Lemma 1. *Constraint (11) is redundant for the integrated crew model*

Proof. Because $0 \leq \sum_{b \in B} \sum_{p \in P_b} b_p^{ij} y_p^b \leq 1$, $-1 \leq \sum_{b \in B} \sum_{p \in P_b} b_p^{ij} y_p^b - 1 \leq 0$. Therefore, constraint (11) is redundant.

We define a nonnegative real vector $(\boldsymbol{\lambda})_{aircraft} = ((\lambda_{ij})_{aircraft})_{(i,j) \in C_A^0}$ that is Lagrangian multiplier of constraint (12) and computed from the integrated aircraft model as described in the next context. Hence, We have the following crew subproblem by the Lagrangian decomposition:

$$\text{ALD-crew } (\boldsymbol{\lambda})_{aircraft} = \min \sum_{b \in B} \sum_{p \in P_b} c_p^b y_p^b$$
$$+ \sum_{(i,j) \in C_A^0} (\lambda_{ij})_{airplane} (\sum_{b \in B} \sum_{p \in P_b} b_p^{ij} y_p^b - 0)$$
$$\text{s.t. } (5), (6), (9), \text{ and } (10).$$

We then obtain an optimal solution $Y^* = (\bar{y}_p^b)_{p \in P_b, b \in B}$ by solving ALD-crew, and fix variables y_p^b in model (1)–(10). The coupling constraint (4) becomes

$$1 - \sum_{a \in A} \sum_{r \in R_a} b_r^{ij} x_r^a \leq 0 \quad (i,j) \in C_B^1 \tag{13}$$

$$0 - \sum_{a \in A} \sum_{r \in R_a} b_r^{ij} x_r^a \leq 0 \quad (i,j) \in C_B^0 \tag{14}$$

where $C_B^1 = \{(i,j) \in C | \sum_{b \in B} \sum_{p \in P_b} b_p^{ij} \bar{y}_p^b = 1\}$ and $C_B^0 = \{(i,j) \in C | \sum_{b \in B} \sum_{p \in P_b} b_p^{ij} \bar{y}_p^b = 0\}$ are the sets of short arcs covered and uncovered by ALD-crew.

Similarly, we also have the following lemma that constraint (14) is redundant for the integrated aircraft model.

Lemma 2. *Constraint (14) is redundant for the integrated aircraft model.*

Proof. Its proof is similar to that of Lemma 1.

We define a nonnegative real vector $(\boldsymbol{\lambda})_{crew} = ((\lambda_{ij})_{crew})_{(i,j)\in C_B^1}$ that is Lagrangian multiplier of constraint (13) and computed from the ALD-crew. We have the following aircraft subproblem by the Lagrangian decomposition:

$$\text{ALD-aircraft } (\boldsymbol{\lambda})_{crew} = \min \sum_{a\in A}\sum_{r\in R_a} c_r^a x_r^a$$

$$+ \sum_{(i,j)\in C_B^1} (\lambda_{ij})_{crew}(1 - \sum_{a\in A}\sum_{r\in R_a} b_r^{ij} x_r^a)$$

$$\text{s.t. } (2), (3), (7), \text{ and } (8).$$

Next, we will describe how to calculate the Lagrange multipliers. In fact, the Lagrange multipliers are computed by the arc reduced costs. We use the formula $\bar{c}_{ij} = \Gamma_j - \Gamma_i$ to compute the arc reduced cost of (i, j), where Γ_j and Γ_i are the minimum costs from the source node to the node j and the node i, respectively. Γ_j and Γ_i are computed using the labeling algorithm when solving the pricing problem. Because there is one pricing problem for each aircraft base or crew base, short arc $(i, j) \in C$ can exist in each pricing problem. Therefore, we first obtain many arc reduced costs on $(i, j) \in C$, and then find the minimum arc reduced cost in these costs. We denote by $(c_{ij})^t_{aircraft}$ the arc reduced cost of $(i, j) \in C_A^0$ computed by the ALD-aircraft at iteration t. Because the Lagrangian multiplier is nonnegative and the arc reduced cost $(c_{ij})^t_{aircraft}$ may be negative at iteration t, we compute the Lagrangian multiplier as the absolute value of $(c_{ij})^t_{aircraft}$ for each $(i, j) \in C_A^0$. Namely, at iteration t, the Lagrangian multiplier for ALD-crew is computed by the following formula:

$$(\lambda_{ij})^t_{aircraft} = |(\bar{c}_{ij})^t_{aircraft}|, \quad (i, j) \in C_A^0.$$

Similarly, at iteration t, the Lagrangian multiplier for ALD-aircraft is computed by the following formula:

$$(\lambda_{ij})^t_{crew} = |(\bar{c}_{ij})^t_{crew}|, \quad (i, j) \in C_B^1,$$

where $(c_{ij})^t_{crew}$ is the arc reduced cost of $(i, j) \in C_B^1$ computed by the ALD-crew.

In addition, we consider these Lagrangian multipliers as dual information. The computed multipliers on each short arc in ALD-crew are projected onto the corresponding arcs in the pricing problems of the crew bases before solving ALD-crew. The ALD-crew then is solved by BP. We have known that $(\lambda_{ij})_{aircraft}$ is positive (i.e., $\sum_{(i,j)\in C_A^0} (\lambda_{ij})_{airplane}(\sum_{b\in B}\sum_{p\in P_b} b_p^{ij} y_p^b - 0)$) in the arc covering term of ALD-crew, so the value $(\lambda_{ij})_{aircraft}$ is added to the arc (i, j) contained in the pricing problems of all crew bases. This doesn't impose ALD-crew to cover the short arcs covered by ALD-aircraft. Secondly, because $(\lambda_{ij})_{crew}$ is negated (i.e., $\sum_{(i,j)\in C_B^1} (\lambda_{ij})_{crew}(1 - \sum_{a\in A}\sum_{r\in R_a} b_r^{ij} x_r^a)$) in the arc covering term of ALD-aircraft, so the value $-(\lambda_{ij})_{crew}$ is added to the arc (i, j) contained in the pricing problems of all aircraft bases. This encourage ALD-aircraft to cover the short arcs covered by ALD-crew.

When the solutions X^* and Y^* are obtained by solving ALD-aircraft and ALD-crew using BP, respectively, we determine whether (X^*, Y^*) violates the coupling constraint (4). If (X^*, Y^*) doesn't violate the coupling constraint (4), then (X^*, Y^*) is a feasible solution for model (1)–(10), and we update the current optimal solution; otherwise, (X^*, Y^*) isn't a feasible solution.

4 Computational Experiments

In this section, we present the results of computational experiments. We consider the first real data sets used in Zeighami et al. [12]. We extracted three four-day data periods from the data set. Because the number of aircraft and aircraft bases aren't given in these data sets, we consider the number of crew and crew bases as the number of aircraft and aircraft bases, respectively. The size of each instance is given in Table 1. The specification of the hardware environment is 13th Gen Intel(R) Core(TM) i9-13900KF central processing unit (CPU) at 3.00 GHz. Our implementations are coded in Python. We make use of the GORUBI solver (version 11.0.0) for solving the RMPs.

Table 1. Data sets

	Instance 1	Instance 2	Instance 3
No. of Flights	136	134	139
No. of Aircraft	33	33	33
No. of Aircraft Bases	3	3	3
No. of Crews	33	33	34
No. of crew Bases	3	3	3

We solve each of these three instances with the integrated and sequential approaches, and compare their performance in Table 2. The integrated (INT) approach solves the integrated model (1)–(10) using the ALD combined with BP. The sequential (SEQ) approach first solves the aircraft routing problem using BP and then solves the crew pairing problem using BP. In addition, for the sequential approach, when solving the crew pairing problem, we neglect the impact of aircraft routing problem, and the sit time must be limited between the minimum and maximum sit time. For the sequential approach, we report the total number of branch-and-bound nodes (Aircraft BB nodes or Crew BB nodes), the total iterations of CG (Aircraft CG iter. or Crew CG iter.), the total number of columns generated (Aircraft Columns or Crew Columns), the total CPU time (in seconds) spent solving the pricing problem (CPU Aircraft SCG or CPU Crew SCG), the total CPU time spent on the CG solution process (CPU Aircraft CG or CPU Crew CG), and the CPU time spent on the BP solution process (CPU Aircraft BP or CPU Crew BP) about aircraft routing or crew pairing. We also report the total solution cost (Cost), and the total CPU time

Table 2. Computational Results for integrated and sequential approaches

	Instance 1	Instance 2	Instance 3
SEQ			
Aircraft BB nodes	1	2	1
Aircraft CG iter.	156	147	149
Aircraft Columns	289	226	221
CPU Aircraft SCG	115.30	64.89	93.80
CPU Aircraft CG	121.30	71.02	100.51
CPU Aircraft BP	121.87	72.29	101.87
Crew BB nodes	1	1	1
Crew CG iter.	144	125	94
Crew Columns	211	205	156
CPU Crew SCG	41.62	43.40	27.57
CPU Crew CG	46.90	49.42	32.40
CPU Crew BP	48.11	50.67	33.68
Cost	411342	572077	476184
CPU Total	169.99	122.97	135.56
INT			
Aircraft BB nodes	4	7	14
Aircraft CG iter.	624	654	1490
Aircraft Columns	1055	1002	2029
CPU Aircraft SCG	359.47	313.71	687.36
CPU Aircraft CG	378.30	339.24	735.07
CPU Aircraft BP	380.59	344.93	741.46
Crew BB nodes	11	15	26
Crew CG iter.	730	867	1776
Crew Columns	1622	1771	3757
CPU Crew SCG	915.21	1090.32	1322.99
CPU Crew CG	944.10	1135.61	1397.43
CPU Crew BP	947.48	1142.82	1406.05
Cost	248636	369669	311497
ALD iter.	3	4	8
CPU Total	1328.11	1487.82	2147.57
INT vs. SEQ			
Cost saving by INT (%)	39.55	35.38	34.58
CPU Total INT/SEQ	7.8	12.1	15.8

(CPU Total). For the integrated approach, we not only report similar statistics for the total performance of solving the ALD-aircraft and ALD-crew by BP, but also provide the total solution cost, the iterations of the alternating Lagrangian decomposition (ALD iter.), and the total CPU time.

For each instance, we observe that the total number of branch-and-bound nodes, the total iterations of CG, the total number of columns generated, the total CPU time spent solving the pricing problem, the total CPU time spent on the CG solution process, and the CPU time spent on the BP solution process of the SEQ are smaller than those of the INT with respect to aircraft or crew, respectively. The main reason is that the ALD iterates between ALD-aircraft and ALD-crew such that these two Lagrangian subproblems need to be solved more than once by BP and considering different Lagrangian multipliers. The last two rows of Table 2 give the percentage of cost savings produced by the INT solutions, and the ratios of the total CPU times of INT and SEQ. The results indicate that the INT approach can yield obvious cost saving, but it yields longer computational time than that derived by SEQ for each instance. In fact, in our experiments, because of the fast convergence of BP solving crew pairing problem, we may obtain a degenerate solution with the large cost. This leads to the degenerate solution containing many artificial variables and the larger total solution cost.

5 Conclusion

In this paper, we have introduced a mathematical formulation and a solution method based on alternating Lagrangian decomposition for solving the integrated aircraft routing and crew pairing problem. Our contributions are as follows: (1) We propose a novel integrated model for the integrated aircraft routing and crew pairing problem. The model integrates the interdependent parts of aircraft routing and crew pairing. (2) An algorithm based alternating Lagrangian decomposition is developed and first applied to solve the integrated aircraft routing and crew pairing problem. Moreover, we also present a heuristic branching strategy in the branch-and-price algorithm. (3) Our computational experiments are conducted on three real instances. The results of the experiments have shown that the integrated approach is better than the sequential approach in terms of cost savings.

References

1. Ahmed, M.B., Mansour, F.Z., Haouari, M.: Robust integrated maintenance aircraft routing and crew pairing. J. Air Transp. Manag. **73**, 15–31 (2018)
2. Bulbul, K.G., Kasimbeyli, R.: Augmented Lagrangian based hybrid subgradient method for solving aircraft maintenance routing problem. Comput. Oper. Res. **132**, 105294 (2021)
3. Cordeau, J.F., Stojković, G., Soumis, F., Desrosiers, J.: Benders decomposition for simultaneous aircraft routing and crew scheduling. Transp. Sci. **35**(4), 375–388 (2001)

4. Crainic, T.G., Frangioni, A., Gendron, B.: Bundle-based relaxation methods for multicommodity capacitated fixed charge network design. Discret. Appl. Math. **112**(1–3), 73–99 (2001)
5. Desaulniers, G., et al.: Crew pairing at air France. Eur. J. Oper. Res. **97**(2), 245–259 (1997)
6. Díaz-Ramírez, J., Huertas, J.I., Trigos, F.: Aircraft maintenance, routing, and crew scheduling planning for airlines with a single fleet and a single maintenance and crew base. Comput. Ind. Eng. **75**, 68–78 (2014)
7. Guignard, M., Kim, S.: Lagrangean decomposition: a model yielding stronger Lagrangean bounds. Math. Program. **39**(2), 215–228 (1987)
8. Haouari, M., Shao, S., Sherali, H.D.: A lifted compact formulation for the daily aircraft maintenance routing problem. Transp. Sci. **47**(4), 508–525 (2013)
9. Irnich, S., Desaulniers, G.: Shortest path problems with resource constraints. In: Desaulniers, G., Desrosiers, J., Solomon, M.M. (eds.) Column Generation, pp. 33–65. Springer, Boston (2005). https://doi.org/10.1007/0-387-25486-2_2
10. Saddoune, M., Desaulniers, G., Soumis, F.: Aircrew pairings with possible repetitions of the same flight number. Comput. Oper. Res. **40**(3), 805–814 (2013)
11. Zeighami, V., Saddoune, M., Soumis, F.: Alternating Lagrangian decomposition for integrated airline crew scheduling problem. Eur. J. Oper. Res. **287**(1), 211–224 (2020)
12. Zeighami, V., Soumis, F.: Combining benders' decomposition and column generation for integrated crew pairing and personalized crew assignment problems. Transp. Sci. **53**(5), 1479–1499 (2019)

Monotone Submodular Meta-learning under the Matroid Constraint

Shufang Gong[1], Bin Liu[1(✉)], Qizhi Fang[1], and Weili Wu[2]

[1] School of Mathematical Sciences, Ocean University of China, Qingdao 266100, China
binliu@ouc.edu.cn

[2] Department of Computer Science, The University of Texas at Dallas, Richardson, TX 75080, USA

Abstract. Meta-learning, or learning to learn, has attracted a lot of attention in the field of artificial intelligence and machine learning in the past few years. The key idea of meta-learning is to utilize prior experience and data to improve performance on new tasks. There are already a large number of meta-learning formulations applied to continuous domains. Recently, Adibi et al. [1] devised a discrete meta-learning framework called submodular meta-learning to reduce the computational cost for new tasks and each task is regarded as a monotone submodular function maximization problem with the cardinality constraint. Motivated by their framework, we study the submodular meta-learning problem with the matroid constraint in this paper. Our method utilizes prior tasks to train an initial solution set that can be rapidly adapted to future tasks. We design a deterministic two-state algorithm using the greedy idea and achieve a 1/2-approximation ratio.

Keywords: Submodular maximization · Matroid constraint · Approximation algorithm · Meta-learning

1 Introduction

Given a ground set of n elements V, a set function $f : 2^V \to \mathbb{R}_+$ is *submodular* if for any subsets $A, B \subseteq V$, it holds

$$f(A) + f(B) \geq f(A \cup B) + f(A \cap B).$$

Another equivalent definition is that for any subsets $S \subseteq T \subseteq V$ and any element $e \in V \setminus T$, it holds

$$f(S \cup \{e\}) - f(S) \geq f(T \cup \{e\}) - f(T).$$

The latter characterizes the *diminishing return property* of the submodular function, that is, the benefit of adding an element to a small set is not less than that of adding it to a larger set. A submodular function is *monotone* (*nondecreasing*) if for any $S \subseteq T \subseteq V$, it holds $f(T) \geq f(S)$. A function is *normalized* if

S. Ghosh and Z. Zhang (Eds.): AAIM 2024, LNCS 15179, pp. 210–221, 2024.
https://doi.org/10.1007/978-981-97-7798-3_18

$f(\emptyset) = 0$. For convenience, we write $f(e|S) = f(S \cup \{e\}) - f(S)$, which quantifies the increase in utility obtained when adding an element e to a set S. Similarly, $f(A|B) = f(A \cup B) - f(B)$. Submodular functions have a wide range of applications in combinatorial optimization as well as in other emerging disciplines such as internet economics, machine learning, and artificial intelligence [15–17,26,29].

We consider such a problem. There is a set of tasks $\Gamma = \{\Gamma_i\}$ where the number of tasks can be infinite. We suppose that every task is independent and identically distributed (i.i.d.) over a probability distribution p. The tasks come randomly according to p and every task Γ_i is a normalized monotone submodular function maximization problem over V under the cardinality constraint, i.e., $\max_{S \subseteq V} f_i(S)$ subject to $|S| \leq k$ where k is the cardinality. For each task Γ_i, it is required to seek a feasible solution set S_i as optimal as possible. Movie recommendation is one of its applications. A silly approach is to solve each task independently. Obviously, this approach is not feasible when computational resources are limited. Therefore, Adibi et al. [1] provided a meta-learning approach that used the visited tasks to reduce the computation load for future tasks and formally summarized the above problem as a *submodular meta-learning problem.*

Meta-learning intends to utilize the visited tasks (or training tasks) to learn model parameters or representations that will work well for future tasks (or testing tasks), especially when computational resources in the testing phase are limited. Suppose that there are sufficient computational resources in the training phase. Meta-learning has drawn increased interest and has been widely applied to many practical problems such as reinforcement learning [9,24] and shot learning [21,27]. So far, there exist a large number of meta-learning formulations. The most famous one is Model-Agnostic Meta-Learning (MAML) applied on the continuous domain proposed by Finn et al. [13] and it is compatible with any model using the gradient descent algorithm. The key idea of the MAML framework is to provide a suitable initialization for the model's parameters so that a future task can be solved rapidly with as few gradient steps as possible. Later, Adibi et al. [1] extended this idea to the discrete domain and proposed the submodular meta-learning framework, where the decision variables are in a discrete set and the goal is to find a suitable initial solution set that can perform well over the entire future tasks.

The submodular meta-learning framework of [1] is divided into two phases. The first phase is to learn a good initial solution set of l ($0 \leq l \leq k$) elements from the training tasks. The second phase is to add the remaining $k - l$ elements to the initial solution set after observing every testing task. Let's consider two extremes: (1) $l = k$, and (2) $l = 0$. The former is that all k elements are selected in the training phase before observing the incoming testing task, implying low quality and high efficiency in the testing phase. The latter is that the k elements are selected after observing the incoming testing task, implying high quality and low efficiency in the testing phase. Hence, the value of l becomes the key to balancing quality and efficiency. According to specific applications, decision-makers can set proper l by themselves, which will not be discussed in this paper.

The focus of this paper is to find a suitable learning strategy for a fixed l to analyze its performance.

Motivated by the work of [1], we study the *submodular meta-learning problem under the matroid constraint*, i.e., each task is a normalized monotone submodular maximization problem under the matroid constraint. Our goal is to find a suitable initial solution set from training tasks that can perform well over the entire testing tasks. In this paper, we devise a computationally efficient deterministic meta-greedy algorithm, which also performs well in approximation ratio.

The organization of this paper is as follows. In Sect. 2, we list the related work of submodular maximization and discrete meta-learning. In Sect. 3, we introduce our problem and some related results. In Sect. 4, we propose a meta-greedy algorithm and give proofs of the theoretical results. Last, in Sect. 5, we summarize our work.

2 Related Work

Submodular Maximization. Most of the submodular function maximization problems are NP-hard, thus it is necessary to design approximation algorithms and analyze performances. For a monotone submodular maximization problem under certain constraints, such as cardinality constraint, knapsack constraint, matroid constraint, and so on, the most classical approach is to design the greedy algorithms and the output solutions generally perform well [19,22]. Later, many prominent scholars improved the previous results in many aspects, such as approximation ratio, query complexity, adaptive complexity, etc., and gave some conclusions on inapproximability [5,6,8,10,12,20]. In the context of massive data, a large number of streaming algorithms have emerged [4,7,14,28].

Discrete Meta-learning. Adibi et al. [1] first introduced the submodular meta-learning problem and designed a two-phase meta-greedy framework. They proposed a deterministic algorithm that achieves a 0.53-approximation ratio and a random algorithm that achieves an almost $(1 - 1/e)$-approximation ratio in expectation. Later, Tang and Yuan extended the problem of [1] to the adaptive setting, i.e., each element in the ground set has a random state drawn from a known prior distribution, and proposed the adaptive submodular meta-learning problem. For this problem, they studied the non-monotone case and devised a two-stage random greedy policy that obtained a $1/32$-approximation ratio [25].

Another approach has been developed to reduce computation called the two-stage submodular optimization framework [2,18,23]. Its goal is to summarize the ground set into a smaller set that can be used during testing. Our framework differs from the two-stage approaches in that it allows for personalization, meaning that a small subset of items can be found at test time specific to the task.

3 Preliminaries

3.1 Problem Formulation

Submodular Meta-learning Problem under the Matroid Constraint. We consider a set of tasks $\Gamma = \{\Gamma_i\}_{i\in I}$ where the number of tasks can be infinite. Assume that every task is i.i.d. over an underlying probability distribution p. The tasks come randomly according to p and every task Γ_i amounts to maximizing a normalized monotone submodular function over V under the matroid constraint, i.e., i-th task:

$$\max_{S\subseteq V} f_i(S)$$
$$\text{s.t.}\quad S\in\mathcal{I},$$

where $i\in I$.

The goal is, for each testing task Γ_i, to find a feasible solution set S_i that is as optimized as possible. To reduce the computational cost in the testing phase, we utilize the idea of meta-learning to find a proper initial solution set $S^{(a)}$ of l elements in the training phase and add the remaining $r-l$ elements $S^{(i)}$ after observing each test task Γ_i in the testing phase, where r is the rank of the matroid $(V,\mathcal{I})$. The rationale behind this is that for monotone case, optimal solutions are bases of the matroid.

Given an initial set $S^{(a)}$ with l elements, its utility on each testing task Γ_i is defined as

$$G_i(S^{(a)}) = \max_{\substack{S^{(a)}\cup S^{(i)}\in\mathcal{I}\\ |S^{(i)}|=r-l}} f_i(S^{(a)}\cup S^{(i)}),$$

and its expected utility on the entire testing tasks is

$$G_{test}(S^{(a)}) = \mathbb{E}_{i\sim p}\Big[\max_{\substack{S^{(a)}\cup S^{(i)}\in\mathcal{I}\\ |S^{(i)}|=r-l}} f_i(S^{(a)}\cup S^{(i)})\Big],$$

where i is the index of testing tasks. Because the occurrence of training tasks and testing tasks obey the same probability distribution p, the expected utility of $S^{(a)}$ on the entire testing tasks can be replaced by the expected utility on the entire training tasks,

$$G_{train}(S^{(a)}) = \mathbb{E}_{i\sim p}\Big[\max_{\substack{S^{(a)}\cup S^{(i)}\in\mathcal{I}\\ |S^{(i)}|=r-l}} f_i(S^{(a)}\cup S^{(i)})\Big],$$

where i is the index of training tasks. Because the underlying distribution p of tasks is often unknown, we usually independently sample a portion of training tasks drawn from p and use the average utility over the sample training tasks to approximate the expected utility over the entire training tasks,

$$G_m(S^{(a)}) = \frac{1}{m}\sum_{i=1}^{m} \max_{\substack{S^{(a)}\cup S^{(i)}\in\mathcal{I}\\ |S^{(i)}|=r-l}} f_i(S^{(a)}\cup S^{(i)}),$$

where m is the number of sample tasks. Stan et al. [23] shows that $G_m(S^{(a)})$ is an arbitrary good approximation of $G_{train}(S^{(a)})$ when the sample size is sufficient. In this paper, we assume that m is a given number in advance.

As a result, we aim to maximize the sample average function

$$\max_{\substack{S^{(a)}\in\mathcal{I}\\|S^{(a)}|=l}} G_m(S^{(a)}) = \max_{\substack{S^{(a)}\in\mathcal{I}\\|S^{(a)}|=l}} \frac{1}{m}\sum_{i=1}^{m} \max_{\substack{S^{(a)}\cup S^{(i)}\in\mathcal{I}\\|S^{(i)}|=r-l}} f_i(S^{(a)}\cup S^{(i)}), \tag{1}$$

where i is the sampled index of training tasks. But this problem is NP-hard, then we turn to seeking an approximate solution, i.e., find an approximate solution $S^{(a)}_{Alg}$ such that $G_m(S^{(a)}_{Alg}) \geq \alpha\cdot \max_{\substack{S^{(a)}\in\mathcal{I}\\|S^{(a)}|=l}} G_m(S^{(a)})$ where $\alpha\in(0,1)$ is called the approximation ratio. However, the objective function $G_m(S^{(a)})$ is not a submodular function although f_i is submodular for any $i=1,2,\cdots,m$, hence this is not a general submodular maximization problem. We overcome this challenge along the following lines. Suppose that $S^{(a)}_{Alg}$ is an approximation solution obtained by executing an approximation algorithm (this approximation algorithm also outputs $\{S^{(i)}_{Alg}\}_{i=1}^m$, where $\{S^{(i)}_{Alg}\}_{i=1}^m$ play a role in approximation ratio analysis). For problem (1), denote by $S^{(a)}_*$ be a set for the outer maximization and $\{S^{(i)}_*\}_{i=1}^m$ be sets for the inner maximization, i.e., optimal sets. Obviously, $|S^{(a)}_{Alg}|=|S^{(a)}_*|=l$, $|S^{(i)}_{Alg}|=|S^{(i)}_*|=r-l$. If we can show that

$$\frac{1}{m}\sum_{i=1}^{m} f_i(S^{(a)}_{Alg}\cup S^{(i)}_{Alg}) \geq \alpha\cdot\frac{1}{m}\sum_{i=1}^{m} f_i(S^{(a)}_*\cup S^{(i)}_*) \tag{2}$$

where $\alpha\in(0,1)$, then we can get that

$$\begin{aligned}
G_m(S^{(a)}_{Alg}) &= \frac{1}{m}\sum_{i=1}^{m} \max_{\substack{S^{(a)}_{Alg}\cup S^{(i)}\in\mathcal{I}\\|S^{(i)}|=r-l}} f_i(S^{(a)}_{Alg}\cup S^{(i)})\\
&\geq \frac{1}{m}\sum_{i=1}^{m} f_i(S^{(a)}_{Alg}\cup S^{(i)}_{Alg})\\
&\geq \alpha\cdot\frac{1}{m}\sum_{i=1}^{m} f_i(S^{(a)}_*\cup S^{(i)}_*)\\
&= \alpha G_m(S^{(a)}_*),
\end{aligned}$$

i.e., $S^{(a)}_{Alg}$ is a α-approximation solution of problem (1). Next, we focus on seeking α in (2).

3.2 Related Content of the Matroid and the Submodular Function

Definition 1. [11] *Given a finite ground set V and a family of subsets $\mathcal{I}\subseteq 2^V$, $(V,\mathcal{I})$ is called a* matroid *if and only if this pair satisfies the following three*

conditions.
(1) $\emptyset \in \mathcal{I}$.
(2) Hereditary property. If $A \subseteq B$ *and* $B \in \mathcal{I}$, *then* $A \in \mathcal{I}$ *for any* $A, B \subseteq V$.
(3) Augmentation property. If $A, B \in \mathcal{I}$ *and* $|B| > |A|$, *then there exists an element* $e \in B \setminus A$ *such that* $A \cup \{e\} \in \mathcal{I}$ *for any* $A, B \subseteq V$.

Lemma 1. [3] *If* $I_1 \in \mathcal{I}$, $I_2 \in \mathcal{I}$ *and* $|I_1| = |I_2|$, *then there exists a bijection* $\phi : I_1 \setminus I_2 \to I_2 \setminus I_1$ *satisfying that* $(I_2 \setminus \{\phi(e)\}) \cup \{e\} \in \mathcal{I}$ *for each* $e \in I_1 \setminus I_2$.

Furthermore, we can obtain an extended bijection $\phi : I_1 \to I_2$ by defining it to be the identity function on $I_1 \cap I_2$.

Lemma 2. *If* $f_i : 2^V \to \mathbb{R}_+$ *is a normalized monotone submodular function for each* $i = 1, 2, \ldots, m$, *then* $\sum_{i=1}^m f_i : 2^V \to \mathbb{R}_+$ *is also a normalized monotone submodular function.*

Due to page limitations, we have included the proof of Lemma 2 in Appendix.

Lemma 3. *Let* $g(\cdot) = f(\cdot|A)$ *for any* $A \subseteq V$. *If* $f : 2^V \to \mathbb{R}_+$ *is a monotone submodular function, then* $g : 2^{V \setminus A} \to \mathbb{R}_+$ *is also a monotone submodular function.*

4 Algorithm and Analysis

Algorithm 1 proceeds in two phases. Phase 1 finds the common initial solution set $S^{(a)}$, and phase 2 finds the remaining elements $S^{(i)}$ for each task Γ_i. In Algorithm 1, due to the monotonicity of the function, it is easy to observe that the output solution $|S^{(a)}| = l$, $|S^{(i)}| = r - l$, and $S^{(a)} \cup S^{(i)}$ is a base of the matroid $(V, \mathcal{I})$ for each i. Each iteration of phase 1 finds the element e^* with the largest marginal contribution to the current solution. If the addition of e^* does not break the matroid constraint, then it is added. After l iterations, the algorithm enters phase 2 and phase 2 can be performed in parallel for m tasks. For each iteration of each task Γ_i, phase 2 finds the element e_i^* from the remaining set V_2 with the largest marginal contribution to the current solution. If the addition of e_i^* does not break the matroid constraint, it is added. Algorithm 1 terminates until V_2 becomes an empty set and outputs sets $S^{(a)}$, $\{S^{(i)}\}_{i=1}^m$. See Algorithm 1 for details.

Since $|S^{(a)}| = |S_*^{(a)}| = l$ and $|S^{(i)}| = |S_*^{(i)}| = r - l$ for each $i = 1, 2, \ldots, m$. By the extension of Lemma 1, there exists a bijection $\phi : S^{(a)} \to S_*^{(a)}$ such that $(S^{(a)} \setminus \{e\}) \cup \{\phi\{e\}\} \in \mathcal{I}$ for any $e \in S^{(a)}$. Phase 1 of Algorithm 1 requires l iterations. For any $t = 1, 2, \ldots, l$, denote by $S_{t-1}^{(a)}$ and $S_t^{(a)}$ the sets at the beginning and end of the t-th iteration of phase 1, respectively. Denote by e_t the element that is added at the t-th iteration of phase 1.

Lemma 4. *For each* $t = 1, 2, \ldots, l$, *it holds*

$$\sum_{i=1}^{m} \left(f_i(S_{t-1}^{(a)} \cup \{e_t\}) - f_i(S_{t-1}^{(a)})\right) \geq \sum_{i=1}^{m} \left(f_i(S_{t-1}^{(a)} \cup \{\phi(e_t)\}) - f_i(S_{t-1}^{(a)})\right).$$

Algorithm 1

Require: the ground set V, the normalized monotone submodular f, the cardinality r, l, m

Initially set $S^{(a)} \leftarrow \emptyset, S^{(i)} \leftarrow \emptyset$ for all $i = 1, 2, \ldots, m$, $V_1 \leftarrow \emptyset$, $V_2 \leftarrow \emptyset$, $V' \leftarrow \emptyset$
$V_1 \leftarrow V$
while $|S^{(a)}| < l$ **do**
  $e^* \leftarrow \arg\max_{e \in V_1} \sum_{i=1}^{m} f_i(e|S^{(a)})$
  if $S^{(a)} \cup \{e^*\} \in \mathcal{I}$ **then**
    $S^{(a)} \leftarrow S^{(a)} \cup \{e^*\}$
  end if
  $V_1 \leftarrow V_1 \setminus \{e^*\}$
end while
$V' \leftarrow V_1$
for $i = 1, 2, \ldots, m$ **do**
  $V_2 \leftarrow V'$
  while $V_2 \neq \emptyset$ **do**
    $e_i^* \leftarrow \arg\max_{e \in V_2} f_i(e|S^{(a)} \cup S^{(i)})$
    if $S^{(a)} \cup S^{(i)} \cup \{e_i^*\} \in \mathcal{I}$ **then**
      $S^{(i)} \leftarrow S^{(i)} \cup \{e_i^*\}$
    end if
    $V_2 \leftarrow V_2 \setminus \{e_i^*\}$
  end while
end for
return $S^{(a)}$ and $\{S^{(i)}\}_{i=1}^{m}$

Proof. When $\phi(e_t) = e_t$, it clearly holds. Assume that $\phi(e_t) \neq e_t$. Because $e_t \notin S_{t-1}^{(a)}$, then $S_{t-1}^{(a)} \cup \{\phi(e_t)\} \subseteq (S^{(a)} \setminus \{e_t\}) \cup \{\phi(e_t)\}$. With $(S^{(a)} \setminus \{e_t\}) \cup \{\phi(e_t)\} \in \mathcal{I}$, by the hereditary property of matroid, we get $S_{t-1}^{(a)} \cup \{\phi(e_t)\} \in \mathcal{I}$. And because ϕ is the identity function on $S^{(a)} \cap S_*^{(a)}$, then $\phi(e_t) \notin S_{t-1}^{(a)}$. Otherwise, suppose that $\phi(e_t) = e_i$ for some $i \in \{1, 2, \ldots, t-1\}$, it holds $\phi(e_t) = e_i = \phi(e_i)$, which contradicts with ϕ is a bijection. Therefore, $\phi(e_t)$ is a candidate for e_t. By the selection rule of Algorithm 1, we get the conclusion. □

We define a transition variable A_t as follows,

$$A_t = (A_{t-1} \setminus \{\phi(e_t)\}) \cup \{e_t\}$$

for each $t = 1, 2, \ldots, l$. Let $A_0 = S_*^{(a)}$. From the construction process, we have $S_t^{(a)} \subseteq A_t$ and $A_l = S^{(a)}$.

Lemma 5. *For each $t = 1, 2, \ldots, l$, we get*

$$\sum_{i=1}^{m} \left(f_i(S_t^{(a)}) - f_i(S_{t-1}^{(a)})\right) \geq \sum_{i=1}^{m} \left(f_i(S_*^{(i)} \cup A_{t-1}) - f_i(S_*^{(i)} \cup A_t\right).$$

Proof. First, we demonstrate that $S_{t-1}^{(a)} \subseteq (S_*^{(i)} \cup A_{t-1}) \setminus \phi(e_t)$. Since $S_{t-1}^{(a)} \subseteq A_{t-1}$, then $S_{t-1}^{(a)} \subseteq (A_{t-1} \cup S_*^{(i)})$. If $\phi(e_t) = e_t$, then $\phi(e_t) \notin S_{t-1}^{(a)}$ because

$e_t \notin S_{t-1}^{(a)}$. If $\phi(e_t) \neq e_t$, then $\phi(e_t) \notin S_{t-1}^{(a)}$. Hence, we get the conclusion that $S_{t-1}^{(a)} \subseteq (S_*^{(i)} \cup A_{t-1}) \setminus \{\phi(e_t)\}$.

Second, we demonstrate that $(S_*^{(i)} \cup A_{t-1}) \setminus \{\phi(e_t)\} \subseteq S_*^{(i)} \cup A_t$. We observe that

$$\begin{aligned}(S_*^{(i)} \cup A_{t-1}) \setminus \{\phi(e_t)\} &\subseteq S_*^{(i)} \cup (A_{t-1} \setminus \{\phi(e_t)\}) \\ &\subseteq S_*^{(i)} \cup ((A_{t-1} \setminus \{\phi(e_t)\}) \cup \{e_t\}) \\ &= S_*^{(i)} \cup A_t.\end{aligned}$$

Now we formally begin to prove our conclusion,

$$\begin{aligned}\sum_{i=1}^{m} (f_i(S_t^{(a)}) - f_i(S_{t-1}^{(a)})) &= \sum_{i=1}^{m} (f_i(S_{t-1}^{(a)} \cup \{e_t\}) - f_i(S_{t-1}^{(a)})) \\ &\geq \sum_{i=1}^{m} (f_i(S_{t-1}^{(a)} \cup \{\phi(e_t)\}) - f_i(S_{t-1}^{(a)})) \\ &= \sum_{i=1}^{m} f_i(\phi(e_t)|S_{t-1}^{(a)}) \\ &\geq \sum_{i=1}^{m} f_i(\phi(e_t)|(S_*^{(i)} \cup A_{t-1}) \setminus \{\phi(e_t)\}) \\ &= \sum_{i=1}^{m} (f_i(S_*^{(i)} \cup A_{t-1}) - f_i((S_*^{(i)} \cup A_{t-1}) \setminus \{\phi(e_t)\})) \\ &\geq \sum_{i=1}^{m} (f_i(S_*^{(i)} \cup A_{t-1}) - f_i(S_*^{(i)} \cup A_t)).\end{aligned}$$

The first inequality utilizes Lemma 4. The second inequality utilizes $S_{t-1}^{(a)} \subseteq (S_*^{(i)} \cup A_{t-1}) \setminus \{\phi(e_t)\}$. The last inequality holds because $(S_*^{(i)} \cup A_{t-1}) \setminus \{\phi(e_t)\} \subseteq S_*^{(i)} \cup A_t$ and the monotonicity of $\sum_{i=1}^{m} f_i$. □

According to the definitions of $S_{t-1}^{(a)}$ and $S_t^{(a)}$ above, specifically, $S_0^{(a)} = \emptyset$ and $S_l^{(a)} = S^{(a)}$.

Lemma 6. *We have*

$$\sum_{i=1}^{m} f_i(S_*^{(i)} \cup S^{(a)}) \geq \sum_{i=1}^{m} f_i(S_*^{(i)} \cup S_*^{(a)}) - \sum_{i=1}^{m} f_i(S^{(a)}).$$

Proof.

$$\begin{aligned}\sum_{i=1}^{m} f_i(S^{(a)}) &= \sum_{i=1}^{m} (f_i(S_l^{(a)}) - f_i(S_0^{(a)})) \\ &= \sum_{i=1}^{m}\sum_{t=1}^{l} (f_i(S_t^{(a)}) - f_i(S_{t-1}^{(a)})) \\ &\geq \sum_{i=1}^{m}\sum_{t=1}^{l} (f_i(S_*^{(i)} \cup A_{t-1}) - f_i(S_*^{(i)} \cup A_t)) \\ &= \sum_{i=1}^{m} (f_i(S_*^{(i)} \cup A_0) - f_i(S_*^{(i)} \cup A_l)) \\ &= \sum_{i=1}^{m} (f_i(S_*^{(i)} \cup S_*^{(a)}) - f_i(S_*^{(i)} \cup S^{(a)})).\end{aligned}$$

The inequality uses Lemma 5 and rearranging the above inequality yields that

$$\sum_{i=1}^{m} f_i(S_*^{(i)} \cup S^{(a)}) \geq \sum_{i=1}^{m} f_i(S_*^{(i)} \cup S_*^{(a)}) - \sum_{i=1}^{m} f_i(S^{(a)}).$$

□

Denote by $S_O^{(i)}$ an optimal solution of the following problem: $\max_{S\subseteq V'} f_i(S|S^{(a)})$ with the constraint $S \cup S^{(a)} \in \mathcal{I}$ where the definition of V' is given in Algorithm 1.

Lemma 7. *For each $i = 1, 2, \ldots, m$, it holds that*

$$f_i(S^{(a)} \cup S^{(i)}) - f_i(S^{(a)}) \geq \frac{1}{2}(f_i(S^{(a)} \cup S_O^{(i)}) - f_i(S^{(a)})).$$

Theorem 1. *For inequality (2), we can get $\alpha = \frac{1}{2}$, i.e.,*

$$\frac{1}{m}\sum_{i=1}^{m} f_i(S^{(a)} \cup S^{(i)}) \geq \frac{1}{2}\Big(\frac{1}{m}\sum_{i=1}^{m} f_i(S_*^{(a)} \cup S_*^{(i)})\Big).$$

Hence, for problem (1), Algorithm 1 is a $\frac{1}{2}$-approximation algorithm.

Proof.

$$\begin{aligned}\sum_{i=1}^{m} (f_i(S^{(a)} \cup S^{(i)}) - f_i(S^{(a)})) &\geq \frac{1}{2}\sum_{i=1}^{m} (f_i(S^{(a)} \cup S_O^{(i)}) - f_i(S^{(a)})) \\ &\geq \frac{1}{2}\sum_{i=1}^{m} (f_i(S^{(a)} \cup S_*^{(i)}) - f_i(S^{(a)})) \\ &\geq \frac{1}{2}\sum_{i=1}^{m} (f_i(S_*^{(a)} \cup S_*^{(i)}) - 2f_i(S^{(a)})).\end{aligned}$$

The first inequality uses Lemma 7. The second inequality uses the definition of $S_O^{(i)}$. The last inequality holds by Lemma 6. Multiplying by $\frac{1}{m}$ in both sides of the above inequality, we get that

$$\frac{1}{m}\sum_{i=1}^{m} f_i(S^{(a)} \cup S^{(i)}) \geq \frac{1}{2} \cdot \frac{1}{m}\sum_{i=1}^{m} f_i(S_*^{(a)} \cup S_*^{(i)}).$$

□

Hence, the output solution $S^{(a)}$ of Algorithm 1 can achieve a 1/2-approximation ratio of our problem (1).

5 Conclusion

In this paper, for the submodular meta-learning problem with matroid constraint, we use the meta-greedy ideas to design a deterministic algorithm that can reduce computation load for future tasks and give a theoretical analysis of the approximation ratio. In the future, we can further consider designing random algorithms to improve the approximation ratio.

Acknowledgments. This work was supported by the National Natural Science Foundation of China (11971447), and the Fundamental Research Funds for the Central Universities (202261097).

Disclosure of Interests. The authors have no competing interests.

References

1. Adibi, A., Mokhtari, A., Hassani, H.: Submodular meta-learning. In: Proceedings of the Thirty-Fourth International Conference on Neural Information Processing Systems, pp. 3821–3832. PMLR Press, Virtual Event (2020)
2. Balkanski, E., Mirzasoleiman, B., Krause, A., Singer, Y.: Learning sparse combinatorial representations via two-stage submodular maximization. In: Proceedings of the Thirty-Third International Conference on Machine Learning, pp. 2207–2216. PMLR Press, New York City, NY, USA (2016)
3. Brualdi, R.-A.: Comments on bases in dependence structures. Bull. Aust. Math. Soc. **1**(2), 161–167 (1969)
4. Badanidiyuru, A., Mirzasoleiman, B., Karbasi, A.: Streaming submodular maximization: massive data summarization on the fly. In: The Proceedings of the Twentieth International Conference on Knowledge Discovery and Data Mining, pp. 671–680. ACM Press, New York (2014)
5. Breuer, A., Balkanski, E., Singer, Y.: The FAST algorithm for submodular maximization. In: Proceedings of the Thirty-Seventh International Conference on Machine Learning, pp. 1134–1143. PMLR Press, Virtual Event (2020)
6. Călinescu, G., Chekuri, C., Pál, M., Vondrák, J.: Maximizing a monotone submodular function subject to a matroid constraint. SIAM J. Comput. **40**(6), 1740–1766 (2011)

7. Chakrabarti, A., Kale, S.: Submodular maximization meets streaming: matchings, matroids, and more. Math. Program. **154**(1), 225–247 (2015)
8. Chekuri, C., Quanrud, K.: Parallelizing greedy for submodular set function maximization in matroids and beyond. In: Proceedings of the Fifty-First International Conference on Symposium on Theory of Computing, pp. 78–89. ACM Press, Phoenix, AZ, USA (2019)
9. Duan, Y., Schulman, J., Chen, X.: Rl^2: fast reinforcement learning via slow reinforcement. ArXiv:1611.02779 (2016)
10. Cui, M., Xu, D., Guo, L., Wu, D.: Approximation guarantees for parallelized maximization of monotone non-submodular function with a cardinality constraint. J. Comb. Optim. **43**(5), 1671–1690 (2022)
11. Edmonds, J.: Submodular functions, matroids, and certain polyhedra. In: Jünger, M., Reinelt, G., Rinaldi, G. (eds.) Combinatorial Optimization—Eureka, You Shrink! LNCS, vol. 2570, pp. 11–26. Springer, Heidelberg (2003). https://doi.org/10.1007/3-540-36478-1_2
12. Ene, A., Nguyen, H.-L.: A nearly-linear time algorithm for submodular maximization with a knapsack constraint. In: Proceedings of the Forty-Sixth International Colloquium on Automata, Languages, and Programming, pp. 53:1–53:12. Schloss Dagstuhl - Leibniz-Zentrum für Informatik Press, Patras, Greece (2017)
13. Finn, C., Abbeel, P., Levine, S.: Model-agnostic meta-learning for fast adaptation of deep networks. In: Proceedings of the Thirty-Fourth International Conference on Machine Learning, pp. 1126–1135. PMLR Press, Sydney, NSW, Australia (2017)
14. Huang, C.-C., Kakimura, N., Yoshida, Y.: Streaming algorithms for maximizing monotone submodular functions under a knapsack constraint. Algorithmica **83**(3), 1006–1032 (2020)
15. Kempe, D., Kleinberg, J., Tardos, É.: Maximizing the spread of influence through a social network. In: Proceedings of the Ninth International Conference on Knowledge Discovery and Data Mining, pp. 137–146. ACM Press, Washington, DC, USA (2003)
16. Krause, A., et al.: Optimizing sensor placements in water distribution systems using submodular function maximization. In: Proceedings of the Eighth Water Distribution Systems Analysis Symposium, pp. 1–17. Cincinnati, Ohio (2006)
17. Liu, M.-Y., Tuzel, O., Ramalingam, S., Chellappa, R.: Entropy-rate clustering: cluster analysis via maximizing a submodular function subject to a matroid constraint. IEEE Trans. Pattern Anal. Mach. Intell. **36**(1), 99–112 (2013)
18. Mitrovic, M., Kazemi, E., Zadimoghaddam, M., Karbasi, A.: Data summarization at scale: a two-stage submodular approach. In: Proceedings of Thirty-Fifth International Conference on Machine Learning. Stockholmsmässan, Stockholm, Sweden (2018)
19. Nemhauser, G.-L., Wolsey, L.-A., Fisher, M.-L.: An analysis of approximations for maximizing submodular set functions-I. Math. Program. **14**(1), 265–294 (1978)
20. Nemhauser, G.-L., Wolsey, L.-A.: Best algorithms for approximating the maximum of a submodular set function. Math. Oper. Res. **3**(3), 177–188 (1978)
21. Ravi, S., Larochelle, H.: Optimization as a model for few-shot learning. In: Proceedings of the Fifth International Conference on Learning Representations. OpenReview.net, Toulon, France (2017)
22. Sviridenko, M.: A note on maximizing a submodular set function subject to a knapsack constraint. Oper. Res. Lett. **32**(1), 41–43 (2004)
23. Stan, S., Zadimoghaddam, M., Krause, A.: Probabilistic submodular maximization in sub-linear time. In: Proceedings of the Thirty-Fourth Conference on Machine Learning, pp. 3241–3250. PMLR Press, Sydney, NSW, Australia (2017)

24. Song, X., Gao, W., Yang, Y., Choromanski, K., Pacchiano, A., Tang, Y.: ES-MAML: simple hessian-free meta learning. In: Proceedings of the Eighth International Conference on Machine Learning. OpenReview.net, Addis Ababa, Ethiopia (2020)
25. Tang, S., Yuan, J.: Non-monotone adaptive submodular meta-learning. In: Proceedings of 2021 Conference on Applied and Computational Discrete Algorithms, pp. 57–65. SAIM Press, Virtual Conference (2021)
26. Trevisan, L., Sorkin, G.-B., Sudan, M., Williamson, D.-P.: Gadgets, approximation, and linear programming. SIAM J. Comput. **29**(6), 2074–2097 (2000)
27. Vinyals, O., Blundell, C., Lillicrap, T., Kavukcuoglu, K., Wierstra, D.: Matching networks for one shot learning. In: Proceedings of the Thirtieth Conference on Neural Information Processing Systems, pp. 3630–3638. PMLR Press, Barcelona, Spain (2016)
28. Yang, R., Xu, D., Guo, L., Zhang, D.: Sequence submodular maximization meets streaming. J. Comb. Optim. **41**(1), 43–55 (2021)
29. Zhang, H., Zhang, H., Kuhnle, A., Thai, M.-T.: Profit maximization for multiple products in online social networks. In: Proceedings of the Thirty-Fifth IEEE International Conference on Computer Communications, pp. 1–9. IEEE Press, San Francisco, CA, USA (2016)

Submodular Participatory Budgeting

Jing Yuan[1] and Shaojie Tang[2(✉)]

[1] Department of Computer Science and Engineering, University of North Texas, Denton, USA

[2] Department of Management Science and Systems, School of Management, University at Buffalo, Buffalo, USA

tangshaojie@gmail.com

Abstract. Participatory budgeting refers to the practice of allocating public resources by collecting and aggregating individual preferences. Most existing studies in this field often assume an additive utility function, where each individual holds a private utility for each candidate project, and the total utility of a set of funded projects is simply the sum of the utilities of all projects. We argue that this assumption does not always hold in reality. For example, building two playgrounds in the same neighborhood does not necessarily lead to twice the utility of building a single playground. To address this, we extend the existing study by proposing a submodular participatory budgeting problem, assuming that the utility function of each individual is a monotone and submodular function over funded projects. We propose and examine three preference elicitation methods and analyze their performances in terms of distortion. Notably, if the utility function is additive, our aggregation rule designed for threshold approval votes achieves a better distortion than the state-of-the-art approach.

Keywords: Participatory Budgeting · Submodular Optimization · Distortion

1 Introduction

Participatory budgeting [5], which allocates public resources by collecting and aggregating individual preferences subject to a budget constraint, has attracted increased attention recently [3]. The input of such a problem consists of a set of alternatives and a group of voters, each voter is allowed to provide its preference over those alternatives-typically through ranking, the goal of this problem is to select a set of winning alternatives that best reflect all individuals' preferences [6]. While most of existing studies in this field overlooks the cost associated with each alterative, [11] extends this study to a setting called budgeted social choice. In their setting, each alterantive has a cost and the selected alterantives must satisfy a budget constraint. Building upon the framework established by [3,9], we can decompose any participatory budgeting approach into two components. The first component is the *input format*, which refers to how each voter's preferences

S. Ghosh and Z. Zhang (Eds.): AAIM 2024, LNCS 15179, pp. 222–234, 2024.
https://doi.org/10.1007/978-981-97-7798-3_19

are elicited. The second component is the *aggregation rule*, which pertains to how these preferences are combined to select a group of alternatives. Some popular input formats include the *ranking-by-values method*, where each voter ranks all alternatives according to their values, and the *approval-based method*, where each voter submits a list of acceptable alternatives. These preferences can be aggregated using variants of classic voting rules, such as the Borda count rule.

Most studies, like [3,9], assume additive utility functions, but this does not always hold, especially with project interactions [10]. For example, two playgrounds in the same neighborhood often do not double the utility. We propose a submodular participatory budgeting problem, capturing negative synergies with monotone submodular utility functions. This generalizes the problem, as all additive functions are submodular. We explore three preference elicitation methods: *ranking-by-marginal-values*, *ranking-by-values*, and *threshold approval votes*, evaluating them using *distortion* [12,13], which measures the worst-case ratio of optimal to achieved social welfare. For ranking based methods, we introduce *ranking-by-marginal-values* for non-additive utilities. Both this and *ranking-by-values* achieve a distortion of $O(\log m\sqrt{m}/(1-c))$, where m is the number of alternatives and $c \in [0,1]$ is the curvature of each submodular utility function. Intuitively, a smaller value of c indicates that a submodular function is closer to an additive function. For additive functions ($c = 0$), our results match the state-of-the-art $O(\log m\sqrt{m})$ [3]. Another contribution is our aggregation rule for *threshold approval votes*, achieving a distortion of $O(\log m/(1-c))$. For additive utility functions ($c = 0$), it improves the current state-of-the-art distortion from $O(\log^2 m)$ to $O(\log m)$.

Additional Related Work. Participatory budgeting is an active field in computational social science [1,14]. The work most relevant to ours is [3], which examines preference elicitation methods assuming additive utility functions. Our study extends this by considering a more general submodular setting. While most studies assume additive settings, some recent papers address project interactions [7]. For example, [10] models synergies between projects through a known partition, assuming complementary or substitution effects. Our study differs in two key aspects: (1) We do not assume access to each voter's utility function, introducing challenges that necessitate using distortion to evaluate methods; (2) We do not assume publicly available interaction structures, but instead use private, voter-specific submodular functions. Our setting captures only negative interactions, leaving positive interactions for future work. Another related work is [6], which selects a set of k alternatives with utility defined as the maximum for any alternative in the subset. Their study is a special case of ours, as their constraints and utility functions are specific instances of the knapsack constraint and monotone submodular functions, respectively.

2 The Model

Preliminaries: Submodular Function and Curvature. Given a function f, define the marginal utility of an item $i \in A$ with respect to a set of items $S \subseteq A$

as $f(i \mid S) \stackrel{\text{def}}{=} f(S \cup \{i\}) - f(S)$. A function f is said to be submodular if and only if for any two sets X and Y such that $X \subseteq Y$ and any item $i \notin Y$, $f(i \mid X) \geq f(i \mid Y)$. Furthermore, a submodular function f is said to have *curvature* $c \in [0, 1]$ if $f(i \mid S) \geq (1 - c)f(\{i\})$ for any $S \subseteq A$ and $i \notin S$ [2]. Curvature intuitively measures how much a given function deviates from a modular function.

The Model. For any integer k, let $[k] = \{1, 2, \cdots, k\}$. The input of our problem is a set of n *voters* $[n]$, and a set A of m *alternatives* (e.g., projects). Each alternative $a \in A$ has a cost of c_a, and the total budget is 1 (after normalization). For any subset of alternatives $S \subseteq A$, we abuse notation and let $c(S) = \sum_{a \in S} c_a$ denote the total cost of S. Define $\mathcal{F} = \{S \subseteq A : c(S) \leq 1\}$ as a collection of all feasible sets of alternatives.

We assume that each voter i holds a utility function $f_i : 2^A \rightarrow \mathbb{R}_{\geq 0}$, here we assume that each utility function f_i is monotone and submodular. To ensure fairness among all voters, we follow the standard setting [3,4] and assume that $f_i(A) = 1$ for all $i \in [n]$. Given the *utility profile* $\hat{f} = \{f_1, f_2, \cdots, f_n\}$, the (utilitarian) social welfare of a set of alternatives $S \subseteq A$ is defined as $\mathsf{sw}(S, \hat{f}) = \sum_{i \in [n]} f_i(S)$.

The utility function f_i of a voter i is not directly accessible, however, it can be accessed through i's vote ρ_i that is induced by f_i. Let $\hat{\rho} = \{\rho_1, \rho_2, \cdots, \rho_n\}$ denote the *input profile*. Let $\hat{f} \triangleright \hat{\rho}$ denote that the utility profile $\hat{f}$ is consistent with the input profile $\hat{\rho}$. In this paper, we examine three preference elicitation methods: *group based ranking-by-marginal-values*, *group based ranking-by-values* and *threshold approval votes*, and analyze their performance through the lens of implicit utilitarian voting.

Group Based Ranking-by-Marginal-Values. The group based ranking-by-marginal-values method asks each voter to rank *partial* alternatives according to a simple greedy algorithm. In particular, building on the idea from [3], we first partition all alternatives into $\log m + 1$ groups based on their costs. Specifically, let $l_0 = 0$ and $u_0 = 1/m$. For $t \in [\log m]$ (for ease of presentation, assume m is a power of 2), define $l_t = 2^{t-1}/m$ and $u_t = 2^t/m$. We partition all alternatives into $\log m + 1$ groups based on their costs: $G_0 = \{a \mid c_a \leq u_0\}$ and $G_t = \{a \in A \mid c_a \in (l_t, u_t]\}$ for $t \in [\log m]$. We choose a group G_t uniformly at random, and asks each voter to rank all alternatives in G_t according to a simple greedy algorithm. This algorithm starts with an empty set $S = \emptyset$. Then for each alternative not yet ranked $a \in G_t \setminus S$, it computes the marginal value $f_i(a \mid S)$, and adds $a_j = \arg\max_{a \in G_t \setminus S} f_i(a \mid S)$ to the ranking list. Here a_j is simply the best alternative that has the highest marginal utility when added to S. The final ranking $\rho_i = (a_1, a_2, \ldots, a_{|G_t|})$ reflects the order in which alternatives were selected.

While the idea of partitioning was adopted from [3], they use this technique solely for analysis; during their preference elicitation stage, voters are required to rank *all* alternatives. In contrast, our method requires voters to rank only the alternatives from a randomly selected group G_t. This design has two advantages: first, it helps to lower the communication cost incurred during the elicitation process compared to the approach that ranks all alternatives; second, and most importantly, it achieves better distortion.

Group Based Ranking-by-Values. Similar to the group based ranking-by-marginal-values method, our second approach, termed as group based ranking-by-values method, asks each voter to rank partial alternatives. The key difference in this method is that voters rank alternatives based on their stand-alone values.

Specifically, we follow the same procedure as described in the previous method to partition all alternatives into $\log m + 1$ groups: $G_0, G_1, \cdots, G_{\lfloor \log m \rfloor}$. We choose a group G_t uniformly at random, and asks each voter to rank all alternatives in G_t according to their stand-alone values. That is, in the final ranking ρ_i from voter $i \in [n]$, for any two alternatives $a \in G_t$ and $b \in G_t$, a is placed ahead of b if and only if $f_i(\{a\}) \geq f_i(\{b\})$.

This method is similar to the classical ranking method designed for additive utility functions [3,4], but it does not require each voter to rank *all* alternatives.

Threshold Approval Votes. The *threshold approval votes* method asks each voter to report all alternatives whose stand-alone utility is above a certain threshold α. Formally, given a threshold α, the approval set for voter i is given by $\rho_i = \{a \in A \mid f_i(\{a\}) \geq \alpha\}$. The term threshold approval votes was first introduced in [3] for additive functions. Our contribution is to develop a novel aggregation rule and use additional techniques to deal with submodular functions. Notably, for the special case of additive functions, our method achieves lower distortion compared to the previous study.

2.1 Problem Formulation

Following the implicit utilitarian voting framework, our goal is to maximize the (utilitarian) social welfare. We introduce the notation of *distortion* [13] to capture the distance between the social welfare achieved by an aggregation rule π and the maximum social welfare. Formally, the distortion of π with respect to a vote profile $\hat{\rho}$ is given by $\mathsf{dist}(\pi, \hat{\rho}) = \sup_{\hat{f}:\hat{f} \triangleright \hat{\rho}} \frac{\max_{T \in \mathcal{F}} \mathsf{sw}(T, \hat{f})}{\mathbb{E}[\mathsf{sw}(\pi(\hat{\rho}), \hat{f})]}$ where $\pi(\hat{\rho})$ denotes the set of alternatives selected by π. In this paper, we focus on a randomized participatory budgeting approach; hence, we are interested in the expected social welfare achieved by an aggregation rule π. The (overall) distortion of a rule π is defined as $\mathsf{dist}(\pi) = \max_{\hat{\rho}} \mathsf{dist}(\pi, \hat{\rho})$. Let π^* denote the optimal (random) aggregation rule that minimizes the distortion. I.e., $\pi^* = \arg\min_{\pi} \mathsf{dist}(\pi)$.

3 Distortion Analysis

3.1 Group Based Ranking-by-Marginal-Values

We first recall the procedure of this approach. Let $l_0 = 0$ and $u_0 = 1/m$. For $t \in [\log m]$ (for ease of presentation, assume m is a power of 2), define $l_t = 2^{t-1}/m$

and $u_t = 2^t/m$. We partition all alternatives into $\log m+1$ groups based on their costs: $G_0 = \{a \in A \mid c_a \leq u_0\}$ and $G_t = \{a \in A \mid c_a \in (l_t, u_t]\}$ for $t \in [\log m]$.

We choose a group G_t uniformly at random, and asks each voter to rank all alternatives in G_t according to a simple greedy algorithm. This algorithm starts with an empty set $S = \emptyset$. Then for each alternative not yet ranked $a \in G_t \setminus S$, it computes the marginal value $f_i(a \mid S)$, and adds $a_j = \arg\max_{a\in G_t \setminus S} f_i(a \mid S)$ to the ranking list. Here a_j is simply the best alternative that has the highest marginal utility when added to S. The final ranking $\rho_i = (a_1, a_2, \dots, a_{|G_t|})$ reflects the order in which alternatives were selected.

We follow the general framework of [3] and introduce some additional notations. Consider a ranking ρ and an alternative $a \in G_t$. We define $\rho(a)$ as the position of a in ρ. For an input profile $\hat{\rho}$, the *harmonic score* of a in $\hat{\rho}$ is given by $\mathsf{sc}(a, \hat{\rho}) = \sum_{i=1}^{n} 1/\rho_i(a)$.

Let $\hat{f}$ denote the *true* utility profile that is consistent with $\hat{\rho}$ and let $S^* = \arg\max_{S\in\mathcal{F}} \mathsf{sw}(S, \hat{f})$ denote the optimal set of alternatives that maximizes the social welfare.

Next, we describe our aggregation rule, which randomly picks a rule from the following two candidates. This aggregation rule was first introduced in [3] to address the case of additive functions; however, additional efforts are required in our analysis to tackle the challenges posed by non-additive functions.

- Rule A: Let G_t^+ denotes the top (up to) $\sqrt{m} \cdot (1/u_t)$ alternatives from G_t that have the largest harmonic scores in $\hat{\rho}$. Select a subset of G_t^+ with a size of $1/u_t$, also chosen uniformly at random.
- Rule B: Choose a single alternative uniformly at random.

For the sake of analysis, we introduce some additional notations. Let $G_t^- = G_t \setminus G_t^+$. Note that because $\sqrt{m} \cdot (1/u_0) = \sqrt{m} \cdot m \geq m$, we have $G_0^+ = G_0$. Define $G^+ = \cup_{t=0}^{\log m} G_t^+$ and $G^- = A \setminus G^+$. We next present two lemmas to bound the social welfare achieved by Rule A and Rule B respectively.

Lemma 1. *The expected social welfare achieved by Rule A is at least* $\mathsf{sw}(G^+ \cap S^*, \hat{f})/(1+\log m)\sqrt{m}$.

Proof. Observe that Rule A chooses each group G_t with probability $1/(1+\log m)$, then it selects each alternative in G_t^+ with probability at least $1/\sqrt{m}$ (this is because $|G_t^+| \leq \sqrt{m} \cdot (1/u_t)$ and Rule A selects a subset of G_t^+ with a size of $1/u_t$). Therefore, Rule A selects each alternative in G^+ with a probability of at least $1/(1+\log m)\sqrt{m}$.

We next present a result that links random sampling to submodular maximization.

Lemma 2. *(Lemma 2.2 of [8]). Consider a monotone submodular set function* $g : 2^U \to \mathbb{R}_{\geq 0}$. *Let* $U(p)$ *denote a subset of* U *where each element is included with a probability of at least* p *(elements are not necessarily independent of each other). The expected value of* $g(U(p))$ *is at least* p *times the value of* $g(U)$. *Formally,* $\mathbb{E}[g(U(p))] \geq p \cdot g(U)$.

Recall that $\mathsf{sw}(S, \hat{f}) = \sum_{i\in[n]} f_i(S)$ and f_i is a monotone submodular function for each $i \in [n]$. Hence, $\mathsf{sw}(\cdot, \hat{f})$ is also a monotone submodular function by the fact that a linear combination of monotone submodular functions is still a monotone submodular function. Recall that Rule A selects each alternative in G^+, with probability at least $1/(1+\log m)\sqrt{m}$. By substituting $g(\cdot)$ with $\mathsf{sw}(\cdot, \hat{f})$ and substituting p with $1/(1+\log m)\sqrt{m}$ in Lemma 2, we have that the returned solution from Rule A has social welfare at least $\mathsf{sw}(G^+, \hat{f})/(1+\log m)\sqrt{m}$, which is no less than $\mathsf{sw}(G^+ \cap S^*, \hat{f})/(1 + \log m)\sqrt{m}$ by the assumption that $\mathsf{sw}(\cdot, \hat{f})$ is a monotone function. □

We next show that the expected social welfare achieved under Rule B is at least $(1-c)\mathsf{sw}(G^- \cap S^*, \hat{f})/2(1+\log m)\sqrt{m}$. While our proof follows the general framework of Theorem 3 in [3], we must make additional efforts to address the challenges posed by non-additive functions, as previously mentioned.

Lemma 3. *Assume the curvature of each function f_i is at least c, the expected social welfare achieved by Rule B is at least* $(1-c)\mathsf{sw}(G^- \cap S^*, \hat{f})/2(1+\log m)\sqrt{m}$.

Proof. Observe that $G_0^- = \emptyset$. Therefore,

$$\mathsf{sw}(G^- \cap S^*, \hat{f}) = \mathsf{sw}(\cup_{t\in[\log m]}(G_t^- \cap S^*), \hat{f}). \tag{1}$$

Fix $t \in [\log m]$ and $a \in G_t^-$. Let H_k be the k-th harmonic number, it is easy to verify that

$$\sum_{b\in G_t} \mathsf{sc}(b, \hat{\rho}) = n \cdot H_{|G_t|} \leq n \cdot H_m. \tag{2}$$

Recall that G_t^+ contains the top $\sqrt{m} \cdot (1/u_t)$ alternatives from G_t that have the largest harmonic scores in $\hat{\rho}$, we have

$$\mathsf{sc}(a, \hat{\rho}) \leq \frac{n \cdot H_m}{\sqrt{m} \cdot (1/u_t)} < \frac{n \cdot (1 + \log m)}{\sqrt{m} \cdot m/2^t}. \tag{3}$$

Meanwhile, for each voter $i \in [n]$ and alternative $b \in G_t$, we have

$$\Delta_i(b) \leq 1/\rho_i(b) \tag{4}$$

where $\Delta_i(b) = f_i(T \cup \{b\}) - f_i(T)$ (here $T = \cup_{a\in G_t:\rho_i(a)\leq\rho_i(b)} a$ denotes b's proceeders in ρ_i) denotes the marginal utility of b over all its proceeders in ρ_i. This is because $f_i(G_t) = \sum_{b\in G_t} \Delta_i(b, G_t) \leq 1$ and $\Delta_i(b, G_t)$ is no larger than the marginal utility brought by any of b's predecessors in ρ_i (this is by the submodularity of f_i and the design of Ranking-by-Marginal-Values).

It follows that for each alternative $a \in G_t^-$,

$$\sum_{i\in[n]} \Delta_i(a, G_t) \leq \sum_{i\in[n]} 1/\rho_i(a) = \mathsf{sc}(a, \hat{\rho}) < \frac{n \cdot (1 + \log m)}{\sqrt{m} \cdot m/2^t} \tag{5}$$

where the first inequality is by inequality (4) and the second inequality is by inequality (3). This, together with the assumption that the curvature of each function f_i is at least c, implies that

$$(1-c)\sum_{i\in[n]} f_i(\{a\}) \leq \sum_{i\in[n]} \Delta_i(a, G_t) < n\cdot(1+\log m)/(\sqrt{m}\cdot m/2^t). \quad (6)$$

Moreover, by the definition of G_t, we have that each alternative $a \in G_t^-$ has a cost of at least $l_t = 2^{t-1}/m$. Therefore, for any $t \in [\log m]$, the "value-to-cost" density of each alternative $a \in G_t^-$ is at most

$$\frac{\mathsf{sw}(a,\hat{f})}{l_t} = \frac{\sum_{i\in[n]} f_i(\{a\})}{2^{t-1}/m} \leq \frac{\frac{n\cdot(1+\log m)}{(1-c)\sqrt{m}\cdot m/2^t}}{2^{t-1}/m} = \frac{2n\cdot(1+\log m)}{(1-c)\sqrt{m}} \quad (7)$$

where the inequality is by inequality (6).

Notice that the above inequality applies to all alternatives in $\cup_{t\in[\log m]}(G_t^- \cap S^*)$, hence, $\sum_{a\in\cup_{t\in[\log m]}(G_t^-\cap S^*)} \mathsf{sw}(a,\hat{f}) \leq \frac{2n\cdot(1+\log m)}{(1-c)\sqrt{m}}$. This is because the "value-to-cost" density of each alternative in G^- is at most $\frac{2n\cdot(1+\log m)}{(1-c)\sqrt{m}}$ and the total cost of $\cup_{t\in[\log m]}(G_t^- \cap S^*)$ is at most one (by the observation that S^* is a budget feasible solution). This, together with inequality (1), implies that

$$\begin{aligned}\mathsf{sw}(G^-\cap S^*,\hat{f}) &= \mathsf{sw}(\cup_{t\in[\log m]}(G_t^-\cap S^*),\hat{f})\\ &\leq \sum_{a\in\cup_{t\in[\log m]}(G_t^-\cap S^*)} \mathsf{sw}(a,\hat{f}) \leq \frac{2n\cdot(1+\log m)}{(1-c)\sqrt{m}}\end{aligned}$$

where the first inequality is by the assumption that $\mathsf{sw}(\cdot,\hat{f})$ is a submodular function.

We are now ready to prove this lemma. Recall that according to Rule B, a single alternative is chosen uniformly at random. Therefore, each alternative has a probability of $1/m$ of being selected. This, together with Lemma 2 and the assumption that $\mathsf{sw}(\cdot,\hat{f})$ is a monotone submodular function, implies that the expected social welfare achieved by Rule B is at least $\mathsf{sw}(A,\hat{f})/m = n/m$. Combining this with inequality (8), we have that the expected social welfare achieved by Rule B is at least $(1-c)\mathsf{sw}(G^-\cap S^*,\hat{f})/2(1+\log m)\sqrt{m}$. □

Now we are in position to present the main theorem of this section.

Theorem 1. *Assume the curvature of each function f_i is at least c, the expected social welfare achieved by our aggregation rule is at least $(1-c)\mathsf{sw}(S^*,\hat{f})/4(1+\log m)\sqrt{m}$. This also implies that the distortion of our approach is at most $O(\log m\sqrt{m}/(1-c))$.*

Proof. Recall that our aggregation rule randomly picks a rule from Rule A and Rule B at uniform, moreover, the expected social welfare of Rule A is at least $\mathsf{sw}(G^+\cap S^*,\hat{f})/(1+\log m)\sqrt{m}$ (by Lemma 1) and the expected social welfare of

Rule B is at least $(1-c)\mathsf{sw}(G^- \cap S^*, \hat{f})/2(1+\log m)\sqrt{m}$ (by Lemma 3). Hence, the expected social welfare of our aggregation rule is at least

$$\frac{\mathsf{sw}(G^+ \cap S^*, \hat{f})/(1+\log m)\sqrt{m} + (1-c)\mathsf{sw}(G^- \cap S^*, \hat{f})/2(1+\log m)\sqrt{m}}{2} \tag{8}$$

$$\geq (1-c)\frac{\mathsf{sw}(G^+ \cap S^*, \hat{f}) + \mathsf{sw}(G^- \cap S^*, \hat{f})}{4(1+\log m)\sqrt{m}} \geq \frac{(1-c)\mathsf{sw}(S^*, \hat{f})}{4(1+\log m)\sqrt{m}} \tag{9}$$

where the second inequality is by the observations that $G^+ \cup G^- = A$ and $\mathsf{sw}(\cdot, \hat{f})$ is a submodular function. □

Remark 1. Note that selecting a rule from Rule A and Rule B uniformly may not be optimal; one can certainly optimize this selection probability to reduce the distortion.

Remark 2. Interestingly, if we require each voter to rank all alternatives instead of only the alternatives from a random group, it might increase the distortion. This can be partially explained by the submodularity of the utility function. For example, ranking all alternatives based on their marginal values makes it harder to disentangle the correlations among alternatives, thereby making it more difficult to assess the "true" value of each individual alternative. This contrasts with our observation in the next method (ranking by stand-alone values), where providing a global ranking is always beneficial.

3.2 Group Based Ranking-by-Values

Recall that this method follows the same procedure as described in the previous method to partition all alternatives into $\log m + 1$ groups: $G_0, G_1, \cdots, G_{|\log m|}$. We choose a group G_t uniformly at random, and asks each voter to rank all alternatives in G_t according to their stand-alone values. That is, in the final ranking ρ_i from voter $i \in [n]$, for any two alternatives $a \in G_t$ and $b \in G_t$, $\rho_i(a) \leq \rho_i(b)$ if and only if $f_i(\{a\}) \geq f_i(\{b\})$.

We next recall some important notations from the previous section. For an input profile $\hat{\rho}$, the *harmonic score* of a in $\hat{\rho}$ is given by $\mathsf{sc}(a, \hat{\rho}) = \sum_{i=1}^{n} 1/\rho_i(a)$. Let $\hat{f}$ denote the *true* utility profile that is consistent with $\hat{\rho}$ and let $S^* = \arg\max_{S \in \mathcal{F}} \mathsf{sw}(S, \hat{f})$ denote the optimal set of alternatives that maximizes the social welfare.

Next, we describe our aggregation rule, which randomly picks a rule from the following two candidates. This, again, aligns with the one introduced in [3].

- Rule A: Let G_t^+ denotes the top (up to) $\sqrt{m} \cdot (1/u_t)$ alternatives from G_t that have the largest harmonic scores in $\hat{\rho}$. Select a subset of G_t^+ with a size of $1/u_t$, also chosen uniformly at random.
- Rule B: Choose a single alternative uniformly at random.

We now present the main theorem, with its proof detailed in our technical report [16].

Theorem 2. *Assume the curvature of each function f_i is at least c, the expected social welfare achieved by our aggregation rule is at least $(1-c)\mathsf{sw}(S^*, \hat{f})/4(1+\log m)\sqrt{m}$. This also implies that the distortion of our approach is at most $O(\log m\sqrt{m}/(1-c))$.*

Remark. While we limit each voter to rank alternatives from a random group, the traditional method, which requires each voter to rank *all* alternatives, achieves at least the same distortion. This is because ranking all alternatives offers more information than ranking only a subset. However, this statement does not apply to ranking-by-marginal-values method.

3.3 Threshold Approval Votes

Recall that the *threshold approval votes* method asks each voter to report all alternatives whose stand-alone utility is above a certain threshold α. Formally, the approval set for voter i is given by $\rho_i = \{a \in A \mid f_i(\{a\}) \geq \alpha\}$.

For $t \in [\log m]$ (for ease of presentation, assume m is a power of 2), define $l_t = 2^{t-1}/m$ and $u_t = 2^t/m$. Let $I_0 = [0, 1/m]$ and $I_t = (l_t, u_t]$ for $t \in [\log m]$.

Let $\hat{f}$ denote the true utility profile that is consistent with $\hat{\rho}$ and let $S^* = \arg\max_{S\in\mathcal{F}} \mathsf{sw}(S, \hat{f})$ denote the optimal set of alternatives that maximizes the social welfare. For $a \in A$ and $t \in \{0, 1, \cdots, \log m\}$, define $d_a^t = |\{i \in [n] \mid f_i(\{a\}) \in I_t\}|$ as the number of voters whose utility of a belongs to the interval I_t. We draw inspiration from [3] and subsequently present two bounds concerning the social welfare of each alternative $a \in A$: $\sum_{t=0}^{\log m} d_a^t \cdot l_t \leq \mathsf{sw}(a, \hat{f}) \leq \sum_{t=0}^{\log m} d_a^t \cdot u_t$.

Moreover, because $I_0 = [0, 1/m]$, the above bound indicates that

$$\sum_{t=0}^{\log m} d_a^t \cdot l_t \leq \mathsf{sw}(a, \hat{f}) \leq \frac{d_a^0}{m} + \sum_{t=1}^{\log m} d_a^t \cdot u_t. \tag{10}$$

Now we are ready to introduce our aggregation rule. Our rule consists of two candidate rules:

- Rule A: The first rule is to compute a set of alternatives through solving a knapsack problem. First, randomly pick a threshold α from $\{l_1, l_2, \cdots, l_{\log m}\}$ at uniform. Then assign a weight w_a to each alternative $a \in A$ such that $w_a = |\{i \in [n] \mid f_i(\{a\}) \geq \alpha\}|$ represents the number of voters who approve alternative a with respect to the threshold α. Then solve the following knapsack problem, which allows a Fully Polynomial Time Approximation Scheme (FPTAS) [15]:

  P.1 $\max_{S\subseteq A} \sum_{a\in S} w_a$ subject to $c(S) \leq 1$.

  This near-optimal solution is provided as the output from Rule A.
- Rule B: Choose a single alternative from A randomly at uniform.

Recall that $\mathsf{sw}(a, \hat{f}) \leq \frac{d_a^0}{m} + \sum_{t=1}^{\log m} d_a^t \cdot u_t$ (by the upper bound in (10)), we have

$$\mathsf{sw}(S^*, \hat{f}) \leq \sum_{a \in S^*} \mathsf{sw}(a, \hat{f}) \leq \sum_{a \in S^*} (\frac{d_a^0}{m} + \sum_{t=1}^{\log m} d_a^t \cdot u_t) \leq \frac{n}{m} + \sum_{t=1}^{\log m} \sum_{a \in S^*} d_a^t \cdot u_t \tag{11}$$

where the first inequality is by the assumption that $\mathsf{sw}(\cdot, \hat{f})$ is a submodular function, and the last inequality is by the observation that $\sum_{a \in S^*} d_a^0 \leq n$. In the following two lemmas, we will prove that the expected social welfare of Rule A is close to $\sum_{t=1}^{\log m} \sum_{a \in S^*} d_a^t \cdot u_t$ and the expected social welfare of Rule B is close to n/m, respectively. Consequently, randomly selecting a rule from the above two candidates yields a bounded distortion.

Lemma 4. *Assume the curvature of each function f_i is at least c, the expected social welfare of Rule A is at least $\frac{(1-\epsilon)(1-c)}{2 \log m} \sum_{t \in [\log m]} \sum_{a \in S^*} d_a^t \cdot u_t$. Here $\epsilon \in [0, 1]$ is a number that is arbitrarily close to zero.*

Proof. Recall that in Rule A, we solve the following problem (approximately), here $w_a = |\{i \in [n] \mid f_i(\{a\}) \geq \alpha\}|$ and α is a random number from $\{l_1, l_2, \cdots, l_{\log m}\}$:

P.1 $\max_{S \subseteq A} \sum_{a \in S} w_a$ subject to $c(S) \leq 1$.

Fix $\alpha = l_t$, we have $w_a = |\{i \in [n] \mid f_i(\{a\}) \geq l_t\}|$. Now consider a different knapsack problem.

P.2 $\max_{S \subseteq A} \sum_{a \in S} d_a^t$ subject to $c(S) \leq 1$.

Let $O_2(\alpha)$ denote an optimal solution of **P.2**. By the observation that S^* is a feasible solution of **P.2** and $O_2(\alpha)$ denotes an optimal solution of **P.2**, we have $\sum_{a \in S^*} d_a^t \leq \sum_{a \in O_2(\alpha)} d_a^t$. This indicates that

$$\sum_{a \in S^*} d_a^t \cdot u_t \leq \sum_{a \in O_2(\alpha)} d_a^t \cdot u_t. \tag{12}$$

Because $d_a^t = |\{i \in [n] \mid f_i(\{a\}) \in (l_t, u_t]\}|$ and $w_a = |\{i \in [n] \mid f_i(\{a\}) \geq l_t\}|$, we have $w_a \geq d_a^t$ for all $t \in A$. Therefore, the value of the optimal solution of **P.1** is no less than the value of the optimal solution of **P.2**. Let $O_1(\alpha)$ denote the optimal solution of **P.1**, we have

$$\sum_{a \in O_1(\alpha)} w_a \geq \sum_{a \in O_2(\alpha)} d_a^t. \tag{13}$$

Given that **P.1** admits a FPTAS, let $V(\alpha)$ denote the solution returned from Rule A, which is a $(1-\epsilon)$-approximation solution of **P.1**. Here $\epsilon \in [0, 1]$ is a number that is arbitrarily close to zero. We have $\sum_{a \in V(\alpha)} w_a \geq (1-\epsilon) \cdot$

$\sum_{a\in O_1(\alpha)} w_a \geq (1-\epsilon)\cdot \sum_{a\in O_2(\alpha)} d_a^t$ where the second inequality follows from inequality (13). This, together with the fact that $l_t \geq u_t/2$ for all $t \in [\log m]$, implies that

$$\sum_{a\in V(\alpha)} w_a \cdot l_t \geq (1-\epsilon)\cdot \sum_{a\in O_2(\alpha)} d_a^t \cdot \frac{u_t}{2}. \tag{14}$$

Moreover, it is easy to verify that $\sum_{i\in[n]} f_i(\{a\}) \geq w_a \cdot l_t$ for all $a \in A$. Hence,

$$\sum_{a\in V(\alpha)} \sum_{i\in[n]} f_i(\{a\}) \geq \sum_{a\in V(\alpha)} w_a \cdot l_t \geq (1-\epsilon)\cdot \sum_{a\in O_2(\alpha)} d_a^t \cdot \frac{u_t}{2} \geq (1-\epsilon)\cdot \sum_{a\in S^*} d_a^t \cdot \frac{u_t}{2} \tag{15}$$

where the second inequality is by inequality (14) and the third inequality is by inequality (12).

Recall that our threshold α is picked randomly from $\{l_1, l_2, \cdots, l_{\log m}\}$ at uniform, we have

$$\mathbb{E}_\alpha[\sum_{a\in V(\alpha)} \sum_{i\in[n]} f_i(\{a\})] = \frac{1}{\log m} \sum_{\alpha\in\{l_1,l_2,\cdots,l_{\log m}\}} \sum_{a\in V(\alpha)} \sum_{i\in[n]} f_i(\{a\})$$
$$\geq \frac{1}{\log m} \sum_{t\in[\log m]} (1-\epsilon)\cdot \sum_{a\in S^*} d_a^t \cdot \frac{u_t}{2} = \frac{1-\epsilon}{2\log m} \sum_{t\in[\log m]} \sum_{a\in S^*} d_a^t \cdot u_t \tag{16}$$

where the inequality is by inequality (15).

Now we are ready to prove this lemma, that is, the expected social welfare of Rule A (i.e., $\mathbb{E}_\alpha[\mathsf{sw}(V(\alpha), \hat{f})]$) is at least $\frac{(1-\epsilon)(1-c)}{2\log m} \sum_{t\in[\log m]} \sum_{a\in S^*} d_a^t \cdot u_t$. Observe that

$$\mathbb{E}_\alpha[\mathsf{sw}(V(\alpha), \hat{f})] = \mathbb{E}_\alpha[\sum_{i\in[n]} f_i(V(\alpha))]$$
$$\geq (1-c)\cdot \mathbb{E}_\alpha[\sum_{a\in V(\alpha)} \sum_{i\in[n]} f_i(\{a\})] \geq \frac{(1-\epsilon)(1-c)}{2\log m} \sum_{t\in[\log m]} \sum_{a\in S^*} d_a^t \cdot u_t$$

where the first inequality is by the assumption that the curvature of each function f_i is at least c, and the second inequality is by inequality (16). □

We next analyze the social welfare achieved by Rule B.

Lemma 5. *The expected social welfare of Rule B is at least n/m.*

Proof. Recall that Rule B chooses a single alternative from A randomly at uniform. Therefore, each alternative has a probability of $1/m$ of being selected. This, together with Lemma 2 and the assumption that $\mathsf{sw}(\cdot, \hat{f})$ is a monotone submodular function, implies that the expected social welfare achieved by Rule B is at least $\mathsf{sw}(A, \hat{f})/m = n/m$. □

Lemmas 4 and 5 imply that randomly choosing between Rule A and Rule B yields an expected social welfare of at least $\frac{1}{2} \cdot (\frac{(1-\epsilon)(1-c)}{2\log m} \sum_{t\in[\log m]} \sum_{a\in S^*} d_a^t \cdot u_t + \frac{n}{m})$.

Observe that

$$\frac{1}{2} \cdot (\frac{(1-\epsilon)(1-c)}{2\log m} \sum_{t\in[\log m]} \sum_{a\in S^*} d_a^t \cdot u_t + \frac{n}{m})$$
$$\geq \frac{(1-\epsilon)(1-c)}{4\log m} \cdot (\sum_{t\in[\log m]} \sum_{a\in S^*} d_a^t \cdot u_t + \frac{n}{m}) \geq \frac{(1-\epsilon)(1-c)}{4\log m} \cdot \mathsf{sw}(S^*, \hat{f})$$

where the second inequality is by inequality (11). This leads to the following main theorem.

Theorem 3. *Assume the curvature of each function f_i is at least c, the expected social welfare achieved by our aggregation rule is at least $(1-\epsilon)(1-c)\mathsf{sw}(S^*, \hat{f})/4\log m$. Here $\epsilon \in [0,1]$ is a number that is arbitrarily close to zero. This also implies that the distortion of our approach is at most $O(\log m/(1-c))$.*

Remark. When $c = 0$ (assuming additive utility functions), our distortion bound improves to $O(\log m)$, surpassing the previous best of $O(\log^2 m)$ [3]. Our method avoids the need to partition alternatives by cost, contributing to this improvement. On the hardness side, [3] showed that the distortion for the threshold approval voting method is at least $\Omega(\log m/\log\log m)$.

References

1. Aziz, H., Lee, B., Talmon, N.: Proportionally representative participatory budgeting: axioms and algorithms. arXiv preprint arXiv:1711.08226 (2017)
2. Balkanski, E., Rubinstein, A., Singer, Y.: The power of optimization from samples. In: Advances in Neural Information Processing Systems, vol. 29 (2016)
3. Benade, G., Nath, S., Procaccia, A.D., Shah, N.: Preference elicitation for participatory budgeting. Manage. Sci. **67**(5), 2813–2827 (2021)
4. Boutilier, C., Caragiannis, I., Haber, S., Lu, T., Procaccia, A.D., Sheffet, O.: Optimal social choice functions: a utilitarian view. In: Proceedings of the 13th ACM Conference on Electronic Commerce, pp. 197–214 (2012)
5. Cabannes, Y.: Participatory budgeting: a significant contribution to participatory democracy. Environ. Urban. **16**(1), 27–46 (2004)
6. Caragiannis, I., Nath, S., Procaccia, A.D., Shah, N.: Subset selection via implicit utilitarian voting. J. Artif. Intell. Res. **58**, 123–152 (2017)
7. Durand, M., Pascual, F.: Detecting and taking project interactions into account in participatory budgeting. arXiv preprint arXiv:2403.19194 (2024)
8. Feige, U., Mirrokni, V.S., Vondrák, J.: Maximizing non-monotone submodular functions. SIAM J. Comput. **40**(4), 1133–1153 (2011)
9. Goel, A., Krishnaswamy, A.K., Sakshuwong, S., Aitamurto, T.: Knapsack voting for participatory budgeting. ACM Trans. Econ. Comput. (TEAC) **7**(2), 1–27 (2019)
10. Jain, P., Sornat, K., Talmon, N.: Participatory budgeting with project interactions. In: IJCAI, pp. 386–392 (2020)

11. Lu, T., Boutilier, C.: Budgeted social choice: from consensus to personalized decision making. In: IJCAI, vol. 11, pp. 280–286 (2011)
12. Mandal, D., Shah, N., Woodruff, D.P.: Optimal communication-distortion tradeoff in voting. In: Proceedings of the 21st ACM Conference on Economics and Computation, pp. 795–813 (2020)
13. Procaccia, A.D., Rosenschein, J.S.: The distortion of cardinal preferences in voting. In: Klusch, M., Rovatsos, M., Payne, T.R. (eds.) CIA 2006. LNCS (LNAI), vol. 4149, pp. 317–331. Springer, Heidelberg (2006). https://doi.org/10.1007/11839354_23
14. Talmon, N., Faliszewski, P.: A framework for approval-based budgeting methods. Proc. AAAI Conf. Artif. Intell. **33**, 2181–2188 (2019)
15. Vazirani, V.V.: Approximation Algorithms, vol. 1. Springer, Heidelberg (2001). https://doi.org/10.1007/978-3-540-71844-4_16
16. Yuan, J., Tang, S.: Submodular participatory budgeting. arXiv preprint arXiv:2406.13586 (2024)

Approximation Algorithms for k-Submodular Maximization Under the Fair Constraints and Size Constraints

Weijia Hu and Bin Liu(✉)

School of Mathematical Sciences, Ocean University of China, Qingdao 266100, People's Republic of China
binliu@ouc.edu.cn

Abstract. Submodular function maximization, a crucial aspect of combinatorial optimization, is extensively applied in various fields like economics and computing. Recently, the focus has shifted towards the generalizations of submodular maximization problems, such as k-submodular function maximization. This approach not only considers the selection of elements from the ground set but also the specific k-set they belong to. It has been proved useful in modeling several machine learning problems, including influence maximization with k topics and sensor placement with k types of sensors. In real-world applications, elements possess certain attributes like gender, age, race, etc., which are used to group elements in the ground set. To maintain fairness, it is desirable to have a balanced number of elements in each group, introducing the concept of group fairness. We divide the ground set V into m groups $(V_1, V_2, ..., V_m)$, each representing items with the same attribute. If the difference in the number of elements between any two groups is less than or equal to α, we say the group fairness constraints are satisfied. Existing k-submodular algorithms do not account for the group fairness constraints, leading to imbalances in group sizes. This paper addresses this gap by studying k-submodular maximization under the fairness constraints. We explore the k-submodular maximization problem under the intersection of the fairness constraints and total size constraint, achieving an approximate ratio of $1/2$ for monotone cases and $1/24$ for non-monotone cases.

Keywords: k-Submodular maximization · Fairness constraints · Approximation algorithm

1 Introduction

Submodular function optimization is an important part of combinatorial optimization. If a set function has the property of diminishing marginal returns, then the function is submodular. Based on this typical property, many practical discrete optimization problems in computing, operations research, and machine learning can be reduced to submodular maximization problems under certain

S. Ghosh and Z. Zhang (Eds.): AAIM 2024, LNCS 15179, pp. 235–246, 2024.
https://doi.org/10.1007/978-981-97-7798-3_20

constraints, such as sample-based clustering [4], document summarization [10], social network analysis [9], etc. The diminishing marginal returns property of submodular functions is described mathematically as follows. A set function $f : 2^N \rightarrow \mathbb{R}_{\geq 0}$ is a submodular function if and only if the set function f satisfies $f_T(e) \leq f_S(e)$ for any subset $S \subseteq T \subseteq N$ and arbitrary element $e \in N \backslash T$, where $f_T(e) = f(T \cup \{e\}) - f(T)$ denotes the marginal benefit from adding e to T.

It has been proved that the monotone submodular function maximization problem is NP-hard [12] which requires the design of an approximation algorithm to obtain the best possible solution. The greedy algorithm can achieve an approximate guarantee of $(1 - 1/e)$ for the submodular maximization problem under cardinality constraint [12], and $(1 - 1/e)$ is the optimal approximate guarantee that the approximation algorithm can achieve in polynomial time [11]. In recent years, many generalized forms of submodular maximization problems have appeared, such as the k-submodular function maximization problem, which has been practically applied in the fields of sensor placement and influence maximization [14,15].

Unlike submodular maximization, which involves selecting a single subset to maximize the objective function, the k-submodular function maximization problem aims to select k disjoint subsets from a family of subsets to achieve this goal. It is known that a k-submodular function reduces to a submodular function when $k = 1$ [7]. However, for $k \geq 2$, the methods used for submodular function maximization cannot be directly applied to the k-submodular function maximization problem. Therefore, it becomes necessary to study approximate algorithms for the k-submodular maximization problem.

What's more, in practical problems, the individuals to be selected often have certain attributes, such as gender, race, age, or other attributes, so that all individuals can be grouped according to a certain attribute. When selecting individuals, it is crucial to prioritize certain attributes to avoid imbalances in attribute proportions in the final results. For instance, based on gender, the "unfairness" phenomenon is that there will be too many or too few female individuals. Recently, with the increasing demand for fairness, the concept of fairness constraint has been gradually applied in various fields, such as the voting system [2], whose goal is to ensure that the election result is a fair preference of voters. Similarly, in data summary [3], there is growing interest in developing fairness algorithms. What's more, group fairness can be conceptualized in various ways, but typically falls into two primary categories by Bronfenbrenner [1]: equity-fairness and equality-fairness. Equity-fairness emphasizes that a specific range limit $[l_i, u_i]$ is given to the number of elements selected in each group V_i. Equality-fairness emphasizes that the difference between the elements selected in each group V_i will not be too large. In this paper, the fairness constraints mentioned all refer to the equality-fairness constraints.

As discussed above, the problem of maximization of submodular functions under fairness constraints and k-submodular functions under various constraints has been extensively studied, but there has been no research on maximizing k-submodular functions under fairness constraints. In this paper, we consider the

k-submodular maximization problem under the intersection constraints of group fairness constraints and total size constraint, We get an approximate ratio of 1/2 for the case of monotone and an approximate ratio of 1/24 for the case of non-monotone.

1.1 Related Work

k-Submodular. The study of k-submodularity was initially proposed by Singh et al. [17], focusing exclusively on the case where $k = 2$, known as bipartite submodularity. Also, since submodular function maximization is NP-hard, it is not hard to find that k-submodular function maximization is also NP-hard. Ward and Zivny [22] introduced a deterministic greedy algorithm to tackle the unconstrained k-submodular function maximization problem. They guaranteed an approximation ratio of 1/3 in the non-monotone case and improved it to 1/2 in the monotone case. Next, Ohsaka and Yoshida [14] explored the monotone k-submodular maximization problem under cardinality constraints, achieving approximation ratios of 1/2 under global constraints and 1/3 under each individual position constraint using greedy methods. For non-monotone k-submodular maximization under knapsack constraints, Xiao et al. [24] achieved an approximation ratio of 0.317 using an alternative greedy algorithm. With the addition of monotonicity constraints on k-submodular functions, the approximation ratio can be improved to 0.432. Under the matroid constraints, Niu et al. [13] used the threshold decline algorithm to give the 1/2 approximation ratio for the monotone k-submodular maximization problem and the 1/3 approximation ratio for the non-monotone k-submodular maximization problem. Iwata et al. [8] proved the inapproximability of maximizing monotone k-submodular, and 1/2 is already asymptotically compact approximation ratio.

Fairness. Celis et al. [2] used the continuous greedy method to solve the problem of maximizing the monotone submodular function under the equity-fairness constraints, and obtained the approximate ratio of $(1-1/e)$. Tang et al. [20] used the method of continuous greedy to solve the problem of maximizing monotone submodular functions under the equity-fairness constraints with the cardinality constraints, and obtained the approximate ratio of $(1 - 1/e)$. Wang et al. [21] studied the submodular maximization problem under the equity-fairness constraints and the cardinality constraints without considering lower bounds. They gave the offline greedy algorithm with 1/2 approximate ratio guarantee and the flow algorithm with $(1/2 - \epsilon)$ approximate ratio guarantee. Halabi et al. [6] introduced the concept of expandable set and gave an approximate ratio of 1/2 for the monotone submodular maximization problem that satisfies both the equity-fairness constraints and the cardinality constraints. Then, Halabi et al. [5] gave an approximate ratio of 1/11.656 for the monotone submodular maximization problem that satisfies both the equity-fairness and the matroid constraints. For maximizing the submodular function under the equality-fairness constraints, Tang [19] obtain a $(1 - 1/e)$ approximate ratio for the case when f is monotone and a 0.045 approximate ratio for the case when f is non-monotone.

1.2 Organization

The rest of this paper is organized as follows. In Sect. 2, we review properties of k-submodular functions and specify the problem we want to study. Sections 3 is devoted to maximizing monotone k-submodular functions and obtaining 1/2 approximation. In Sect. 4, we consider maximizing non-monotone k-submodular functions and obtain a 1/24 approximation for the case f is non-monotone. Then we conclude our results and talk about the future work in Sect. 5.

2 Preliminaries and Problem Statement

Given an integer k and a finite set V with its partition $V_1, ..., V_m$ satisfied $V = \bigcup_{r\in[m]} V_r$ and $V_i \cap V_j = \emptyset$ for any $i, j \in [m]$. For any $\mathbf{x} = (X_1, X_2, ..., X_k), \mathbf{y} = (Y_1, Y_2, ..., Y_k) \in (k+1)^V$, we define $supp(\mathbf{x}) = \bigcup_{j\in[k]} X_j$. If $e \in X_j$, we write $\mathbf{x}(e) = j$; if $e \notin supp(\mathbf{x})$, we write $\mathbf{x}(e) = 0$. Adding $e \notin supp(\mathbf{x})$ to X_j can be represented by $\mathbf{x} \sqcup (e, j)$, adding $X \cap supp(\mathbf{x}) = \emptyset$ to X_j can be represented by $\mathbf{x} \sqcup (X, j)$. We denote by $\mathbf{x} \sqsubseteq \mathbf{y}$ if and only if $X_j \subseteq Y_j$ for all $j \in [k]$. For $e \in V, j \in [k]$, $\triangle f((e, j)|\mathbf{x}) = f(\mathbf{x} \sqcup (e, j)) - f(\mathbf{x})$.

Definition 1. *(k-**submodular**) A function $f : (k+1)^V \mapsto \mathbb{R}_{\geq 0}$ is k-**submodular** if for any $\mathbf{x} = (X_1, X_2, ..., X_k)$ and $\mathbf{y} = (Y_1, Y_2, ..., Y_k)$, we have*

$$f(\mathbf{x}) + f(\mathbf{y}) \geq f(\mathbf{x} \sqcap \mathbf{y}) + f(\mathbf{x} \sqcup \mathbf{y}),$$

where $\mathbf{x}\sqcap\mathbf{y} = (X_1\cap Y_1, ..., X_k\cap Y_k)$ and $\mathbf{x}\sqcup\mathbf{y} = (X_1\cup Y_1 \setminus (\bigcup_{i\neq 1} X_i\cup Y_i), ..., X_k\cup Y_k \setminus (\bigcup_{i\neq k} X_i \cup Y_i))$.

A function $f : (k+1)^V \to \mathbb{R}_{\geq 0}$ is *monotone* if $f(\mathbf{x}) \leq f(\mathbf{y})$, for any $\mathbf{x} = (X_1, ..., X_k)$ and $\mathbf{y} = (Y_1, ..., Y_k)$ with $X_i \subseteq Y_i \subseteq V$ for each $i \in \{1, ..., k\}$, namely $\mathbf{x} \sqsubseteq \mathbf{y}$. A function $f : (k+1)^V \to \mathbb{R}_{\geq 0}$ is *normalized* if $f(\{\emptyset, \emptyset, ..., \emptyset\}) = 0$. In this paper, we assume that f is always normalized.

Proposition 1. ***(Properties of** k**–submodular function)***

- *A function $f : (k+1)^V \to \mathbb{R}_{\geq 0}$ has **orthant submodularity** iff*

$$\triangle f((e, j)|\mathbf{x}) \geq \triangle f((e, j)|\mathbf{y}),$$

  for any $\mathbf{x}, \mathbf{y} \in (k+1)^V$ satisfied $\mathbf{x} \sqsubseteq \mathbf{y}, e \notin supp(\mathbf{y})$ and $j \in [k]$.
- *When $k \geq 2$, a function $f : (k+1)^V \to \mathbb{R}_{\geq 0}$ has **pairwise monotonicity** iff*

$$\triangle f((e, i)|\mathbf{x}) + \triangle f((e, j)|\mathbf{x}) \geq 0,$$

  for any $\mathbf{x} \in (k+1)^V, e \notin supp(\mathbf{x})$ and $i, j \in [k]$ with $i \neq j$.
- *When $k \geq 2$, the function $f : (k+1)^V \to \mathbb{R}_{\geq 0}$ is k-submodular iff f is orthant submodular and pairwise monotone [23].*

Definition 2. ***(Fairness constraints)*** *Given a ground set* $V = \{1, 2, ..., n\}$ *and its partition* $V_1, ..., V_m$ *satisfy* $V = V_1 \cup ... \cup V_m$, $V_i \cap V_j = \emptyset$ *for any* i *and* j. *We say* $\mathbf{s} \in \{0, ..., k\}^n$ *satisfying the fairness constraints* $\alpha \in [0, n]$ *if*

$$|supp(\mathbf{s}) \cap V_i| - |supp(\mathbf{s}) \cap V_j| \leq \alpha, \forall i, j \in [m].$$

Based on the above definition of the fairness constraints, the purpose of this paper is to find an approximate solution of k-submodular maximization problem under the total size constraint and the fairness constraints, formally stated as follows.

FTSk-SM Problem: Given a ground set V and a k-submodular function $f : \{0, ..., k\}^n \rightarrow \mathbb{R}_{\geq 0}$ together with a value oracle $\mathcal{O}_f$, the k-submodualr maximization problem under the fairness constraints $\alpha \in [0, n]$ and total size constraint can be formulated mathematically as follows.

P1

$$\max_{\mathbf{s} \in \{0,...,k\}^n} \quad f(\mathbf{s})$$
$$s.t. \quad |supp(\mathbf{s}) \cap V| \leq b,$$
$$|supp(\mathbf{s}) \cap V_i| - |supp(\mathbf{s}) \cap V_j| \leq \alpha, \forall i, j \in [m].$$

For the sake of further discussion, we will refer to the above problem as **P1**. And in the following section, we will discuss two cases: one where f is monotone, and the other where f is non-monotone.

3 Maximizing Monotone k-Submodular Functions

In this section, we assume that the k-submodular function f is monotone. Then we consider the **FTSk-SM Problem**, namely **P1** when f is monotone. In order to solve **P1**, similar with [19], our main idea is that transforming **P1** into k-submodular maximization problem under a matroid constraint step by step. And there are rich theoretical results for the k-submodular maximization problem under the matroid constraints [15,16,18], then we can invoke the corresponding algorithm to obtain an approximate solution. The specific process is as follows.

Let $\mathbf{o}$ be an optimal solution for **P1**, $z = \min_{i \in [m]} |supp(\mathbf{o}) \cap V_i|$. Note although we do not know $\mathbf{o}$, we can get the value of z by enumeration (no more than b) then for each z discussed and choose the best one. Thus in this section assumes that z is known and then we can reduce **P1** to **P1.1**.

P1.1

$$\max_{\mathbf{s} \in \{0,...,k\}^n} \quad f(\mathbf{s})$$
$$s.t. \quad |supp(\mathbf{s}) \cap V| \leq b,$$
$$z \leq |supp(\mathbf{s}) \cap V_i| \leq z + \alpha, i \in [m].$$

Let $\mathbf{o}^{P11}$ be the optimal solution for **P1.1**. It is not difficult to find that $\mathbf{o}$ is also a feasible solution for **P1.1**, then we have $f(\mathbf{o}^{P11}) \geq f(\mathbf{o})$. Moreover, we can verify that every feasible solution of **P1.1** must be feasible to **P1**. Therefore, solving **P1.1** suffices to solve **P1**.

In order to solve **P1.1**, we introduce **P1.2** as follows, which relaxes the constraints of **P1.1**. EI Halabi et al. [6] proved the constraints of **P1.2** is a matroid constraint.

P1.2

$$\begin{aligned} \max_{\mathbf{s}\in\{0,\ldots,k\}^n} \quad & f(\mathbf{s}) \\ s.t. \quad & \textstyle\sum_{i\in[m]} \max\{z, |supp(\mathbf{s}) \cap V_i|\} \leq b, \\ & |supp(\mathbf{s}) \cap V_i| \leq z + \alpha, i \in [m]. \end{aligned}$$

The following lemma discusses the relationships between the optimal solutions of **P1.1** and **P1.2**.

Lemma 1. *Let $\mathbf{o}^{P11}$ and $\mathbf{o}^{P12}$ be the optimal solution to **P1.1** and **P1.2**, then we have $f(\mathbf{o}^{P12}) \geq f(\mathbf{o}^{P11})$.*

The proof of Lemma 1 is trivial, because **P1.2** is a relaxation of **P1.1**. From Lemma 1 we know $f(\mathbf{o}^{P12}) \geq f(\mathbf{o}^{P11})$. Note **P1.2** is a monotone k-submodular maximization problem under the matroid constraints, and there exists a 1/2 approximation algorithm for this problem [16], denoted $\mathcal{A}_1$. Let $\mathbf{s}^{P12}$ be the output obtained by the algorithm $\mathcal{A}_1$ for **P1.2**, then we have $f(\mathbf{s}^{P12}) \geq \frac{1}{2}f(\mathbf{o}^{P12}) \geq \frac{1}{2}f(\mathbf{o}^{P11}) \geq \frac{1}{2}f(\mathbf{o})$. But note that $\mathbf{s}^{P12}$ is not necessarily a feasible solution to **P1**, and we will discuss this in more detail.

When $\mathbf{s}^{P12}$ is a feasible solution for **P1.1**, then let $\mathbf{s}^{final} = \mathbf{s}^{P12}$ be the final solution. When $\mathbf{s}^{P12}$ is not a feasible solution for **P1.1**, it means that there exists some $i \in [m]$ such that $|supp(\mathbf{s}^{P12}) \cap V_i| < z$. For these i, we can select $z - |supp(\mathbf{s}^{P12}) \cap V_i|$ elements of V_i and add to any set of k-sets to make $|supp(\mathbf{s}^{P12}) \cap V_i| \geq z$ holds. After these operations, we let the new solution be $\mathbf{s}^{final}$, it's easy to verify that $\mathbf{s}^{final}$ is a feasible solution of **P1.1**. And because of the monotonicity of f, we know $f(\mathbf{s}^{final}) \geq f(\mathbf{s}^{P12}) \geq \frac{1}{2}f(\mathbf{o})$. Combined with the above process of constructing $\mathbf{s}^{final}$, the following Theorem 1 has been proved and the proof will not be repeated here.

Theorem 1. *Let $\mathbf{o}$ be the optimal solution of **P1**, $\mathbf{s}^{final}$ be a solution obtained from the above discussion, then $\mathbf{s}^{final}$ is feasible solution of **P1** and $f(\mathbf{s}^{final}) \geq \frac{1}{2}f(\mathbf{o})$.*

4 Maximizing Non-monotone k-Submodular Functions

In this section, we discuss the case that f is non-monotone. To solve **P1** when f is non-monotone, similar to the case where f is monotone, we transform **P1** into a k-submodular maximization problem under the matroid constraint. Firstly we introduce the Algorithm 1 to construct $\mathbf{u}$, which is useful in Lemma 2. With Lemma 2, we can achieve the above transformation.

Note that the input of Algorithm 1 is $\mathbf{o}$, but we do not know $\mathbf{o}$ in advance. In fact, Algorithm 1 is used solely to establish quantitative relationships between the constructed $\mathbf{u}$ and the optimal solution $\mathbf{o}$, facilitating the transformation of the original problem **P1**. We do not need to know the elements of $\mathbf{u}$ when solving **P1**, so we can use $\mathbf{o}$ as the input of Algorithm 1.

Algorithm 1 CONSTRUCT $(\mathbf{o}, \mathbf{u})$

Input: $\mathbf{o} \in (k+1)^V$.
Output: $\mathbf{u} \in (k+1)^V$.
Let $\mathbf{u} = \{\emptyset, \emptyset, ..., \emptyset\}$;
while $\exists i \in [m], s.t. |supp(\mathbf{u}) \cap V_i| < \left\lfloor \frac{|supp(\mathbf{o}) \cap V_i|}{2} \right\rfloor$ **do**
  $(e^*, \mathbf{o}_{e^*}) \leftarrow arg\max_{(e,\mathbf{o}_e) \in \mathbf{o} \setminus \mathbf{u}} f((e, \mathbf{o}_e)|\mathbf{u})$;
  Assume $e^* \in V_i$
  if $|supp(\mathbf{u}) \cap V_i| < \left\lfloor \frac{|supp(\mathbf{o}) \cap V_i|}{2} \right\rfloor$ **then**
    $\mathbf{u} \leftarrow \mathbf{u} \sqcup (e^*, \mathbf{o}_{e^*})$
  else
    $\mathbf{o} \leftarrow \mathbf{o} \setminus (e^*, \mathbf{o}_{e^*})$
  end if
end while
return $\mathbf{u}$

For $\mathbf{u}$ constructed by Algorithm 1, we have the following properties. Lemma 2 describes the relationships between $\mathbf{u}$ and $\mathbf{o}$.

Lemma 2. *Assume that* $\mathbf{u} \in (k+1)^V$ *is returned by Algorithm 1, then* $\mathbf{u}$ *have the following properties.* (1) *For each* $i \in [m]$, $|supp(\mathbf{u}) \cap V_i| = \left\lfloor \frac{|supp(\mathbf{o}) \cap V_i|}{2} \right\rfloor$. (2) $f(\mathbf{u}) \geq \frac{\lambda}{1+\lambda} f(\mathbf{o})$, *where* $\lambda = \min_{i \in [m]} \frac{|supp(\mathbf{u}) \cap V_i|}{|supp(\mathbf{o}) \cap V_i|}$. (3) *Let* $z' = \min_{i \in [m]} |supp(\mathbf{u}) \cap V_i|$, *then for each* $i \in [m]$ *satisfies* $z' \leq |supp(\mathbf{u}) \cap V_i| \leq z' + \alpha$, *and* $|supp(\mathbf{u})| \leq b$.

Combined with Lemma 2, when f is non-monotone, we can also reduce **P1** to get the **P1.3**. Note $z' = \min_{i \in [m]} |supp(\mathbf{u}) \cap V_i| \leq \left\lfloor \frac{|supp(\mathbf{o}) \cap V_i|}{2} \right\rfloor$, similarly, we can get the value of z' by enumeration (no more than $\frac{b}{2}$), then for each z' discussed and choose the best one.

P1.3

$$
\begin{aligned}
\max_{\mathbf{s} \in \{0,\ldots,k\}^n} \quad & f(\mathbf{s}) \\
s.t. \quad & |supp(\mathbf{s}) \cap V| \leq b, \\
& z' \leq |supp(\mathbf{s}) \cap V_i| \leq z' + \alpha, i \in [m].
\end{aligned}
$$

What's more, from Lemma 2 we know that $\mathbf{u}$ is also a feasible solution of **P1.3** and hence $f(\mathbf{o}^{P13}) \geq f(\mathbf{u})$. The following Lemma 3 introduces the relationships between **P1** and **P1.3**.

Lemma 3. *Every feasible solution to **P1.3** must be feasible to **P1**. Let* $\mathbf{o}^{P13}$ *be the optimal solution to **P1.3**, we have that* $f(\mathbf{o}^{P13}) \geq \frac{\lambda}{1+\lambda} f(\mathbf{o})$, *where* $\lambda = \min_{i \in [m]} \frac{|supp(\mathbf{u}) \cap V_i|}{|supp(\mathbf{o}) \cap V_i|}$.

Proof. The discussion of feasibility is trivial, and is not discussed here. According to the construction of $\mathbf{u}$ and Lemma 2, we know $f(\mathbf{u}) \geq \frac{\lambda}{1+\lambda} f(\mathbf{o})$. Moreover, $\mathbf{u}$ is also a feasible solution to **P1.3**, so we have $f(\mathbf{o}^{P13}) \geq f(\mathbf{u}) \geq \frac{\lambda}{1+\lambda} f(\mathbf{o})$.

Combined with Lemma 3, we know that a β approximate solution to **P1.3** is also a $\frac{\beta\lambda}{1+\lambda}$ approximate solution to **P1**, so the next part will focus on how to solve **P1.3**. In order to solve **P1.3**, we introduce **P1.4** after relaxing the constraints of **P1.3** as follows.

P1.4

$$\max_{\mathbf{s} \in \{0,\ldots,k\}^n} \quad f(\mathbf{s})$$
$$s.t. \quad \sum_{i \in [m]} \max\{z', |S \cap V_i|\} \leq b,$$
$$|supp(\mathbf{s}) \cap V_i| \leq z' + \alpha, i \in [m].$$

The following Lemma 4 introduces the relationships between **P1.3** and **P1.4**.

Lemma 4. *Every feasible solution of **P1.3** is a feasible solution of **P1.4**. Let* $\mathbf{o}^{P14}$ *be the optimal solution of **P1.4**, we have* $f(\mathbf{o}^{P14}) \geq f(\mathbf{o}^{P13})$.

Proof. It is easy to verify that **P1.4** is a relaxation of **P1.3**, it means that every feasible solution of **P1.3** must also be feasible to **P1.4**. Hence we have $f(\mathbf{o}^{P14}) \geq f(\mathbf{o}^{P13})$. But the reverse is not necessarily true, because the feasible solution of **P1.4** may not satisfy the lower bound constraint in **P1.3**.

As discussed in Lemma 4, we have $f(\mathbf{o}^{P14}) \geq f(\mathbf{o}^{P13}) \geq \frac{\lambda}{1+\lambda} f(\mathbf{o})$. It is worth noting that **P1.4** is a k-submodular maximization problem under matroid constraints [6], for which we already know an approximation algorithm with the 1/3 approximation ratio [18], denoted $\mathcal{A}_2$.

Note the key point is that any feasible solutions to **P1.4** is not necessarily a feasible solution to **P1.3**. Since the solution of **P1.4** may not satisfy the lower bound constraint of **P1.3**, if this is satisfied, the feasibility of **P1.3** is also satisfied. If we can convert a feasible solution of **P1.4** to a feasible solution of **P1.3** with some loss, then we can solve **P1.3** by using the approximation algorithm of **P1.4**, followed by solving **P1**. We introduce Algorithm 2 to describe how to convert a feasible solution of **P1.4** to a feasible solution of **P1.3** when $\min_{i \in [m]} |\mathbf{o}_i| > 1$.

Algorithm 2 Obtain $\mathbf{s}^{final}$ for **P1**($\min_{i\in[m]} |\mathbf{o}_i| > 1$)

Input: An algorithm $\mathcal{A}_2$ for non-monotone k-submodular maximization under the matroid constraint.
Output: $\mathbf{s}^{final} \in (k+1)^V$.

```
Call A_2 to solve P1.4 and obtain a 1/3 approximation solution s^{P14}
Initialize s^1 = s^2 = s^{P14}, choose j ∈ [k] at random
for i ∈ [m] do
    if |supp(s^{P14}) ∩ V_i| ≥ z' then
        continue
    else
        Picks X_i, Y_i ⊆ V_i \ supp(s^{P14}) such that
        |X_i| = |Y_i| = z' − |supp(s^{P14}) ∩ V_i| and X_i ∩ Y_i = ∅.
        let s^1 = (S^1_1, ..., S^1_j ∪ X_i, ..., S^1_k), s^2 = (S^2_1, ..., S^2_j ∪ Y_i, ..., S^2_k)
    end if
end for
s^{final} = arg max {f(s^1), f(s^2)}
return s^{final}
```

The $\mathbf{s}^{final}$ obtained by Algorithm 2 satisfies the lower bound constraints of **P1.3** by adding elements to $\mathbf{s}^{P14}$. The following Theorem 2 provides an approximate ratio guarantee for **P1**.

Theorem 2. *Let* $\mathbf{o}$ *denote the optimal solution of* ***P1****,* $\mathbf{s}^{final}$ *is the solution returned by Algorithm* 2*, then* (1) $\mathbf{s}^{final}$ *is a feasible solution to* ***P1.3*** *and hence* ***P1****.* (2) *When* $\min_{i\in[m]} |supp(\mathbf{o}) \cap V_i| > 1$*, we have* $f(\mathbf{s}^{final}) \geq \frac{1}{24} f(\mathbf{o})$*.*

Proof. We will discuss the feasibility and approximate ratio analysis respectively.

Feasibility:
Observe that $z' = \min_{i\in[m]} |supp(\mathbf{u}) \cap V_i| = \min_{i\in[m]} \left\lfloor \frac{|supp(\mathbf{o})\cap V_i|}{2} \right\rfloor$. Although we do not know $\mathbf{o}$, we know $\min_{i\in[m]} \left\lfloor \frac{|supp(\mathbf{o})\cap V_i|}{2} \right\rfloor \leq \frac{b}{2}$. Then we can enumerate all possibilities of z' in the range of $\left[0, \frac{b}{2}\right]$, and return the best one as the final solution.

Let $L = \{i \in [m] : |supp(\mathbf{s}^{P14}) \cap V_i| < z'\}$. For any $i \in [m] \setminus L$, $|supp(\mathbf{s}^{final}) \cap V_i| = |supp(\mathbf{s}^{P14}) \cap V_i| \geq z'$, then $z' \leq |supp(\mathbf{s}^{final}) \cap V_i| \leq z' + \alpha$ holds. Consider each $i \in L$, because $|X_i| = |Y_i| = z' - |supp(\mathbf{s}^{P14}) \cap V_i|$, thus $|supp(\mathbf{s}^1) \cap V_i| = |supp(\mathbf{s}^2) \cap V_i| = z'$. Thus for any $i \in L$, $z' = |supp(\mathbf{s}^{final}) \cap V_i| \leq z' + \alpha$. What's more, we have

$$\begin{aligned}\sum_{i\in[m]} \max\left\{|supp(\mathbf{s}^{final}) \cap V_i|, z'\right\} &= \sum_{i\in L} z' + \sum_{i\in[m]\setminus L} |supp(\mathbf{s}^{P14}) \cap V_i| \\ &\leq \sum_{i\in[m]} \max\left\{|supp(\mathbf{s}^{P14}) \cap V_i|, z'\right\} \qquad (1) \\ &\leq b.\end{aligned}$$

Thus, $\sum_{i\in[m]} \left|supp(\mathbf{s}^{final}) \cap V_i\right| \leq \sum_{i\in[m]} \max\left\{\left|supp(\mathbf{s}^{final}) \cap V_i\right|, z'\right\} \leq b$. From what has been discussed above, we know $\mathbf{s}^{final}$ is a feasible solution.

Approximation Ratio: In the following we discuss the approximation guarantee when $\min_{i\in[m]} |\mathbf{o}_i| > 1$. Firstly we know

$$\begin{aligned} f(\mathbf{s}^{P14}) &\geq \frac{1}{3} f(\mathbf{o}^{P14}) \\ &\geq \frac{1}{3} f(\mathbf{o}^{P13}) \\ &\geq \frac{\lambda}{3(\lambda+1)} f(\mathbf{o}). \end{aligned} \tag{2}$$

The first inequality is due to the fact that the k-submodular maximization problem under the matroid constraints can be solved approximately with an approximation guarantee of 1/3 in [18]. The second and third inequality is due to Lemma 3 and Lemma 4. Then we have

$$\begin{aligned} f(\mathbf{s}^{final}) &\geq \frac{f(\mathbf{s}^1) + f(\mathbf{s}^2)}{2} \\ &= \frac{f(\mathbf{s}^{P14} \sqcup (\bigcup_{i\in L} X_i, j)) + f(\mathbf{s}^{P14} \sqcup (\bigcup_{i\in L} Y_i, j))}{2} \\ &\geq \frac{f(\mathbf{s}^{P14})}{2} \\ &\geq \frac{\lambda}{6(\lambda+1)} f(\mathbf{o}). \end{aligned} \tag{3}$$

The first inequality is due to the choose of $\mathbf{s}^{final}$ in Algorithm 2. The second inequality is due to the k-submodular and $X_i \cap Y_j = \emptyset$ for any $i, j \in [m]$. The last inequality is by Eq. (2). Note that $\lambda = \min_{i\in[m]} \frac{|supp(\mathbf{u})\cap V_i|}{|supp(\mathbf{o})\cap V_i|} = \min_{i\in[m]} \frac{\left\lfloor \frac{|supp(\mathbf{o})\cap V_i|}{2} \right\rfloor}{|supp(\mathbf{o})\cap V_i|} \geq \frac{1}{3}$ when $\min_{i\in[m]} |supp(\mathbf{o}) \cap V_i| > 1$, then we have $\frac{\lambda}{\lambda+1} \geq \frac{1}{4}$. Combined with Eq. (3), we have $f(\mathbf{s}^{final}) \geq \frac{1}{24} f(\mathbf{o})$.

Notice that Theorem 2 only discusses the approximation ratio when $\min_{i\in[m]} |supp(\mathbf{o}) \cap V_i| > 1$. In fact, this is the approximation ratio for **P1**, because we want to find the worst-case approximation ratio to be the final approximation ratio for **P1**. And we can verity that for both cases of $\min_{i\in[m]} |supp(\mathbf{o}) \cap V_i| = 1$ and $\min_{i\in[m]} |supp(\mathbf{o}) \cap V_i| = 0$, the approximation ratio is better than the case of $\min_{i\in[m]} |supp(\mathbf{o}) \cap V_i| > 1$.

5 Conclusion and Future Work

For the k-submodualr maximization problem under the fairness constraints and total size constraint, we design the discrete optimization algorithms with an approximate ratio of 1/2 for the monotone case and an approximate ratio of 1/24($\approx$ 0.041) for the non-monotone case.

Fairness is needed in various optimization fields, including k-submodular. We know that there are various forms of fairness constraints. In this paper, we only consider the equality-fairness constraints, another common fairness constraints is the equity-fairness constraints. In the future, we can consider the design of approximation algorithms for k-submodular maximization problems under the equity-fairness constraints.

Disclosure of Interests. The authors declare that they have no conflict of interest.

References

1. Bronfenbrenner, M.: Equality and equity. Ann. Am. Acad. Pol. Soc. Sci. **409**(1), 9–23 (1973)
2. Celis, L.E., Huang, L., Vishnoi, N.K.: Multiwinner voting with fairness constraints. In: Proceedings of the Twenty-Seventh International Joint Conference on Artificial Intelligence, pp. 144–151. IJCAI Press, Stockholm, Sweden (2018)
3. Celis, L.E., Keswani, V., Straszak, D., Deshpande, A., Kathuria, T., Vishnoi, N.K.: Fair and diverse DPP-based data summarization. In: Proceedings of the 35th International Conference on Machine Learning, vol. 80, pp. 715–724. PMLR Press, Stockholm, Sweden (2018)
4. Dueck, D., Frey, B.J.: Non-metric affinity propagation for unsupervised image categorization. In: Proceedings of the 11th International Conference on Computer Vision, pp. 1–8. IEEE Computer Society press, Rio de Janeiro, Brazil (2007)
5. Halabi, M.E., Fusco, F., Norouzi-Fard, A., Tardos, J., Tarnawski, J.: Fairness in streaming submodular maximization over a matroid constraint. In: Proceedings of the 15th International Conference on Machine Learning, vol. 202, pp. 9150–9171. PMLR press, Honolulu, Hawaii, USA (2023)
6. Halabi, M.E., Mitrovic, S., Norouzi-Fard, A., Tardos, J., Tarnawski, J.: Fairness in streaming submodular maximization: algorithms and hardness. In: Advances in Neural Information Processing Systems, abs/2010.07431 (2020)
7. Huber, A., Kolmogorov, V.: Towards minimizing k-submodular functions. In: Mahjoub, A.R., Markakis, V., Milis, I., Paschos, V.T. (eds.) ISCO 2012. LNCS, vol. 7422, pp. 451–462. Springer, Heidelberg (2012). https://doi.org/10.1007/978-3-642-32147-4_40
8. Iwata, S., Tanigawa, S., Yoshida, Y.: Improved approximation algorithms for k-submodular function maximization. In: Proceedings of the 27th Annual ACM-SIAM Symposium on Discrete Algorithms, pp. 404–413. SIAM Press, Arlington, VA, USA (2016)
9. Kempe, D., Kleinberg, J.M., Tardos, É.: Maximizing the spread of influence through a social network. Theoret. Comput. Sci. **11**, 105–147 (2015)
10. Lin, H., Bilmes, J.: A class of submodular functions for document summarization. In: Proceedings of the 49th Annual Meeting of the Association for Computational Linguistics: Human Language Technologies, pp. 510–520. The Association for Computer Linguistics Press, Portland, Oregon, USA (2011)
11. Nemhauser, G.L., Wolsey, L.A.: Best algorithms for approximating the maximum of a submodular set function. Math. Method Oper. Res. **3**, 177–188 (1978)

12. Nemhauser, G.L., Wolsey, L.A., Fisher, M.L.: An analysis of approximations for maximizing submodular set functions - I. Math. Program. **14**, 265–294 (1978)
13. Niu, S., Liu, Q., Zhou, Y., Li, M.: Fast algorithms for k-submodular maximization subject to a matroid constraint. CoRR abs/2307.13996 (2023)
14. Ohsaka, N., Yoshida, Y.: Monotone k-submodular function maximization with size constraints. In: Advances in Neural Information Processing Systems, vol. 28 (2015)
15. Rafiey, A., Yoshida, Y.: Fast and private submodular and k-submodular functions maximization with matroid constraints. In: Proceedings of the 37th International Conference on Machine Learning, pp. 7887–7897. PMLR Press, Virtual Event (2020)
16. Sakaue, S.: On maximizing a monotone k-submodular function subject to a matroid constraint. Discret. Optim. **23**, 105–113 (2017)
17. Singh, A.P., Guillory, A., Bilmes, J.A.: On bisubmodular maximization. In: Proceedings of the 15th International Conference on Artificial Intelligence and Statistics, vol. 22, pp. 1055–1063. JMLR Press, La Palma, Canary Islands, Spain (2012)
18. Sun, Y., Liu, Y., Li, M.: Maximization of k-submodular function with a matroid constraint. In: Du, D.Z., Du, D., Wu, C., Xu, D. (eds.) Proceedings of the Theory and Applications of Models of Computation - 17th Annual Conference, vol. 13571, pp. 1–10. Springer, Cham (2022). https://doi.org/10.1007/978-3-031-20350-3_1
19. Tang, S., Yuan, J.: Group equality in adaptive submodular maximization. INFORMS J. Comput. **36**(2), 359–376 (2023)
20. Tang, S., Yuan, J., Mensah-Boateng, T.: Achieving long-term fairness in submodular maximization through randomization, pp. 161–173 (2023)
21. Wang, Y., Fabbri, F., Mathioudakis, M.: Fair and representative subset selection from data streams. In: Proceedings of the 30th Web Conference 2021, pp. 1340–1350. ACM Press, Ljubljana, Slovenia (2021)
22. Ward, J., Zivný, S.: Maximizing bisubmodular and k-submodular functions. In: Proceedings of the 25th Annual ACM-SIAM Symposium on Discrete Algorithms, pp. 1468–1481. SIAM Press, Portland, Oregon, USA (2014)
23. Ward, J., Živnỳ, S.: Maximizing k-submodular functions and beyond. ACM Trans. Algorithms **12**(4), 1–26 (2016)
24. Xiao, H., Liu, Q., Zhou, Y., Li, M.: Approximation algorithms for k-submodular maximization subject to a knapsack constraint. CoRR arXiv: abs/2306.14520 (2023)

Systematic Review of Cybersecurity Disclosure Research

Hongmin W. Du[1(✉)], Jiahui Xu[2], and Miklos A. Vasarhelyi[2]

[1] School of Business, Clark University, Worcester, MA 01610-1477, USA
hongmindu@gmail.com

[2] Accounting and Information Systems Department, Rutgers University, Piscataway, NJ 08854, USA
miklosv@business.rutgers.edu

Abstract. This literature review examines the field of cybersecurity disclosure, highlighting key research themes and identifying gaps in the literature. It covers topics including the impact of regulatory frameworks on disclosure practices, financial consequences of cybersecurity breaches, and the effectiveness of communication strategies in the disclosure process. The review also considers the role of emerging technologies and social media in shaping perceptions. Through analysis of empirical and exploratory studies from recent years, the review explores the dynamics between cybersecurity incidents, regulatory responses, and corporate disclosure strategies. It emphasizes the importance of transparent and timely disclosures for maintaining stakeholder trust and protecting organizational interests amid increasing cyber threats. The paper concludes by proposing future research directions to enhance understanding and effectiveness of cybersecurity disclosures, calling for new methodologies and tools to manage the dissemination of cybersecurity information.

Keywords: Cybersecurity Disclosure Practices · Information Security · Risk Management · Compliance and Regulations · Data Protection · Security Breaches · Vulnerability Disclosure · Threat Intelligence · Legal Frameworks in Cybersecurity

1 Introduction

The increase in the use of fast-advancing technologies among companies to facilitate their businesses has also brought more attention to cybersecurity. Cybersecurity has gathered much attention throughout the years due to increased threats of cybersecurity attacks along with the increase in reliance of digital technologies among companies. Recent events such as the Covid-19 pandemic have exacerbated the growing reliance on technologies; thereby, increasing the concern among companies about their cybersecurity risks. Due to these concerns, there has been an increase over the years of SEC governmental regulation on what is required by companies to disclose in their 10-K statements. One of these is related to cybersecurity disclosures.

S. Ghosh and Z. Zhang (Eds.): AAIM 2024, LNCS 15179, pp. 247–262, 2024.
https://doi.org/10.1007/978-981-97-7798-3_21

The purpose of this paper is to advance the literature on cybersecurity research in the cybersecurity disclosure domain by investigating how well recent literature addresses the concerns for disclosure behaviors among companies. We compile cybersecurity research in the cybersecurity disclosure context into different categories to inform researchers about what has been done currently and provide opportunities for future research in this area.

There are two main issues within cybersecurity disclosure research. Some papers think that cybersecurity disclosure research is about measuring the quantity of breach information companies disclose, and some others think that it is about measuring the potential risk of incurring a breach. There seems to be a misconception between what is considered cybersecurity disclosure in the literature. In this literature review, research papers examining any type of disclosure a firm makes with a reference to cybersecurity are considered a form of cybersecurity disclosure.

In this paper, we identify two main streams of cybersecurity disclosure research, cybersecurity risk disclosure and cybersecurity breach disclosure. Cybersecurity breach disclosure can further be broken down into breach disclosures found in the 8-K statements filed by companies and breach disclosures found in social media platforms. Cybersecurity risk disclosures disclose what may happen in the future whereas, cybersecurity breach disclosures disclose what has already happened. One discloses the potential in the future, while the latter discloses what happened in the past; therefore, both are important for the objectives of this literature review.

This literature review differentiates and contributes to prior literature review studies in cybersecurity research, by concentrating more in depth on cybersecurity disclosure behaviors and trends. Prior literature reviews generally focus on a wide range of topic within cybersecurity; therefore, not covering any of the topics in particular detail leaving out many details regarding a research topic. Since cybersecurity is a vast topic, it is difficult for one literature review to cover every single topic within cybersecurity thoroughly. Due to this, the research question that is addressed in this paper include the following: "What is the most up to date trends in cybersecurity disclosure research?", "What tools can improve the current state of cybersecurity disclosure research?", and "What are future research opportunities within cybersecurity disclosure research?". Cybersecurity disclosure related papers are important because it provides practitioners an idea of whether what is currently being disclosed is sufficient for investors or if more detailed disclosures are necessary.

This literature review has three primary objectives. The first is to provide a comprehensive and systematic overview of the current trends of research on cybersecurity disclosure and to provide a set of categories into which these studies fall under. The second objective is to identify the various types of cybersecurity disclosure discussed in the literature and to identify key topics and issues that are common among the current topics. Finally, the third objective is to identify areas in the literature which are the most promising for future research. This literature review has significant contributions for research and practice by detailing the pros and cons of cybersecurity disclosure found among various research papers. The remainder of this paper is organized as follows. Section two provides a background of cybersecurity and the evolution of cybersecurity disclosure. Section three contains the methodology of how we select our literatures for this review. Section four provides the organizing framework. Section five provides a overall summary of current cybersecurity disclosure research from the selected literatures. Section six provides future research opportunities as well as tools that can improve current cybersecurity disclosure research. Finally, section seven concludes the literature review.

2 Cybersecurity Disclosure Background

2.1 Cybersecurity Risk Disclosures

In 2005 the SEC shifted from allowing voluntary disclosure of risks to mandatory disclosure of risk factors. The SEC mandated that firms needed to include risk factor disclosures in their quarterly and annual filings. To reflect this change in policy, the SEC added Item 1A to Form 10-Q and 10-K [1]. The new mandate included cybersecurity risk within the Item 1A section. Cybersecurity risk was only one part of the overall risk in determining firm risk and did not seem to be as important as other risks until cyber-attacks became more frequent. As the occurrence of cyber-attacks increased due to the increased reliance of technology in day-to-day business activities, businesses including the SEC began to see cybersecurity risk as a significant risk in determining overall firm risk. In response, the SEC implemented new guidance on disclosing cybersecurity risk and incidents in 2011. The guidelines implemented by the SEC in 2011 were recommendations for companies when disclosing cybersecurity risk disclosure. Due to the lack of regulation on cybersecurity disclosure, the informativeness of cybersecurity disclosures began to be questioned by practioners, regulators, and stakeholders. In most cases, it was found that the majority of cybersecurity disclosures were boilerplate, using common languages and not very informative to stakeholder in making financial decisions. In response, the SEC implemented new cybersecurity guidelines in 2018 to regulate the disclosure of cybersecurity risks in firms. Although the SEC attempted to regulate the disclosure of cybersecurity risk, the guidelines were vague and still gave companies the leisure to choose what to disclose in their quarterly and annual filings for cybersecurity disclosure. Specifically, some of the key takeaways from the 2018 guidelines stated that firms would need to disclose cybersecurity risk or incidents of cybersecurity breaches if it was material to the firm and would "result in reputational, financial or operational harm" [2]. These key takeaways from the 2018 guidelines also included what the criteria for determining materiality would be and the timing of disclosures related to a cybersecurity incident. Besides these two main takeaways, the other items discussed in the guidelines were once again mere suggestions for companies when preparing their cybersecurity disclosures; however, more detailed than the guidelines introduced in 2011. Since determination of materiality is different for each company, it was solely up to the company to determine if a risk or incident was material. In 2023, the Securities and Exchange Commission (SEC) updated its cybersecurity guidelines, building on the framework established in 2018. Despite these advancements, concerns about the effectiveness of these guidelines persist, highlighting the need for ongoing monitoring. The revised guidelines now mandate that companies disclose material cybersecurity incidents within four business days of determining their materiality, a stipulation that introduces greater stringency compared to the less specific requirements of the 2018 guidelines [3]. Additionally, the new regulations require companies to provide comprehensive descriptions of the material aspects of cybersecurity incidents, encompassing their nature, scope, timing, and significant impact on the company [3]. While these changes are expected to enhance the quality of cybersecurity disclosures significantly, there remains a critical need for robust methodologies to evaluate the effectiveness of these disclosures and to track improvements over time.

2.2 Cybersecurity Breach Disclosures

Cybersecurity breach disclosures are essential in information security management. They involve notifying stakeholders about security breaches that could compromise data integrity and confidentiality. Effective breach disclosure supports corporate transparency and assists in managing the aftermath of a breach. The disclosure of cybersecurity incidents was also regulated by the 2018 SEC cybersecurity guidelines. In the guidelines, it was stated that publicly traded companies would need to disclose cybersecurity breach incidents that were material [1]. Companies were required to file a form 8-K within four days of a material event. The determination of materiality once again varies based on the company. The disclosure of form 8-K includes the disclosure of cybersecurity breaches that were large enough to be considered material to a company. The way a breach is disclosed can significantly affect the company's reputation and stakeholder trust. Strategic communication involves timing the disclosure appropriately, ensuring the accuracy of the information, and being transparent about the extent of the breach and its repercussions. It's also crucial to tailor the communication to the needs of different stakeholders, including customers, partners, regulators, and the public. One main challenge in disclosing cybersecurity breaches is balancing transparency and security. Companies must provide enough detail to inform affected parties without revealing sensitive information that could aid further attacks. Another challenge is managing legal risks, as premature or inaccurate disclosures can lead to regulatory penalties and damage claims. As cyber threats evolve, so do the standards and practices around breach disclosures. There is a move towards more standardized, automated reporting mechanisms and greater inter-organizational cooperation to combat threats. Future trends may also involve the use of technology to provide real-time updates about the status of data breaches and their resolution.

2.3 Cybersecurity Social Media Disclosures

Aside from official SEC filings, companies may also choose to disclose certain events or incidents on their social media platforms. Along with the reliance of technology, the rise and reliance on social media platforms also followed. Due to this, social media became a crucial channel for companies to quickly communicate cybersecurity incidents to stakeholders. This method aligns with trends towards immediate corporate communication. Social media offers opportunities and challenges for cybersecurity disclosures. Platforms like Twitter, LinkedIn, and Facebook enable companies to rapidly share information about security breaches. This quick communication can control misinformation and shape the incident narrative. Social media also allows for interactive engagement, where stakeholders can inquire, comment, and interact directly with the company, fostering transparency and trust. Contrary to the benefits of social media, the rapid spread of information can also lead to challenges, such as the release of unverified details, potentially leading to public confusion. Companies must balance quick disclosure with the responsibility to ensure information is accurate and thorough. The informal nature of social media complicates adherence to regulatory requirements for incident reporting. Therefore, companies need to establish comprehensive guidelines that specify the timing, content, and responsibilities for disclosures. These guidelines should ensure that communications are legally compliant and align with the company's broader communication strategy. Training employees on appropriate social media use during incidents is essential to maintain consistency and prevent the release of unauthorized information. As guidelines for cybersecurity disclosures continue to evolve, the role of social media may become more defined and regulated. Companies must stay updated with these changes to ensure their social media strategies comply with legal standards and best practices. As stakeholders grow accustomed to receiving information via digital platforms, companies need to refine their engagement strategies to communicate effectively. In summary, while social media provides a valuable tool for disclosing cybersecurity information, it requires careful strategy and operational management to leverage its advantages and minimize risks.

2.4 Motivation

Although the regulations for cybersecurity disclosure over the years have been increasing, there still seems to be a lack of control and consistency among what companies disclose. Due to this, we see a rise in research papers addressing the informativeness of cybersecurity disclosures and whether what is disclosed is useful and informative. Since stakeholders and other users of the financial statements rely on these documents to make decisions, it is important that the contents of the cybersecurity disclosure provide meaningful information to these users. Companies who acknowledge more cybersecurity risks and disclose them should also in turn be able to better address these risks. There are multiple arguments of whether more disclosure in cybersecurity is needed or if less is needed. Also, arguments of whether too much disclosure will lead to negative results and whether too little disclosure will become meaningless. Regardless, it is interesting to study how differences in cybersecurity disclosure affects user decisions in various ways when it is just a small subset of a firm's overall risk. To address these concerns on the variation of cybersecurity disclosure, we focus our literature review on the cybersecurity disclosure area of cybersecurity research and look at the following research questions: "What is the most up to date trends in cybersecurity disclosure research?", "What tools can improve the current state of cybersecurity disclosure research?", and "What are future research opportunities within cybersecurity disclosure research?"

3 Methodology and Literature Selection

3.1 Existing Literature Review

We found three literature reviews in the existing cybersecurity literature that covers cybersecurity topics: Ali et al. [6], Haapamaki and Sihvonen [4], and Walton et al. [5]. Ali et al. [6] does a literature review focusing on cybersecurity papers looking at breach impacts on stock market response, long-term performance, internal control reports, and audit fees. The primary focus of the authors paper is to look at the economic impact of data breaches on these factors. Our paper is different in that we look at papers that look at the impact of cybersecurity disclosures including risk disclosure and breach disclosure rather than solely focusing on breaches that occurred as well as the disclosures impact on certain market variables. Haapamaki and Sihvnonent [4] does a literature review on research papers within the accounting and auditing field. The authors of this literature review provide significant implication for research and practice by only looking at cybersecurity papers related to accounting and auditing with an emphasis on the role of internal auditing and controls to improve a firm's cybersecurity. Our review also differs in that we look at all cybersecurity related papers in all fields related to cybersecurity disclosure. Finally, Walton et al. [5] did an integrative review for all of the topics to date in cybersecurity research up to 2019. The author collects multiple papers for each topic within cybersecurity and addresses the current state of research in those topics, but they do not collect all the papers for each topic. It seems that they collect just a small sample of research papers for each topic to discuss what is currently being done in those topics. The author also provides future research opportunities within each of those cybersecurity topics. Cybersecurity disclosure is one of the subtopics discussed by this author; however, the details of cybersecurity disclosure aren't discussed thoroughly. In their review, they only talk about the determinant and impact of cybersecurity disclosure and fail to go further than that. Our paper differs from their paper in that we focus on one specific area of cybersecurity research providing more concrete information and organization of all relevant papers up to date (2022) in the cybersecurity disclosure field. Our paper also organizes the cybersecurity disclosure papers differently in our organizing framework and the selection of papers is also different. Our selection of research papers is intended to include only papers that have been published after 2019 or are currently working papers with the intention of updating prior literature with a focus on cybersecurity disclosures. The motivation for writing this literature review is because the last review paper on cybersecurity research stopped at year 2019. This review is not only meant to be an update to prior literature review paper but also provide a more detailed view on cybersecurity disclosure research.

3.2 Literature Selection

We first define a keyword list which include the following: cybersecurity risk disclosure, cybersecurity breach disclosure, 8-K disclosure, 10-K disclosure, breach disclosure, cyber-attack disclosure, information security risk disclosure, information security breach disclosure, information breach disclosure, data breach disclosure, data breach risk disclosure, data theft disclosure, security breach disclosure, security risk disclosure. Using this keyword list, we search “ssrn.com”, “sciencedirect.com”, google scholar, and the American Accounting Association Journals for cybersecurity disclosure related papers. We restrict our search for papers published in 2020 to 2022 as well as current working papers. After compiling and removing duplicate papers, we find a total of 26 unique papers from SSRN and ScienceDirect including non-published working papers. Additionally, we find 3 additional unique papers from the American Accounting Association journals via a manual search through each edition from 2020 to 2022. In total, we find 29 unique and distinct papers related to cybersecurity disclosure across all fields.

Table 1 contains the distribution of the literature gathered for further investigation. From the distribution of papers in the literature, the majority of cybersecurity disclosure papers are published in business journals including information system, accounting and finance journals. In comparison to prior years, more emphasis on cybersecurity disclosure has been placed in the business context, and less in other journals. We also notice that the majority of the papers publishing cybersecurity disclosure related content are empirical papers looking at historical data. From this analysis, we find that there is a diverse selection of journals that are interested in cybersecurity disclosure related articles. Another trend that is found is that there are significantly less papers published in the year 2019 than in the subsequent two years after. This is likely due to the Covid-19 impact impacting publication during that time. Overall, the number of published and new papers in this area are relatively low compared to other research topics in the accounting and business domain.

4 Organizing Framework

We organize our literature review around three main types of cybersecurity disclosures that businesses may choose to disclose cybersecurity related events (Cybersecurity Risk Disclosure, Cybersecurity Breach Disclosure, and Cybersecurity Social Media Disclosure). Table 2 provides a classification based on the

Table 1. Summary of Academic Journal Publications on Cybersecurity Disclosures

Academic Journal	Authors
Journal of Accounting Research	Badawy and Hebatallah [34]
Critical Perspectives on Accounting	Andrew, Jane, Max Baker, and Casey Huang [33]
The Accounting Review	Ashraf, Musaib, and Jayanthi Sunder [35]
Journal of Accounting and Public Policy	Barry, Thomas, Jonathan Jona, and Naomi Soderstrom [32]; Calderon, Thomas G., and Lei Gao [8]
Journal of Law, Technology & Policy	Kesari, Aniket [30]
International Journal of Information Management Data Insights	Barati, Mehdi, and Benjamin Yankson [40]
Information and Management	Chen, Rui, Dan J. Kim, and H. Raghav Rao [16]
Journal of Business Ethics	Chen, Jing, Elaine Henry, and Xi Jiang [9]
Data in Brief	Rosati, Pierangelo, and Theo Lynn [38]
The Review of Financial Studies	Florackis, Chris, et al. [10]
The Journal of Finance and Data Science	Ashraf, Musaib, John Xuefeng Jiang, and Isabel Yanyan Wang [35]; Nissim, Doron [25]
International Journal of Accounting Information Systems	Gao, Lei, Thomas G. Calderon, and Fengchun Tang [31]; Smith, Thomas, Amanuel F. Tadesse, and Nishani Edirisinghe Vincent [21]; Wang, Tawei, Ju-Chun Yen, and Kyunghee Yoon [11]
Artificial Intelligence in Accounting	Haapamäki, Elina, and Jukka Sihvonen [4]
SMU Cox School of Business Research Paper	Lattanzio, Gabriele, and Yue Ma [29]
Auditing: A Journal of Practice & Theory	Li, He, Won Gyun No, and J. Efrim Boritz [28]
Journal of Financial Markets	Lin, Zhaoxin, et al. [13]
European Financial Management	Liu, Jinyu, and Xiaoran Ni [15]
Research in Economics	Masoud, Najeb, and Ghassan Al-Utaibi [7]
Computers and Security	McLeod, Alexander, and Diane Dolezel [26]; Roumani, Yaman [22]; Thomas, Louise, et al. [14]
Global Finance Journal	Piccotti, Louis R., and Heng Wang [23]
International Review of Financial Analysis	Tosun, Onur Kemal [19]
Finance Research Letters	Wang, Heng Emily et al. [12]
Risk Management	Uddin, Md, Md Ali, and Mohammad Kabir Hassan [18]
Working Papers	Ashraf, Musaib [36]; Binkley, Marcy Ruthann [37]; Liang, Yangfan, and Rahul Telang [27]

items studied within each paper. We further classify each paper into the methodology they use to complete their study and provide a description of their results if significant enough.

4.1 Cybersecurity Risk Disclosure

Cybersecurity risk disclosure is the process of identifying, assessing, and communicating potential cybersecurity risks to relevant stakeholders. Effective cybersecurity risk disclosure is critical for promoting transparency, accountability,

Table 2. Organization of Research Papers in Cybersecurity Disclosures

Authors	Type of Paper	Classification
Liang, Yangfan, and Rahul Telang [27]	Empirical	Cybersecurity Breach Disclosure
Barry, Thomas, Jonathan Jona, and Naomi Soderstrom [32]	Empirical	Cybersecurity Risk Disclosure
Li, He, Won Gyun No, and J. Efrim Boritz [28]	Empirical	Cybersecurity Breach Disclosure
Binkley, Marcy Ruthann [37]	Empirical	Cybersecurity Risk Disclosure
Liu, Jinyu, and Xiaoran Ni [15]	Empirical	Cybersecurity Breach Disclosure
Ashraf, Musaib, and Jayanthi Sunder [35]	Empirical	Cybersecurity Breach Disclosure
Lattanzio, Gabriele, and Yue Ma [29]	Empirical	Cybersecurity Risk Disclosure
Ashraf, Musaib [36]	Empirical	Cybersecurity Risk Disclosure
Uddin, Md, Md Ali, and Mohammad Kabir Hassan [18]	Literature Review	Cybersecurity Risk Disclosure
Chen, Jing, Elaine Henry, and Xi Jiang [9]	Empirical	Cybersecurity Risk Disclosure
Florackis, Chris, et al. [10]	Empirical	Cybersecurity Risk Disclosure
Badawy and Hebatallah [34]	Experiment	Cybersecurity Risk Disclosure
Kesari, Aniket [30]	Exploratory	Cybersecurity Risk Disclosure
Haapamäki, Elina, and Jukka Sihvonen [4]	Literature Review	Cybersecurity Risk Disclosure
Piccotti, Louis R., and Heng Wang [23]	Empirical	Cybersecurity Breach Disclosure
Tosun, Onur Kemal [19]	Empirical	Cybersecurity Breach Disclosure
Thomas, Louise, et al. [14]	Design Science	Cybersecurity Breach Disclosure
Barati, Mehdi, and Benjamin Yankson [40]	Exploratory	Cybersecurity Breach Disclosure
Gao, Lei, Thomas G. Calderon, and Fengchun Tang [31]	Exploratory	Cybersecurity Risk Disclosure
Smith, Thomas, Amanuel F. Tadesse, and Nishani Edirisinghe Vincent [21]	Empirical	Cybersecurity Breach Disclosure
Roumani, Yaman [22]	Exploratory	Cybersecurity Breach Disclosure
Masoud, Najeb, and Ghassan Al-Utaibi [7]	Empirical	Cybersecurity Risk Disclosure
Wang, Heng Emily et al. [12]	Empirical	Cybersecurity Social Media Disclosure
Wang, Tawei, Ju-Chun Yen, and Kyunghee Yoon [11]	Exploratory	Cybersecurity Risk Disclosure
Lin, Zhaoxin, et al. [13]	Empirical	Cybersecurity Social Media Disclosure
Chen, Rui, Dan J. Kim, and H. Raghav Rao [16]	Exploratory	Cybersecurity Social Media Disclosure
Andrew, Jane, Max Baker, and Casey Huang [33]	Exploratory	Cybersecurity Breach Disclosures
Ashraf, Musaib, John Xuefeng Jiang, and Isabel Yanyan Wang [35]	Empirical	Cybersecurity Breach Disclosure

and informed decision-making in cybersecurity management. In this section, we examine the key themes and findings from cybersecurity risk disclosure research papers. The research papers on cybersecurity risk disclosures identify several key themes essential for understanding how organizations manage and communicate cybersecurity risks.

The literature extensively explores the factors influencing firms to disclose cybersecurity risks. Notable among these is the study by Masoud and Ghassan [7], which identifies regulatory environment, perceived severity of cybersecurity threats, and industry-specific risks as key determinants. Their findings suggest that stricter regulatory environments and higher perceived threats lead to more comprehensive risk disclosures. This alignment with regulatory expectations indicates a proactive approach to risk management disclosure.

The influence of regulatory bodies significantly shapes disclosure practices. Calderon and Lei [8] investigate how SEC comment letters prompt changes in the way firms report cybersecurity risks. Their findings illustrate that regulatory scrutiny compels firms to enhance transparency in their disclosures, which not only aligns with legal obligations but also strategically modifies firm behavior in risk communication.

Evaluating the informativeness of cybersecurity risk disclosures is critical for stakeholders, such as investors and clients. Chen et al. [9] assess whether disclosures following a data breach provide meaningful information. Their analysis concludes that detailed disclosures, which elaborate on risk management and the impacts of incidents, are significantly beneficial in shaping stakeholder perceptions and preserving firm value.

Another important theme is the financial market's reaction to cybersecurity risk disclosures. Florackis et al. [10] delve into how the announcement of cybersecurity risks influences investor behavior and stock prices. The study highlights that accurate and timely risk disclosures are crucial, as they directly affect market confidence and can lead to substantial financial implications for the firm.

The literature on cybersecurity risk disclosures shows a complex interaction between regulatory requirements, corporate disclosure practices, and stakeholder reactions. It emphasizes the need for effective disclosure practices that satisfy regulatory demands and meet the informational needs of stakeholders.

4.2 Cybersecurity Breach Disclosure

Cybersecurity breach disclosure is a vital component of cybersecurity management. It involves the disclosure of information related to a cybersecurity breach, including the nature of the breach, the extent of the damage, and the steps taken to mitigate the impact of the breach. Cybersecurity breach disclosure is important for organizations and individuals to protect themselves against further harm and to hold responsible parties accountable. In this section, we examine the key themes and findings from cybersecurity breach disclosure research papers.

A consistent theme is the influence of regulatory frameworks on breach disclosure practices. For instance, Wang et al. [11] analyze how companies respond to SEC comment letters regarding their cybersecurity disclosures. This analysis highlights the pressure regulatory bodies exert on firms to comply and the subsequent adjustments in disclosure practices to meet these requirements.

The impact of breach disclosures on financial markets is a notable theme. Studies like those by Lin et al. [13] and Wang et al. [12] focus on the trading behaviors that precede and follow breach announcements. These studies provide insights into investor reactions and market dynamics, demonstrating that breach disclosures can significantly influence stock values and trading patterns.

Effective communication strategies during and after a breach are critical for managing stakeholder relationships. Thomas et al. [14] discuss the importance of a structured communication framework in breach disclosures. Their findings emphasize that how information is communicated impacts stakeholder trust and the company's ability to manage the crisis.

A more subtle theme is the unintended consequences that can arise from breach disclosures. Liu and Xiaoran [15] explore how disclosures, while necessary, can sometimes lead to manipulations in firm activities or misinterpretations of the data provided. This aspect highlights the delicate balance companies must maintain between transparency and the risk of providing too much strategic information to potential adversaries.

Overall, The literature on cybersecurity breach disclosures identifies themes related to regulatory compliance, market impact, communication strategies, and the potential unintended consequences of disclosures. These themes illustrate the challenges firms face in effectively managing and communicating cybersecurity breaches. As threats evolve and regulatory environments change, continued examination of these themes is necessary to develop better disclosure practices that protect both firms and their stakeholders.

4.3 Cybersecurity Social Media Disclosure

Cybersecurity social media disclosure refers to the disclosure of information related to cybersecurity incidents, vulnerabilities, and threats through social media platforms. Social media has become an important channel for cybersecurity disclosure due to its widespread use, ease of access, and real-time nature. In recent years, cybersecurity social media disclosure has been the subject of a significant amount of research. In this section, we examine the key themes and findings from cybersecurity social media disclosure research papers.

One prominent theme is how social media facilitates compliance with regulatory requirements for timely and effective disclosures. The study by Ong and Sandy [17] explores Hong Kong's data breach notification scheme from stakeholders' perspectives, highlighting how social media platforms are used to meet regulatory demands for swift public notification.

Social media's immediacy and broad reach significantly impact stakeholder perceptions. Chen et al. [16] examine social networking site usage from a security and privacy threat assessment perspective. Their findings suggest that the way information is disclosed on social media can quickly alter stakeholder trust and engagement levels, either mitigating or exacerbating the perceived severity of a cybersecurity incident.

Another theme is the challenge of strategic communication through social media. Due to the informal and unstructured nature of social media, firms face difficulties in maintaining a consistent and professional tone while meeting the urgency of disclosure requirements. The need for rapid response can sometimes lead to premature or incomplete information dissemination, which can further complicate the management of a cybersecurity incident.

The themes identified in the literature on cybersecurity social media disclosures highlight the integral role of social media in modern cybersecurity strategies. These platforms offer powerful tools for rapid communication and regulatory compliance but also pose significant challenges in strategic communication and stakeholder management. As social media continues to evolve as a primary communication medium, understanding its dynamics in the context of cybersecurity disclosures becomes increasingly crucial for effective crisis management and stakeholder engagement.

5 Summary of Current Cybersecurity Disclosure Research

In summary, we find that the majority of the papers in the literature focus around examining the affect of cybersecurity risk disclosure on breach or the affect of public behavior in the event of a breach. There seems to be much less research on the disclosure of breach from 8-K statements and more of reliance of breached firms from already established breach data sets such as the ones obtainable from Privacy Right Clearing House. However, there is far less research using social media data in the cybersecurity disclosure literature. We see common research themes in each of the three main types of cybersecurity disclosures.

A significant portion of the research examines how regulatory frameworks influence cybersecurity disclosure practices. Studies such as those by Wang et al. [11] detail the adjustments firms make in response to regulatory bodies like the SEC. Compliance with these regulations is shown to impact not only disclosure practices but also market reactions, as investors and other stakeholders respond to the perceived adequacy and timeliness of the information disclosed.

The financial implications of cybersecurity breaches and their disclosures are well-documented, with research indicating that how and when breaches are disclosed can significantly affect a company's stock price and investor trust. Research by Lin et al. [13], and Wang et al. [12] provides insights into market behavior before and after breach announcements, underscoring the economic stakes involved in disclosure practices.

Effective communication strategies are critical in managing cybersecurity incidents, especially in terms of maintaining public trust and managing stakeholder relationships. The literature emphasizes the need for clear, accurate, and timely communication, with studies like those by Thomas et al. [14] suggesting frameworks for improving data privacy and security accountability in communications. The role of social media in disseminating information quickly and broadly is also highlighted as both a tool and a challenge, given the rapid spread of information and the potential for misinformation.

The use of social media for cybersecurity disclosures is a relatively new area of study, focusing on how these platforms can both aid and complicate the disclosure process. Studies by Chen et al. [16] and Ong and Sandy [17] explore the strategic use of social media to meet regulatory requirements and manage public perception, highlighting the need for careful planning and execution in using these platforms.

Current cybersecurity disclosure research covers a broad spectrum of themes, each intertwined with the next. The collective findings emphasize the complexity of managing cybersecurity risks and the critical role of strategic disclosures in protecting organizational interests and maintaining stakeholder trust. As cyber threats evolve and new technologies emerge, ongoing research will be essential to adapt disclosure practices to meet changing regulatory, technological, and social demands.

6 Future Research Opportunities

There seems to be plenty of research opportunities in the cybersecurity disclosure literature. Although a lot has been done in this topic over the recent years, there is still room to redo some previous research with updated results. Some research streams that can be further looked into could include looking at the effectiveness of current SEC cybersecurity disclosure regulations: This research could evaluate the extent to which the current SEC regulations have improved the quality of cybersecurity disclosures made by public companies. It could also assess whether the regulations are effectively meeting the needs of investors in terms of understanding the cybersecurity risks and incidents of companies. Although this research has been done before in a similar fashion, there is new regulations coming out with new research opportunities to explore the affect of the new regulations.

Another research opportunity would be to explore options to develop new metrics for evaluating the quality of cybersecurity disclosures: This research could involve the development of objective metrics that can be used to assess the quality of cybersecurity disclosures made by public companies. This could help investors better understand the risks associated with investing in companies with poor cybersecurity practices.

Researchers can also investigate the impact of cybersecurity incidents on the financial performance of public companies. This research could examine the relationship between cybersecurity incidents and the financial performance of public companies. It could evaluate whether companies with poor cybersecurity practices are more likely to experience financial losses following a cyber attack. Although there is plenty of papers on this topic in the literature, there is still research potential to update the results with more current data.

One of the trending themes in the literature was emphasizing the importance of communication methods to disclose cybersecurity disclosure. So another interesting research opportunity would be to look at the effectiveness of different communication methods for disclosing cybersecurity incidents. This research could examine the different ways in which companies disclose information about cybersecurity incidents to their stakeholders, such as investors, employees, and customers. For example, researchers could assess the effectiveness of different communication methods, such as press releases, regulatory filings, and social media posts, in conveying the severity of the incident and the actions being taken to mitigate the damage.

Another research opportunity could include looking at the impact of cybersecurity disclosures on a company's reputation: This research could examine the effect of cybersecurity disclosures on a company's reputation, both in the short and long term. It could evaluate whether investors are more likely to avoid investing in companies that have suffered cybersecurity incidents or whether customers are more likely to stop using a company's products or services following a cybersecurity incident. Although there are several papers that already look at this, there can still be room for update in the current literature as most of these papers are outdated.

Finally, one of the biggest opportunities for research is to explore the use of technology to improve disclosures. There is immense potential to explore the use of technology to improve the effectiveness of cybersecurity disclosures. For example, researchers could investigate the use of artificial intelligence or blockchain to ensure the accuracy and transparency of disclosures, or explore the use of data visualization techniques to make disclosures more accessible and readable to non-experts.

7 Conclusion and Remarks

In conclusion, cybersecurity disclosure is a critical topic that has attracted attention from scholars, policymakers, and industry practitioners. This literature review highlights the complexities and challenges associated with cybersecurity disclosures, including the lack of consensus on what constitutes material cybersecurity information, concerns about the potential for disclosures to create additional vulnerabilities, and the impact of disclosure on firm value and reputation. In this paper, we provide a summary of the current research themes within cybersecurity disclosure research, as well as a view of the type of journals interested in this type of research. The review also identifies several opportunities for future research, such as examining the effectiveness of different types of disclosures in mitigating cyber risks, the impact of cybersecurity disclosure on consumer behavior and trust, and the development of standardized metrics and frameworks for assessing and reporting cybersecurity risks. As cybersecurity threats continue to evolve and become more sophisticated, the need for effective cybersecurity disclosures will only grow in importance, making it a critical area for continued research and innovation.

References

1. Commission Statement and Guidance on Public Company Cybersecurity Disclosure. https://www.sec.gov/files/rules/interp/2018/33-10459.pdf. Accessed 9 June 2024
2. Key Takeaways from the SEC's 2018 Cybersecurity Guidance. https://www.kirkland.com/siteFiles/Publications/Key_Takeaways_from_the_SEC's_2018_Cybersecurity_Guidance.pdf. Accessed 3 Dec 2023

3. SEC's new cyber disclosure rule. https://www.pwc.com/us/en/services/consulting/cybersecurity-risk-regulatory/sec-final-cybersecurity-disclosure-rules.html. Accessed 4 Jan 2024
4. Haapamäki, E., Sihvonen, J.: Cybersecurity in accounting research. In: Artificial Intelligence in Accounting, pp. 182–214 (2022)
5. Walton, S., Wheeler, P.R., Zhang, Y., Zhao, X.: An integrative review and analysis of cybersecurity research: current state and future directions. J. Inf. Syst. **35**(1), 155–186 (2021)
6. Ali, S.E.A., Lai, F.-W., Dominic, P.D.D., Brown, N.J., Lowry, P.B.B., Ali, R.F.: Stock market reactions to favorable and unfavorable information security events: a systematic literature review. Comput. Secur. **110**, 102451 (2021)
7. Masoud, N., Al-Utaibi, G.: The determinants of cybersecurity risk disclosure in firms' financial reporting: empirical evidence. Res. Econ. **76**(2), 131–140 (2022)
8. Calderon, T.G., Gao, L.: Changes in corporate cybersecurity risk disclosures after SEC comment letters. J. Account. Public Policy **41**(5), 106993 (2022)
9. Chen, J., Henry, E., Jiang, X.: Is cybersecurity risk factor disclosure informative? Evidence from disclosures following a data breach. J. Bus. Ethics **187**(1), 199–224 (2023)
10. Florackis, C., Louca, C., Michaely, R., Weber, M.: Cybersecurity risk. Rev. Financ. Stud. **36**(1), 351–407 (2023)
11. Wang, T., Yen, J.-C., Yoon, K.: Responses to SEC comment letters on cybersecurity disclosures: an exploratory study. Int. J. Account. Inf. Syst. **46**, 100567 (2022)
12. Wang, H.E., Wang, Q.E., Wu, W.: Short selling surrounding data breach announcements. Finance Res. Lett. **47**, 102690 (2022)
13. Lin, Z., Sapp, T.R.A., Ulmer, J.R., Parsa, R.: Insider trading ahead of cyber breach announcements. J. Financ. Mark. **50**, 100527 (2020)
14. Thomas, L., Gondal, I., Oseni, T., Firmin, S.S.: A framework for data privacy and security accountability in data breach communications. Comput. Secur. **116**, 102657 (2022)
15. Liu, J., Ni, X.: Ordeal by innocence in the big-data era: intended data breach disclosure, unintended real activities manipulation. Eur. Financ. Manag. **30**(1), 129–163 (2024)
16. Chen, R., Kim, D.J., Rao, H.R.: A study of social networking site use from a three-pronged security and privacy threat assessment perspective. Inf. Manage. **58**(5), 103486 (2021)
17. Ong, R., Sabapathy, S.: Hong Kong's data breach notification scheme: from the stakeholders' perspectives. Comput. Law Secur. Rev. **42**, 105579 (2021)
18. Hamid, U.M., Ali, M.H., Hassan, M.K.: Cybersecurity hazards and financial system vulnerability: a synthesis of literature. Risk Manage. **22**(4), 239–309 (2020)
19. Tosun, O.K.: Cyber-attacks and stock market activity. Int. Rev. Financ. Anal. **76**, 101795 (2021)
20. Taufick, R.D.: The underdeterrence, underperformance response to privacy, data protection laws. Technol. Soc. **67**, 101752 (2021)
21. Smith, T., Tadesse, A.F., Vincent, N.E.: The impact of CIO characteristics on data breaches. Int. J. Account. Inf. Syst. **43**, 100532 (2021)
22. Roumani, Y.: Detection time of data breaches. Comput. Secur. **112**, 102508 (2022)
23. Piccotti, L.R., Wang, H.: Informed trading in the options market surrounding data breaches. Global Finance J. **56**, 100774 (2022)

24. Ogbanufe, O., Kim, D.J., Jones, M.C.: Informing cybersecurity strategic commitment through top management perceptions: the role of institutional pressures. Inf. Manage. **58**(7), 103507 (2021)
25. Nissim, D.: Big data, accounting information, and valuation. J. Finance Data Sci. **8**, 69–85 (2022)
26. McLeod, A., Dolezel, D.: Information security policy non-compliance: can capitulation theory explain user behaviors? Comput. Secur. **112**, 102526 (2022)
27. Liang, Y., Telang, R.: Customer response to adverse security events: an empirical study. Available at SSRN 3523788 (2020)
28. Li, H., No, W.G., Boritz, J.E.: Are external auditors concerned about cyber incidents? Evidence from audit fees. Auditing J. Pract. Theory **39**(1), 151–171 (2020)
29. Lattanzio, G., Ma, Y.: Corporate innovation in the cyber age. SMU Cox School of Business Research Paper 20-04 (2021)
30. Kesari, A.: Predicting cybersecurity incidents with machine learning and mandatory disclosure regulation. Illinois J. Law Technol. Policy, 57 (2022)
31. Gao, L., Calderon, T.G., Tang, F.: Public companies' cybersecurity risk disclosures. Int. J. Account. Inf. Syst. **38**, 100468 (2020)
32. Barry, T., Jona, J., Soderstrom, N.: The impact of country institutional factors on firm disclosure: cybersecurity disclosures in Chinese cross-listed firms. J. Account. Public Policy **41**(6), 106998 (2022)
33. Andrew, J., Baker, M., Huang, C.: Data breaches in the age of surveillance capitalism: do disclosures have a new role to play? Crit. Perspect. Account. **90**, 102396 (2023)
34. Badawy, H.A.E.S.: The impact of assurance quality and level on cybersecurity risk management program on non-professional Egyptian investors' decisions: an experimental study. Alexandria J. Account. Res. **5**(3) (2021)
35. Ashraf, M., Jiang, J.X., Wang, I.Y.: Are there trade-offs with mandating timely disclosure of cybersecurity incidents? Evidence from state-level data breach disclosure laws. J. Finance Data Sci. **8**, 202–213 (2022)
36. Ashraf, M.: Potentially unintended consequences of the SEC restricting managerial discretion: evidence from peer data breaches and cyber risk factors. Available at SSRN 3807487 (2021)
37. Binkley, M.R.: Can you see it coming? How disclosure and corporate social responsibility activity predict cybersecurity breach. Diss., The University of North Carolina at Charlotte (2021)
38. Rosati, P., Lynn, T.: A dataset for accounting, finance and economics research on US data breaches. Data Brief **35**, 106924 (2021)
39. Ashraf, M., Sunder, J.: Can shareholders benefit from consumer protection disclosure mandates? Evidence from data breach disclosure laws. Account. Rev. **98**(4), 1–32 (2023)
40. Barati, M., Yankson, B.: Predicting the occurrence of a data breach. Int. J. Inf. Manage. Data Insights **2**(2), 100128 (2022)

Revisiting the Stretch Factor of Delaunay Triangulations of Points in Convex Position

Xuehou Tan[1,2](✉), Rong Chen[1], and Qing Jiang[2]

[1] Dalian Maritime University, Linghai Road 1, Dalian, China
[2] Tokai University, 4-1-1 Kitakaname, Hiratsuka 259-1292, Japan
th7625@tokai.ac.jp

Abstract. Let S be a set of n points in the plane, and let $DT(S)$ be the planar graph of the Delaunay triangulation of S. For a pair of points $a, b \in S$, denote by $|ab|$ the Euclidean distance between a and b. Denote by $DT(a,b)$ the shortest path in $DT(S)$ between a and b, and let $|DT(a,b)|$ be the total length of $DT(a,b)$. $DT(S)$ can be used to approximate the complete graph of S in the sense that the stretch factor $\frac{|DT(a,b)|}{|ab|}$ is upper bounded by a constant, independent of S and n. The currently known best factor for a set of planar points is 1.998.

In this paper, we prove that for a set S of points in convex position (i.e., they form the vertices of a convex polygon), the stretch factor of $DT(S)$ is at most 1.77. This improves upon the previously known factor 1.84 in the convex case.

1 Introduction

Let S be a set of n points in the plane. For a pair of points p, q in the plane, denote by pq the line segment connecting p and q, and $|pq|$ the Euclidean distance between p and q. Let $G(S)$ be such a graph that each vertex corresponds to a point in S and the weight of an edge is the Euclidean distance between its two endpoints. Denote by $P(a,b)$ the shortest path in $G(S)$ between two points $a, b \in S$, and $|P(a,b)|$ the total length of path $P(a,b)$. The graph $G(S)$ is said to approximate the complete graph of S if $\frac{|P(a,b)|}{|ab|}$, called the *stretch factor* of $G(S)$, is upper bounded by a constant, independent of S and n. As compared with $O(n^2)$ edges of the complete graph of S, it is desirable to identify classes of graphs that approximate complete graphs well and have only $O(n)$ edges; these graphs have potential applications in geometric network design problems [7,8].

Denote by $DT(S)$ the planar graph of the Delaunay triangulation of S. Dobkin *et al.* [6] gave a stretch factor $(1+\sqrt{5})/2)\pi \approx 5.08$ of Delaunay triangulations with respect to complete graphs, which was later improved to $2\pi/\ (3\cos(\pi/6)) \approx 2.42$ by Keil and Gutwin [9], and further to 1.998 by Xia [12]. Determining the smallest stretch factor of $DT(S)$ has been a long-standing open problem in computational geometry. On the other hand, Xia and Zhang [13] gave a lower bound of 1.5932 on the stretch factor of $DT(S)$.

This work was partially supported by JSPS KAKENHI Grant Number 20K11683.

S. Ghosh and Z. Zhang (Eds.): AAIM 2024, LNCS 15179, pp. 263–273, 2024.
https://doi.org/10.1007/978-981-97-7798-3_22

Cui *et al.* [5] have also studied the stretch factor of $DT(S)$ for the points in convex position. A set of points is said to be in *convex position*, if all points form the vertices of a convex polygon. Amani *et al.* [1] have constructed a planar graph, whose vertices are in convex position, such that its stretch factor is 1.88. (The planar graph studied by Amani *et al.* is not the Delaunay triangulation of the given point set.) The currently known best factor in this special situation is 1.84, due to a work of Tan et al. on $DT(S)$ [11]. On the other hand, Bose et al. [4] have shown a lower bound 1.5810 on the stretch factor of $DT(S)$ in the convex case.

There is a large work on stretch factors of other graphs (e.g., [1,3]). Particularly, the study of stretch factors of planar graphs is important, as planar graphs (including Delaunary triangulations) have $O(n)$ edges, in comparison with $O(n^2)$ edges of complete graphs.

In this paper, we revisit the stretch factor of $DT(S)$ for points in convex position, and show that it is less than 1.77. This improves upon the previously known factor 1.84 [11]. A deep, important observation made in this paper is that except for a special case, the path used to calculate the stretch factor for two given points can be enclosed by a chain of four circular arcs. In the exception case, the stretch factor is $\pi/2$. Our work also shows a large possibility of reducing the stretch factor of $DT(S)$ in the convex case to 1.73 ($< \sqrt{3}$), or even to a smaller factor close to its lower bound 1.5932.

2 Preliminaries

For a point p in the plane, denote the coordinates of p by $p.x$ and $p.y$, respectively. Without loss of generality, assume that no four points of S are cocircular in the plane, and no three points of S are collinear. The *Voronoi diagram* for S, denoted by *Vor*(S), is a partition of the plane into regions, each containing exactly one point in S, such that for each point $p \in S$, every point within its corresponding region, denoted by *Vor*(p), is closer to p than to any other point of S [2]. The boundaries of these Voronoi regions form a planar graph.

The *Delaunay triangulation* of S, denoted by $DT(S)$, is the straight-line dual of the Voronoi diagram for S; that is, we connect a pair of points in S if and only if they share a Voronoi boundary. Since $DT(S)$ is a planar graph, it has $O(n)$ edges.

For a pair of points $a, b \in S$, denote by $DT(a,b)$ the shortest path in $DT(S)$ between a and b, and $|DT(a,b)|$ the total length of path $DT(a,b)$. The *stretch factor* of $DT(S)$ to the complete graph of S is then the maximum ratio $\frac{|DT(a,b)|}{|ab|}$ among all point pairs (a,b) [6,7].

Suppose that two points $a, b \in S$ are on a horizontal line, with $a.x < b.x$. Let $a = a_0, a_1, \ldots, a_m = b$ be the sequence of points of S, whose Voronoi regions intersect segment ab, see Fig. 1.[1] The path obtained in this way is called the *direct path from a to b* [6]. For ease of presentation, denote by $dp(a,b)$ the direct path from a to b in $DT(S)$.

Path $dp(a,b)$ is said to be *one-sided* if all points of the path are to the same side of the line through a and b.

[1] In this section, we discuss on $DT(S)$ for sets S of points in general position.

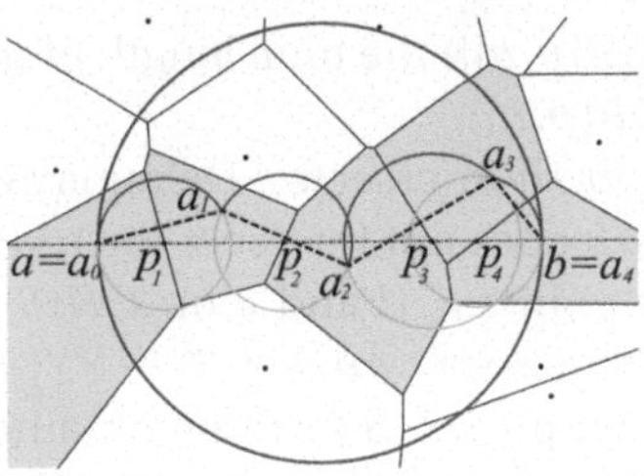

Fig. 1. An instance of path $dp(a,b)$.

Lemma 1. *(Dobkin et al. [6]) If path $dp(a,b)$ is one-sided, then it has length at most $\pi|ab|/2$.*

Let p_i be the intersection point of ab with the Voronoi edge between $Vor(a_{i-1})$ and $Vor(a_i)$, for $1 \le i \le m$. From the definition of $Vor(S)$, the circle passing through a_{i-1} and a_i, centered at p_i, does not contain any point of S in its *interior.* All points a_i are then contained in the circle of diameter ab, no matter whether or not $dp(a,b)$ is one-sided (Fig. 1).

A more general result than Lemma 1 is the following.

Lemma 2. *(Dobkin et al. [6]) Let $C_1, C_2, \ldots, C_k$ be the circles centered on the same line such that $U = \bigcup_{1\le i\le k} C_i$ is connected. The boundary of U has length at most $\pi|ab|$ and is contained in the circle of diameter ab, where a and b are two extreme endpoints of U on the line.*[2]

3 The Main Result

Denote by C the circle of diameter ab, and o the center of C. In the rest of this paper, assume that the set S of given points is in convex position. Let $CH(S)$ be the convex hull of S, i.e., the boundary of the smallest convex polygon containing all points of S.

Most of the notations in this paper are borrowed from that of [11]. Let a and b be two arbitrary points of S. Assume without loss of generality that a and b are on a horizontal line, with $a.x < b.x$. Denote by $SA[a,b]$ (resp. $SB[a,b]$) the sequence of vertices of $CH(S)$, which are above (resp. below) the line through a and b. For a point $p \in SA[a,b]$, denote by $SA[a,p]$ and $SA[p,b]$ the intervals of $SA[a,b]$ from a to p and from p to b, respectively. Analogously, $SB[a,q]$ (resp. $SB[q,b]$), $q \in SB[a,b]$, represents the interval of $SB[a,b]$ from a to q (resp. from q to b). Also, denote by $SA(a,b)$ and $SB(a,b)$ the open intervals of $SA[a,b]$ and $SB[a,b]$ respectively, excluding the extreme points a and b.

Suppose that all points of $SA[a,b]$ (resp. $SB[a,b]$) are ordered from a to b. Sometimes, $SA[a,b]$ or $SB[a,b]$ also refers to as the convex chain formed by its points.

---

[2] From the proof [6, Lemma 2], the boundary of U is also contained in the circle of diameter ab.

Denote by $|SA[p,q]|$ (resp. $|SB[p,q]|$) the total length of the convex chain formed by all points of $SA[p,q]$ (resp. $SB[p,q]$).

Segment ab is said to *properly* intersects a Delaunay triangle if it goes across the interior of the triangle (i.e., ab does not intersect the triangle only at a vertex). Suppose that three consecutive points c, d and e on $CH(S)$ form a Delaunay triangle $\triangle_{c,d,e}$ that does not properly intersect ab. Let $S' \leftarrow S \setminus \{d\}$. From the convexity of S and the planarity of $DT(S)$, graph $DT(S')$ can be obtained from $DT(S)$ by deleting two edges cd and de. Path $DT(a,b)$ in $DT(S)$, by its optimality, is the same as that in $DT(S')$. Since the above operation of deleting a point from the set of present points, if possible, can be performed repeatedly, the following observation can be made.

Observation 1. *([11]) For any two points $a, b \in S$, one can assume that segment ab properly intersects all triangles of $DT(S)$, in calculating the value of $\frac{|DT(a,b)|}{|ab|}$.*

The previous work of Tan et al. [11] showed that either $SA[a,b]$ or $SB[a,b]$ is enclosed by a convex chain that consists of a circular arc and a few line segments, which leads to the stretch factor 1.84 for $DT(S)$. In this paper, we make a deep observation that excluding a special case, one of two chains $SA[a,b]$ and $SB[a,b]$ can be enclosed by a chain of four circular arcs (see the proofs of Lemmas 3 and 4). In the exception case, $|DT(a,b)|/|ab| \leq \pi/2$.

Let us describe our result in detail. If the first and last points of $SA(a,b)$ (resp. $SB(a,b)$) are both contained in C, then we can prove that chain $SA[a,b]$ (resp. $SB[a,b]$) is enclosed by a chain of four circular arcs. In this case, the stretch factor of $DT(S)$ is 1.77.

Lemma 3. *Suppose that both the first and last points of $SA(a,b)$ are contained in C. Then, $|DT(a,b)|/|ab| \leq |SA[a,b]|/|ab| < 1.77$.*

Notice that if $dp(a,b)$ intersects ab an even number of times, then the first and last turning points of path $dp(a,b)$ are to the same side of the line through a and b. So, they are contained in C and belong to either of $SA(a,b)$ and $SB(a,b)$. Lemma 3 can be thus applied, so as to give the stretch factor 1.77.

The remaining situation is that $dp(a,b)$ intersects ab an odd number of times. If the condition stated in Lemma 3 holds for $SA(a,b)$ or $SB(a,b)$, then the stretch factor of $DT(S)$ is also 1.77 (even if $dp(a,b)$ intersects ab an odd number of times). Otherwise, we prove that except for a special case, Lemma 3 can still apply to either of $SA[a,b]$ and $SB[a,b]$. In the exception case, $|DT(a,b)|/|ab| \leq \pi/2$. In conclusion, we obtain that the stretch factor of $DT(S)$ is 1.77 (Theorem 1).

Lemma 4. *Suppose that $dp(a,b)$ properly intersects segment ab an odd number of times. Then, $|DT(a,b)|/|ab| < 1.77$.*

Theorem 1. *Suppose that the set S of given points is in convex position, and a, b are two points of S. In the Delaunay triangulation of S, there is a path from a to b such that its length is less than $1.77|ab|$.*

As will be seen, Lemma 3 or an analog of Lemma 3 is a key in our study of the stretch factor of $DT(S)$.

4 Proofs

Assume that segment ab is on the x-axis. The *bisector* of two points p and q in the plane, denoted by $B_{p,q}$, is the perpendicular line through the middle point of segment pq. For an arc $\widehat{pq}$ of a circle clockwise from p to q, denote its length by $|\widehat{pq}|$.

4.1 Proof of Lemma 3

Assume that $SA[a,b]$ is not wholly contained in C; otherwise, $|DT(a,b)|/|ab| \leq |SA[a,b]|/|ab| \leq \pi/2$. From the convexity of $SA[a,b]$, there is an edge uv in $SA[a,b]$ such that u is outside of C, v is inside C, and $SA[v,b]$ is contained in C. Also, there is an edge $u'v'$ in $SA[a,b]$ such that u' is inside C, v' is outside of C, and $SA[a,u']$ is contained in C. See Fig. 2(a). Note that the slope of uv (resp. $u'v'$) is negative (resp. positive).

Let l_1 and l_2 be the left and right intersection points of C with the horizontal line through the topmost Voronoi vertex (i.e., its y-coordinate is the maximum among those of all Voronoi vertices), respectively. See Fig. 2(a). Denote by D the circle of diameter l_1l_2, and o' the center of D. Then, o' is on $B_{a,b}$ and vertically above the x-axis, and two points a, b are outside of D. From Observation 1, segment l_1l_2 intersects regions $Vor(r)$, for all points $r \in SA[u',v]$. Since u' (resp. v) lies in C, the leftmost (resp. rightmost) point of the circumcircle of the Delaunay triangle having an edge $u'v'$ (resp. uv) is contained in C; in the case that $u'v'$ (resp. uv) intersects l_1l_2, the circle centered on l_1l_2 and passing through v' (resp. u) and that intersection point is used instead, and it is thus to the right of l_1 (resp. left of l_2). Lemma 2 can be then applied to the collection of the circles, which are centered on l_1l_2 and pass through every pair of consecutive points of $SA[u',v]$. Therefore, all points of $SA[u',v]$ are contained in D, see Fig. 2(a).

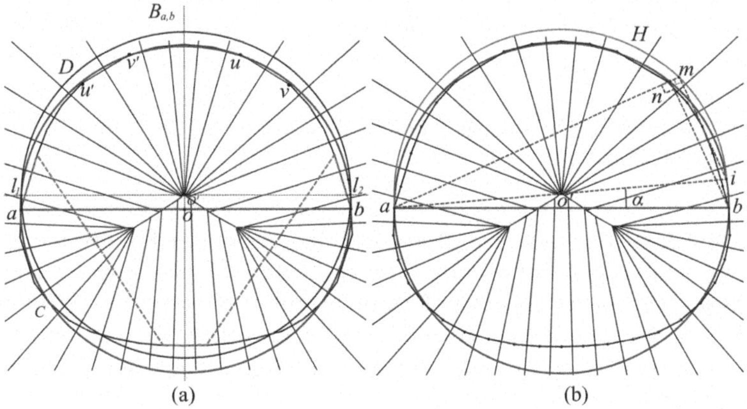

Fig. 2. Illustration of the proof of Lemma 3

Let i ($\neq a$) and j ($\neq b$) be the intersection points of C with the lines containing ao' and bo', respectively. (Clearly, $i.y = j.y$.) Since l_1l_2 intersects region $Vor(u)$, point

i (on C) is outside of $CH(S)$, no matter whether v or not is vertically above the line through a and i. Since all points of $SA[a,u']$ are contained in C, as described above, all points of $SA[a,v]$ are contained in the circle of diameter ai. Analogously, j is outside of $CH(S)$, and all points of $SA[u',b]$ are contained in the circle of diameter bj.

Denote by H the *semi*-circle of diameter ai, whose *interior* points are all vertically above the line through a and i. See Fig. 2(b). As discussed above, $SA[a,b]$ is contained in the region bounded by ab, H and the arc $\widehat{ib}$ of C. See Fig. 3(a). Let H' be the semi-circle of diameter bj, whose interior points are vertically above the line through b and j. Also, $SA[a,b]$ is contained in the region bounded by ab, H' and the arc $\widehat{ja}$ of C.

Let $\angle iab = \alpha$. Then, $\angle jba = \alpha$, too. As in [11], we can show that $\alpha < \pi/6$. Draw a tangent from i to the portion of $SA[a,b]$, which is vertically above ai. Let m be the intersection point of the tangent with H, and let n be the intersection point of C with am. See Fig. 2(b). Since $\angle anb = \angle ami = \pi/2$, two segments im and bn are parallel. Since im intersects C, $\angle mai > \alpha$ holds. As $m.y > i.y$ $(= j.y)$, we have $\angle iab + \angle jba + \angle nai < \pi/2$. By noticing the fact that $\angle nai = \angle mai$, we obtain $\alpha < \pi/6$.

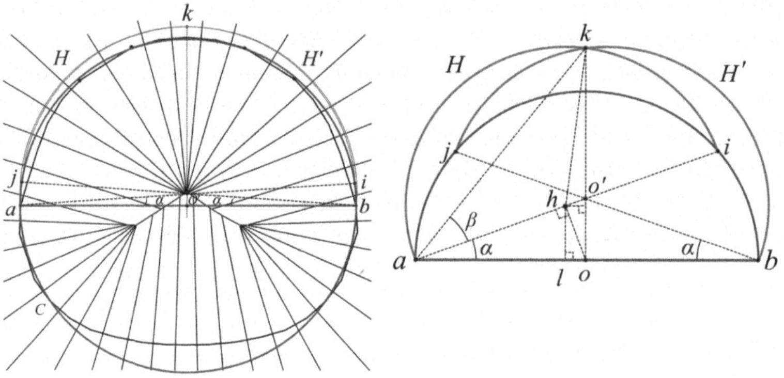

Fig. 3. $SA[a,b]$ is contained in the region bounded by arcs $\widehat{aj}$ (of C), $\widehat{jk}$ (of H'), $\widehat{ki}$ (of H), $\widehat{ib}$ (of C), and segment ab.

Let us now give an analysis on the stretch factor obtained in this work. Denote by k the intersection point of H and H'. Since $\angle iba = \angle jab$, we have $k.x = o.x$. See Fig. 3. As described above, all points of $SA[u',v]$ are contained in the region bounded by H and ai (resp. H' and bj). Since $SA[a,u']$ and $SA[v,b]$ are contained in C, all points of $SA[a,b]$ are thus contained in the region bounded by arcs $\widehat{aj}$ (of C), $\widehat{jk}$ (of H'), $\widehat{ki}$ (of H), $\widehat{ib}$ (of C), and segment ab. See Fig. 3(b).

Denote by h the center of H, and t the point on ao such that $\angle ath = \pi/2$. Then, $|ai| = |ab|\cos(\alpha)$. Since $\angle aho = \pi/2$, $|ah| = |hk| = |ab|\cos(\alpha)/2$, and $|at| = \cos(\alpha)|ah| = |ab|\cos^2(\alpha)/2$. So, $|to| = |ao| - |at| = |ab|/2 - |ab|\cos^2(\alpha)/2 = |ab|\sin^2(\alpha)/2$, and $|ht| = \sin(\alpha)|ah| = |ab|\sin(2\alpha)/4$. Since $|oh| = \sin(\alpha)|ab|/2$, $|ok| = \sqrt{|oh|^2 - |to|^2} + \sqrt{|hk|^2 - |to|^2} = (\sin(2\alpha)/2 + \sqrt{\cos^2(\alpha) - \sin^4(\alpha)})|ab|/2$.

Let $\beta = \angle kai$. Then, $\tan(\beta+\alpha) = |ok|/|ao| = \sin(2\alpha)/2 + \sqrt{\cos^2(\alpha) - \sin^4(\alpha)}$, and $\beta = \arctan(\sin(2\alpha)/2 + \sqrt{\cos^2(\alpha) - \sin^4(\alpha)}) - \alpha$. Note that $\widehat{jk}$ (of H') is of the same length as $\widehat{ki}$ (of H). Since the points of $SA[a,b]$ are in convex position, a simple argument (as in [10, Lemma 3.6]) can then show that $|SA[a,b]| \le |\widehat{aj}| + |\widehat{jk}| + |\widehat{ki}| + |\widehat{ib}|$. Hence, $|DT(a,b)|/|ab| \le |SA[a,b]|/|ab| \le 2\alpha + 2\cos(\alpha)(\arctan(\sin(2\alpha)/2 + \sqrt{\cos^2(\alpha) - \sin^4(\alpha)}) - \alpha)$.

Let $f(\alpha) = 2\alpha + 2\cos(\alpha)(\arctan(\sin(2\alpha)/2 + \sqrt{\cos^2(\alpha) - \sin^4(\alpha)}) - \alpha)$. It seems difficult to formally prove a better upper bound on $f(\alpha)$, $0 < \alpha < \pi/6$. For this reason, we simplify function $f(\alpha)$ a little bit. Since $\sin(\alpha) < \alpha$ and $\sin(2\alpha) < \sqrt{3}/2$, $0 < \alpha < \pi/6$, let $f_1(\alpha) = 2\alpha + 2\cos(\alpha)(\arctan(\alpha + \cos(\alpha)) - \alpha)$, $0 < \alpha \le 10\pi/73$, and let $f_2(\alpha) = 2\alpha + 2\cos(\alpha)(\arctan(\sqrt{3}/4 + \cos(\alpha)) - \alpha)$, $10\pi/73 \le \alpha < \pi/6$. Then, $f_1'(\alpha) = 2 - 2\sin(\alpha)(\arctan(\alpha + \cos(\alpha)) - \alpha) + 2\cos(\alpha)(\frac{1-\sin(\alpha)}{1+(\alpha+\cos(\alpha))^2} - 1)$. It can be shown (in Sect. 4.4) that $f_1'(\alpha) > 0$, $0 < \alpha \le 10\pi/73$. Then, we have $f_1(\alpha) \le f_1(10\pi/73) < 1.77$. On the other hand, $f_2'(\alpha) = 2 - 2\sin(\alpha)(\arctan(\sqrt{3}/4 + \cos(\alpha)) - \alpha) + 2\cos(\alpha)(\frac{-\sin(\alpha)}{1+(\sqrt{3}/4+\cos(\alpha))^2} - 1) < 0$, see Sect. 4.4. Hence, we have $f_2(\alpha) \le f_2(10\pi/73) < 1.77$, $10\pi/73 \le \alpha < \pi/6$.

Finally, since $\sqrt{\cos^2(\alpha) - \sin^4(\alpha)} < \cos(\alpha)$, we have $f(\alpha) < f_1(\alpha)$, $0 < \alpha \le 10\pi/73$, and $f(\alpha) < f_2(\alpha)$, $10\pi/73 \le \alpha < \pi/6$. The proof is thus complete.[3]

4.2 Proof of Lemma 4

Suppose that $dp(a,b)$ is not one-sided, otherwise, the stretch factor of $DT(S)$ is $\pi/2$ (Lemma 1). Since path $dp(a,b)$ intersects ab an odd number of times, it consists of at least three line segments. Denote by ac and bd the first and last segments of path $dp(a,b)$, as viewed from a, respectively. Assume that c is below and d is above ab. From the definition of $dp(a,b)$, both c and d are contained in C.

Denote by p (resp. q) the point of $SA[a,b]$ (resp. $SB[a,b]$) immediately after a (resp. before b). Assume that both p and q are outside of C; otherwise, Lemma 3 applies to $SA(a,b)$ or $SB(a,b)$, and thus, the stretch factor of $DT(S)$ is 1.77.[4] Let p' (resp. q') be the intersection point of $B_{b,d}$ (resp. $B_{a,c}$) with the line containing ap (resp. bq), and d' (resp. c') the midpoint of bd (resp. ac). See Fig. 4.

The stretch factor of $DT(S)$ can be given according to the following situations.

Case 1. Either $\angle ap'd' \ge \pi/2$ or $\angle bq'c' \ge \pi/2$ holds.

Suppose that $\angle ap'd' \ge \pi/2$. Denote by s and t two consecutive vertices on path $dp(a,b)$ such that s is below and t is above ab. (In the instance of Fig. 4, point s is identical to c.) Segment st then intersects ab, at a point, say, k. Let k_1 and k_2 be the intersection points of the line through s and t with the lines containing ap and bd, respectively. See Fig. 4.

Let us extend bd until it touches C at a point, say, l ($\ne b$). Then, $\angle alb = \pi/2$. Since the chain consisting of $SB[a,s]$ and segment st (or sk_2) is convex, the orthogonal projection of s onto lb is to the left of point k_2. See Fig. 4. Thus, $\angle sk_2l < \pi/2$ and $\angle kk_2b > \pi/2$. Moreover, since $\angle ld'p' = \pi/2$, $B_{b,d}$ (it contains $p'd'$) intersects

[3] The proof of $f_1'(\alpha) > 0$ and $f_2'(\alpha) < 0$ is provided in Sect. 4.4.

[4] Lemma 3 is independent of the number of intersection points between $dp(a,b)$ and ab.

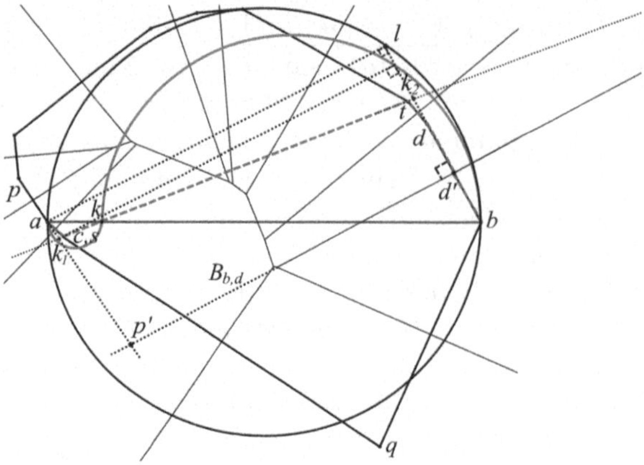

Fig. 4. Illustration of Case 1.

the line containing st at a point that is vertically above the x-axis (Fig. 4). Thus, $\angle ak_1k > \angle ap'd' \geq \pi/2$. Hence, the convex chain consisting of $SB[a,s]$ and segment sk is enclosed by a semi-circle of diameter ak, and the convex chain consisting of segment kt and $SA[t,b]$ is enclosed by a semi-circle of diameter kb. See Fig. 4. Since $st \in DT(S)$, we obtain $|DT(a,b)|/|ab| \leq \pi/2$.

Case 2. Both $\angle ap'd' < \pi/2$ and $\angle bq'c' < \pi/2$ hold.

Recall that db is the last segment of $dp(a,b)$. From the assumption that no three points are collinear, we have $d.y > 0$. Since d is contained in C, the slope of bd is negative. Analogously, the slope of ac is also negative. Assume below that the slope of bd is smaller or equal to that of ac. Since $\angle ap'd' < \pi/2$ and $\angle p'd'd = \pi/2$, two lines containing ap (or pp') and db intersect at a point that is above the x-axis. Since $a.x < b.x$, the slope of ap, if *negative*, is smaller than that of bd. As will be seen, this makes it possible to apply Lemma 3 again. (In the case that the slope of bd is larger than that of ac, the other inequality $\angle bq'c' < \pi/2$ is employed).

Let us extend bd until it intersects C at a point, say, b' ($\neq b$). Then, b' is outside of $CH(S)$. Let a' be the (other) intersection point of C with the line through b' and o. Since aa' is parallel to $b'b$ and the (negative) slope of bd is smaller or equal to that of ac, point a' is outside of $CH(S)$, too. See Figs. 5 and 6. We further distinguish two different situations depending on the slope of ap.

Case 2.1 The slope of ap is negative.

Denote by e the last vertex of $SA[a,b]$ such that e is outside of C, and f the vertex of $SA[a,b]$ immediately after e. Assume that f is on C (Fig. 5(a)); otherwise, the intersection point of C with ef can be considered (as f). Hence, f is to the left of b' (i.e., $f.x < b'.x$).

Denote by L the line through a' and b'. Find the smallest circle D, whose center is on $B_{a',b'}$ and which contains all points of $SA[a,f]$ and the portion of path $dp(a,b)$ that is to the left of L. Denote by o' the center of D, see Fig. 5. Since the points asked to be contained in D are all to the left of L, and since at least one point of $SA[a,f]$ is outside of C, point o' is to the left of L (Fig. 5(a)).

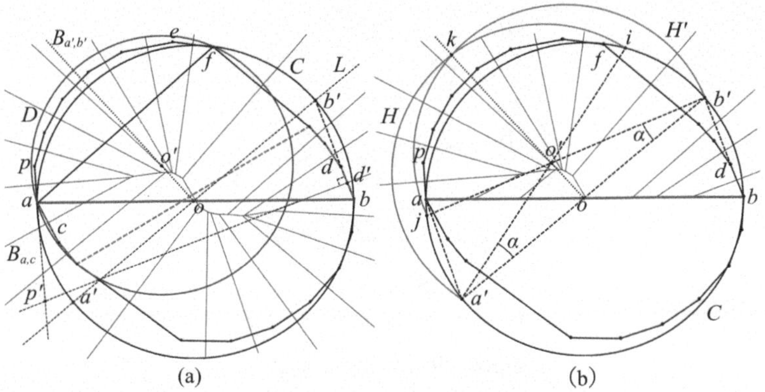

Fig. 5. Illustrating Case 2.1.

Again, let i and j be the intersection points of C with the lines containing $a'o'$ and $b'o'$, respectively (Fig. 5(b)). Assume that both i and j are outside of $CH(S)$; otherwise, it can be done by moving the center o' towards o directly, while keeping chain $SA[a, f]$ and the portion of path $dp(a, b)$, which is to the left of L, contained in the enlarged circle, until the new position of o' is vertically below both lines containing $a'f$ and ab'.

Let H (resp. H') be the semi-circle of diameter $a'i$ (resp. jb'), which is to the left of the line containing $a'i$ (resp. jb'). See Fig. 5(b). By establishing the one-to-one correspondence for pairs $(a'b', ab)$, $(ap, u'v')$ and (ef, uv), as discussed in Sect. 4.1, all points of $SA[a, f]$ are contained in the region bounded by $a'i$ and H (resp. $b'j$ and H'). Since $a'a$ and $b'b$ are of the same length, $|SA[a, b]| \leq |SA[a, f]| + |\widehat{fb'}| + |b'b| = |a'a| + |SA[a, f]| + |\widehat{fb'}|$. See Fig. 5. Since the slope of ap is smaller than that of bd, the chain consisting of segment $a'a$ and $SA[a, f]$ is *convex*. From the proof of Lemma 3, the stretch factor of $DT(S)$ is then less than 1.77.

Case 2.2 The slope of ap is positive.

In this case, the slopes of ap and bd have different signs. Obviously, the chain consisting of $a'a$ and $SA[a, f]$ is convex, where points a' and f are defined in the same manner as in Case 2.1. See Fig. 6. As argued in Case 2.1, the stretch factor of $DT(S)$ is less than 1.77, too.

The above cases enumerate all possible situations. Thus, the proof is complete.

4.3 The Proof of Theorem 1

Assume that path $dp(a, b)$ is not one-sided; otherwise, the stretch factor of $DT(S)$ is $\pi/2$ (Lemma 1). If path $dp(a, b)$ intersects ab an even number of times, then its first and last segments are contained in C, i.e., the first and last points of either $SA(a, b)$ or $SB(a, b)$ are contained in C. Thus, the stretch factor of $DT(S)$ is 1.77 (Lemma 3). In the case that $dp(a, b)$ intersects ab an odd number of times, the stretch factor of $DT(S)$ is 1.77, too (Lemma 4).

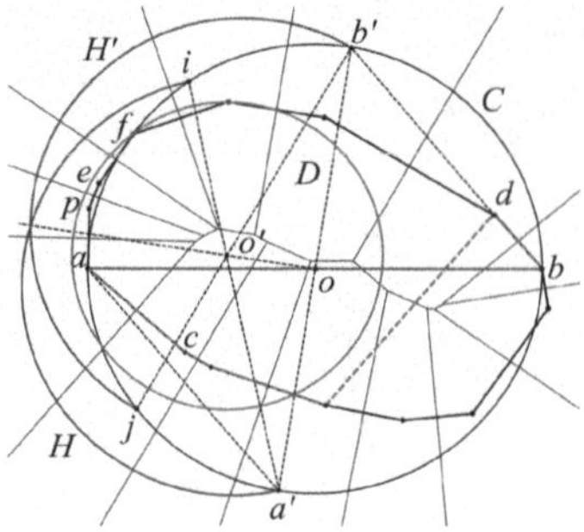

Fig. 6. Illustrating Case 2.2.

4.4 The Proof of $f_1'(\alpha) > 0$ and $f_2'(\alpha) < 0$

Recall that $f_1'(\alpha) = 2 - 2\sin(\alpha)(\arctan(\alpha + \cos(\alpha)) - \alpha) + 2\cos(\alpha)(\frac{1-\sin(\alpha)}{1+(\alpha+\cos(\alpha))^2} - 1))$, $0 < \alpha \le 10\pi/73$. Let $g(\alpha) = \alpha + \cos(\alpha)$, $0 < \alpha \le 10\pi/73$. Since $g'(\alpha) = 1 - \sin(\alpha) > 0$, we have $g(\alpha) \le 10\pi/73 + \cos(10\pi/73)$. Thus, $\arctan(\alpha + \cos(\alpha)) \le \arctan(10\pi/73 + \cos(10\pi/73)) < 1$, and $1 + (\alpha + \cos(\alpha))^2 \le 1 + (10\pi/73 + \cos(10\pi/73))^2 < 2.8$. Hence, $f_1'(\alpha) > 2 - 2\sin(\alpha) + 2\alpha\sin(\alpha) - \frac{2}{2.8}\cos(\alpha)\sin(\alpha) - \frac{3.6}{2.8}\cos(\alpha) = 2 - 2\sin(\alpha) + 2\alpha\sin(\alpha) - \frac{\sin(2\alpha)}{2.8} - \frac{3.6}{2.8}\cos(\alpha)$. Since $(2 - 2\sin(\alpha) + 2\alpha\sin(\alpha) - \frac{\sin(2\alpha)}{2.8} - \frac{3.6}{2.8}\cos(\alpha))' = -2\cos(\alpha) + 2\sin(\alpha) + 2\alpha\cos(\alpha) - \frac{\cos(2\alpha)}{1.4} + \frac{3.6}{2.8}\sin(\alpha) < -2\cos(\alpha) + 3.3\sin(\alpha) + 2\alpha\cos(\alpha) - \frac{\cos(2\alpha)}{1.4} = (-0.9\cos(\alpha) + 2\alpha\cos(\alpha)) + (-1.1\cos(\alpha) + 2.3\sin(\alpha)) + (-\frac{\cos(2\alpha)}{1.4} + \sin(\alpha)) < 0$, $0 < \alpha \le 10\pi/73$, we have $(2 - 2\sin(\alpha) + 2\alpha\sin(\alpha) - \frac{\sin(2\alpha)}{2.8} - \frac{3.6}{2.8}\cos(\alpha)) \ge 2 - 2\sin(10\pi/73) + 20\pi\sin(10\pi/73)/73 - \frac{\sin(20\pi/73)}{2.8} - \frac{3.6}{2.8}\cos(10\pi/73) > 0$. Hence, $f_1'(\alpha) > 0$, $0 < \alpha \le 10\pi/73$.

Note that $f_2'(\alpha) = 2 - 2\sin(\alpha)(\arctan(\sqrt{3}/4 + \cos(\alpha)) - \alpha) + 2\cos(\alpha)(\frac{-\sin(\alpha)}{1+(\sqrt{3}/4+\cos(\alpha))^2} - 1)$. Since $10\pi/73 \le \alpha < \pi/6$, we have $\arctan(\sqrt{3}/4 + \cos(\alpha)) \ge \arctan(\sqrt{3}/4 + \cos(\pi/6)) > 0.9$ and $1 + (\sqrt{3}/4 + \cos(\alpha))^2 \ge 1 + (\sqrt{3}/4 + \cos(\pi/6))^2 > 2.6$. Thus, $f_1'(\alpha) < 2 - 1.8\sin(\alpha) + 2\sin(\alpha)\alpha - \frac{2}{2.6}\cos(\alpha)\sin(\alpha) - 2\cos(\alpha) = 2\sin(\alpha)(\alpha - 0.9) + (2 - \frac{\sin(2\alpha)}{2.6} - 2\cos(\alpha))$. Since $(\alpha - 0.9) \le \pi/6 - 0.9 < 0$ and $(2 - \frac{\sin(2\alpha)}{2.6} - 2\cos(\alpha)) \le 2 - \frac{\sin(20\pi/73)}{2.6} - 2\cos(\pi/6) < 0$, we obtain $f_2'(\alpha) < 0$, $10\pi/73 \le \alpha < \pi/6$.

5 Concluding Remarks

1. It is an interesting work to further improve the result given in Lemma 3. Although a strict proof cannot be given, we believe that an upper bound 1.73 ($< \sqrt{3}$) on function $f(\alpha) = 2\alpha + 2\cos(\alpha)(\arctan(\sin(2\alpha)/2 + \sqrt{\cos^2(\alpha) - \sin^4(\alpha)}) - \alpha)$, $0 < \alpha < \pi/6$, can be obtained. Figure 7 shows an experimental result on $f(\alpha)$, which is produced using *gunplot.*
2. We have shown that the stretch factor of the Delaunay triangulation of a set of points in convex position is less than 1.77. We believe that the same stretch factor also holds for the set of points in general position. Whether or not Lemma 3 can

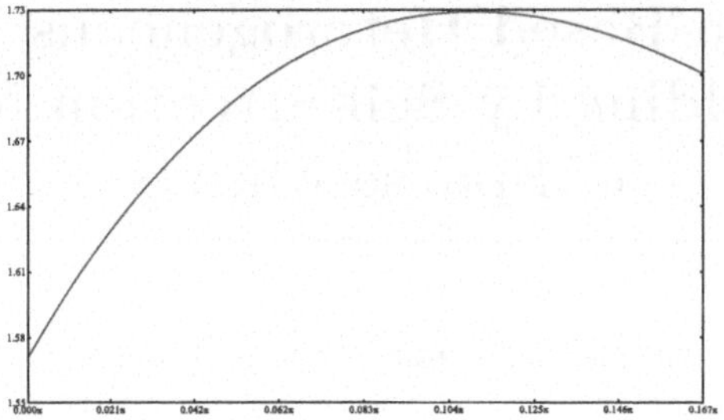

Fig. 7. An illustration of the function $f(\alpha)$, $0 < \alpha < \pi/6$.

be generalized to the set of points in general position is clearly an interesting open problem.

References

1. Amani, M., Biniaz, A., Bose, P., De Carufel, J.-L., Maheshwair, A., Smid, M.: A plane 1.88-spanner for points in convex position. J. Comput. Geom. **7**, 520–539 (2016)
2. de Berg M., van Kreveld M., Overmars M., Schwarzkopf O.: Computational Geometry: Algorithms and Applications. Springer, Cham (2008). https://doi.org/10.1007/978-3-540-77974-2
3. Bose, P., Smid, M.: On planar geometric spanners: a survey and open problems. Comput. Geom. **46**, 818–830 (2013)
4. Bose, P., Devroye, L., Loffler, M., Snoeyink, J., Verma, V.: Almost all Delaunay triangulations have stretch factor larger than $\pi/2$. Comput. Geom. **44**, 121–127 (2011)
5. Cui, S., Kanj, I.A., Xia, G.: On the stretch factor of Delaunay triangulations of points in convex position. Comput. Geom. **44**, 104–109 (2011)
6. Dobkin, D.P., Friedman, S.T., Supowit, K.J.: Delaunay graphs are almost as good as complete graphs. Discrete Comput. Geom. **5**, 399–407 (1990)
7. Eppstein, D.: Spanning trees and spanners. In: Sack, J.-R., Urrutia, J. (eds.) Handbook of Computational Geometry, pp. 425–462, North-Hollard Press (2000)
8. Narasimhan, G., Smid, M.: Geometric Spanner Networks. Cambridge University Press, Cambridge (2007)
9. Keil, J.M., Gutwin, C.A.: The Delaunay triangulation closely approximates the complete graph. Discrete Comput. Geom. **7**, 13–28 (1992)
10. Hershberger, J., Suri, S.: Practical methods for approximating shortest paths on a convex polytope in R^3. Comput. Geom. **10**, 31–46 (1998)
11. Tan, X., Sakthip, C., Jiang, B., Liu, S.: Improved stretch factor of Delaunay triangulations of points in convex position. J. Combin. Opt. **45**(3), 1–12 (2023)
12. Xia, G.: The stretch factor of the Delaunay triangulation is less than 1.998. SIAM J. Comput. **42**, 1620–1659 (2013)
13. Xia, G., Zhang, L.: Towards the tight bound of the stretch factor of Delaunay triangulations. In: Proceedings of CCCG 2011, pp. 175–180 (2011)

Transition-Based Heterogeneous Network Embedding by Self-attention Model of Sparse Matrix

Keqin Chen and Zhipeng Jiang(✉)

School of Mathematical Sciences, University of Chinese Academy of Sciences, Beijing 100049, China
chenkeqin22@mails.ucas.ac.cn, jiangzhipeng@ucas.ac.cn

Abstract. Heterogeneous network embedding aims to embed the networks into low dimensional spaces, in which each vertex is represented as a low-dimensional vector. Compared with homogeneous graphs, heterogeneous graphs increase the complexity of representation learning due to their heterogeneity. In this paper, we distinguish different types of edges in heterogeneous networks and project the nodes into different relational spaces. In addition, all previous works on heterogeneous graph representation ignored the imbalance in heterogeneous networks. Inspired by the TranSparse method used in the field of knowledge graph embeddings, we extend the idea of this method to heterogeneous networks to handle the imbalance. We consider replacing the transition matrix used in projection with sparse-matrices-correlation matrix, where the sparse matrices used are adaptive. For the two sparse patterns (structured and unstructured), we propose to combine them using a self-attention model, which effectively integrates the information extracted from the two patterns. Experiments are conducted on three heterogeneous networks to evaluate our algorithm on link prediction tasks and the experimental results show that our algorithm outperforms other baseline algorithms for heterogeneous network embeddings.

Keywords: Heterogeneous network embedding · Sparse pattern · Self-attention model

1 Introduction

Heterogeneous information network (heterogeneous graph) contains various types of nodes or edges. Compared with homogeneous graphs, heterogeneous graphs model real-world systems as interactions between a large number of multimodal and multi-type objects, providing a better characterization of the real world. As for heterogeneous network embedding, the low-dimensional vectors learned by network embedding methods can capture rich structural and semantic information in the graph. Moreover, the methods of encoding nodes in the graph into continuous vector spaces also exhibit strong robustness, providing

S. Ghosh and Z. Zhang (Eds.): AAIM 2024, LNCS 15179, pp. 274–285, 2024.
https://doi.org/10.1007/978-981-97-7798-3_23

effective and scalable approaches for modeling graphs. At the same time, the heterogeneity poses more challenges for learning.

In this paper, we complement the non-network method for structural preservation with a network-based approach using a self-attention model. Firstly, using Euclidean distance to measure the similarity of nodes directly may lead to logical errors due to the different semantics of different types of edges. For example, in a heterogeneous network, the triangle inequality cannot be satisfied, which mathematically constrains it geometrically and leads to an indeterminate algebraic system as it attempts to fit each pair of connected nodes to the same point in a low-dimensional space, but each node may have many neighbors with different relationships [1]. So the idea of projected metric embedding is introduced into our method.

Furthermore, when doing statistics on the datasets, we find that the issue of the imbalance exists widely in heterogeneous networks, which manifests in two ways: (1) imbalance in the number of head and tail nodes. There may be a large number of "1-N" and "N-1" edges in heterogeneous networks. For example, in a publication network, a "venue" as a tail node receives submissions from multiple "paper" nodes, and an "author" node as a head node participates in writing multiple "paper"; (2) imbalance in the number of different types of edges in the graph, which is likely to occur. Previous works related to heterogeneous network embedding learning have not considered the issue of imbalance. We will exploit sparse matrix to characterize the imbalance through the difference in sparsity.

For the two sparse patterns proposed by Yousef Saad [2], including structured sparse pattern and unstructured sparse pattern. These two patterns are also applied in the TranSparse [3] method of knowledge graph embedding, but the integration of the two has not been achieved. In this paper, we propose a self-attention model to combine the two patterns, integrating the advantages of both.

The basic idea of our method is illustrated in Fig. 1. The sparse matrices of the two sparse patterns of each node on the edge are input into the self-attention model. Then the resulting matrix serves as the transition matrix for the projection, and thereby the node embeddings are optimized for the next iteration based on the distance between the projected vectors.

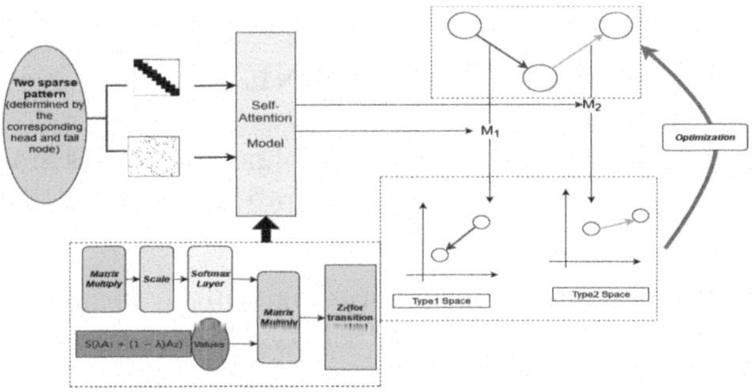

Fig. 1. Simple illustration of TSAS-HetNE

The contributions of this paper are: (1) To the best of our knowledge, we are the first to extend the TranSparse approach from knowledge graph embedding to a projected metric embedding method for heterogeneous graphs, effectively addressing the imbalance inherent in heterogeneous graphs; (2) We propose the fusion of two sparse patterns, which is not considered in previous methods, combining the respective advantages using a self-attention model and demonstrate the effectiveness mathematically; (3) Link predictions are conducted on three large-scale heterogeneous networks: Amazon, Twitter, and Youtube. Compared with the state-of-the-art methods, our method TSAS-HetNE demonstrates the advantages.

2 Related Work

Common methods of heterogeneous graph embedding learning can be roughly categorized into non-network-based methods and network-based methods.

Non-network-based methods adhere to the principle of structural preservation, attempting to maintain consistency with the graph in the embedding vector space as much as possible during the embedding process. These methods consider the different structures of heterogeneous graphs, including edges, metapaths, and subgraphs. Some representative works are: PME [1] and EOE [4]. PME maps edge types to different metric spaces separate from the node representation space. On the other hand, EOE differs from PME in that it characterizes interaction information using inner products for same-type nodes while introducing a collaborative embedding matrix to handle nodes of different types.

Network-based methods utilize the information propagation mechanism of neural networks to aggregate and propagate information in the graph including heterogeneous graph neural networks like HetGNN [5] and heterogeneous graph attention network [6]. HetGNN consists of five components: random walk sampling, content aggregation, neighbor aggregation, type aggregation, and construction of the objective function and training. Besides, heterogeneous graph attention network employs a hierarchical attention mechanism to characterize the importance of nodes. Network-based methods generalize through induction and inference (i.e., learn representations of nodes outside the training samples).

3 Proposed Method: TSAS-HetNE

In this section, we first explain the definitions of heterogeneous graphs and graph representation (or embedding). Then we elaborate on our method TSAS-HetNE.

Definition 1. *A heterogeneous information network is defined in a graph* $G = \{V, E\}$. *In a homogeneous network, there is only one node type and one edge type, while in a heterogeneous network, there are multiple types of nodes or edges in the graph.*

We formally define our problem for representation learning on networks(graph).

Definition 2. *Graph representation (also known as network embedding) learns a mapping* $\Phi : V \rightarrow \mathbb{R}^d$*, embedding the nodes of the graph into a low-dimensional Euclidean space, where* $d \ll |V|$.

The framework of TSAS-HetNE involves projecting the nodes in the graph into their respective semantic space based on the types of edges they belong to. Instead of using a general matrix as the projection transition matrix, we utilize the matrix obtained from the self-attention model which uses sparse matrices of two patterns for each node with the aim of alleviating the imbalance. Additionally, the loss function used in training penalizes negative samples through defining a parameterized scoring function and a margin-based training objective.

3.1 Sparse Matrix

We focus on the issue of imbalance existing in heterogeneous networks. Two types of imbalance are tackled: imbalance in the number of head and tail nodes and imbalance in the number of different types of edges in the graph. There is a distinction between complex and simple relationships in heterogeneous graphs. Modeling each relationship in the same way may lead to overfitting of simple relationships or underfitting of complex relationships [3]. Inspired by the TranSparse method used in the field of knowledge graph embedding for handling imbalance, we extend the idea of this method to heterogeneous networks. We consider replacing the transition matrix used in projection with a sparse-matrices-correlation matrix, where the sparse matrices used are adaptive.

It is worth noting that when a sparse matrix participates in the calculation of a weight matrix, in order to maintain the sparsity, the positions and values of zero elements remain unchanged, while only the non-zero elements are considered as parameters. This means that the sparsity of the matrices will control the degrees of freedom of the learning objects. Specifically, in our method, the sparse-matrices-correlation matrix is used as the transition matrix. Because complex relationships connect more head and tail nodes, making them more meaningful than simple relationships, complex relationships require more parameters to fully learn, while simple relationships require fewer parameters for appropriate learning when determining the sparse matrix. This is manifested in the fact that there are fewer learning variables in the transition matrix of simple relationship, i.e., lower degree of freedom, fewer non-zero elements, and greater sparsity. Therefore, the sparse matrix of simple relationship is sparser than that of complex relationship.

Let $A(\theta_r)$ be the sparse matrix with edge type r and sparsity θ_r. N_r denotes the number of edge type r in the heterogeneous network, and let r^* be the edge type with the highest number of occurrences. At this point, the corresponding sparse matrix $A(\theta_{r^*})$ has the minimum sparsity, denoted as θ_{min} $(0 \leq \theta_{min} < 1)$. According to the above idea, the following equation should hold:

$$\frac{1-\theta_r}{1-\theta_{min}} = \frac{N_r}{N_{r^*}} \tag{1}$$

That is:

$$\theta_r = 1 - (1-\theta_{min})\,\frac{N_r}{N_{r^*}} \tag{2}$$

The distinction between head nodes and tail nodes can be considered, implying that the head and tail node on the same edge are likely to have different sparsity. Similarly, their respective sparsity can be calculated using the following equation:

$$\theta_r^l = 1 - (1-\theta_{min})\,\frac{N_r^l}{N_{r^*}^{l^*}}(l = h, t) \tag{3}$$

where θ_r^l represents the sparsity of the sparse matrix with edge type r at position l, and N_r^l represents the number of nodes with edge type r at position l.

Additionally, we introduce two sparse patterns proposed by Yousef Saad [2]: structured sparse pattern and unstructured sparse pattern, as shown in Fig. 2. In the structured sparse pattern, non-zero elements are concentrated along and around the diagonal, while in the unstructured sparse pattern, non-zero elements are randomly distributed in the matrix. This two kinds of sparse patterns have different advantages. The structured sparse pattern provides fixed and local linear combinations, whereas the unstructured sparse pattern provides flexible and global linear combinations.

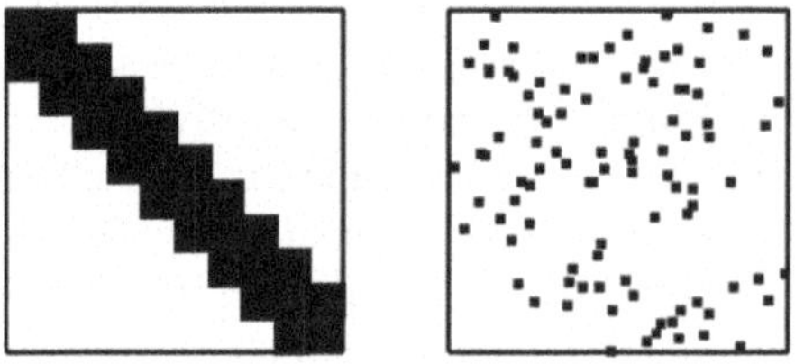

Fig. 2. Structured sparse pattern and unstructured sparse pattern. The black portions within the square borders indicate the positions occupied by non-zero elements.

3.2 Self-attention Model

Assuming a specific edge e(with edge type r), the sparse matrix of the structured pattern connected by edge e is denoted as A_1, while the sparse matrix of the unstructured pattern is denoted as A_2, $A_1, A_2 \in \mathbb{R}^{d_r \times d_v}$. For the same node, A_1 and A_2 have the same sparsity. TSAS-HetNE employs scaled dot-product attention to learn the embedding of nodes on the edge e. Let $S \in \mathbb{R}^{T \times d_r}$ be the input trainable parameter matrix and the corresponding output matrix is $Z_r \in \mathbb{R}^{T \times d_v}$. The mapping relationships of the input matrix are as follows:

$Q = SA_1$ for queries, $K = SA_2$ for keys, and $V = S(\lambda A_1 + (1-\lambda)A_2)$ for values. The self-attention model is defined as:

$$Z_r(\theta_r) = \Gamma V = softmax(\frac{(SA_1)(SA_2)^T}{\sqrt{d_v}})[S(\lambda A_1 + (1-\lambda)A_2)] \tag{4}$$

here, Γ is the importance matrix, and λ is an adjustable parameter. Then, the transition matrix $Z_r(\theta_r)$ for projection is obtained.

Below, we demonstrate the effectiveness of self-attention model mathematically. That means, the property, that the complexity of the relationships in the graph can be depicted by the sparsity, is preserved by the self-attention model above. Firstly, a relevant definition and a lemma are provided, and then a theorem is introduced.

Definition 3. *Let A and B be two $n \times m$ matrices. Then*

$$A \leq B \tag{5}$$

if by definition, $a_{ij} \leq b_{ij}$ for $1 \leq i \leq n, 1 \leq j \leq m$.

The binary relation "$\leq$" imposes only a partial order on $\mathbb{R}^{n\times m}$ since two arbitrary matrices in $\mathbb{R}^{n\times m}$ are not necessarily comparable by this relation.

Lemma 1. *Let A, B, C be nonnegative matrices, with $A \leq B$. Then*

$$AC \leq BC \text{ and } CA \leq CB \tag{6}$$

if by definition, $a_{ij} \leq b_{ij}$ for $1 \leq i \leq n, 1 \leq j \leq m$.

Proof. Same as the proof of Proposition 1.7 in [2].

Theorem 1. *The input matrices with low sparsity capture more information of parameter update than matrices with relatively high sparsity in self-attention model above.*

Proof. For the convenience of formulation, we still use the notations above. For comparison, we consider a complex relationship with sparse matrices $A_1, A_2 \in \mathbb{R}^{d_r \times d_v}$ for structured pattern and unstructured pattern respectively, and the structured pattern and unstructured pattern corresponding to a simple relationship are $\widetilde{A_1}, \widetilde{A_2} \in \mathbb{R}^{d_r \times d_v}$ respectively. The parameter matrix is $S \in \mathbb{R}^{T \times d_r}$. From the previous analysis, it is known that the sparsity of $\widetilde{A_1}, \widetilde{A_2}$ is greater than A_1, A_2 and $S, A_1, A_2, \widetilde{A_1}, \widetilde{A_2}$ are set nonnegative.

Firstly, for the matrices of unstructured sparse pattern $A_2, \widetilde{A_2}$, it is trivial that the sparse matrix with more non-zero elements will capture more information of parameters through the computation of keys.

Secondly, according to the construction process of the structured sparse pattern and the Definition 3, $\widetilde{A_1} \leq A_1$ is obtained. Therefore, for the same parameter matrix S, according to the Lemma 1, $S\widetilde{A_1} \leq SA_1$. We consider the elements arranged above and below the diagonal. They have the properties of translating and adding together the elements in S during the process of column transformation. Therefore, the matrix of structured pattern with lower sparsity corresponding to the complex relationship extracts information from the parameter matrix S more frequently and will be more perceptive of parameter variations.

Similar analysis is applicable to the calculation of values, and obviously, the operations of scaling and softmax do not affect the difference of the information captured. To sum up, we can draw the conclusion.

3.3 Transition-Based Embedding

In view of the heterogeneity, distinct latent spaces are introduced to model objects to alleviate the potential geometrical inflexibility of existing metric learning approaches [1].

We project the head nodes and tail nodes connected by edges of different types into different semantic spaces (specific spaces corresponding to relationships). In different semantic spaces, we can more reasonably measure the first-order and second-order similarity of neighbors using Euclidean distance, especially as manifested by two nodes connected by an edge being closer in the relational space corresponding to that edge.

With the sparse matrices as input to the self-attention model, the projected vectors of the head and tail nodes of the specific edge e in their corresponding semantic space for edge type r are:

$$h_p = Z_r(\theta_r)h, t_p = Z_r(\theta_r)t \tag{7}$$

When the differences between the head and tail nodes are considered and their respective sparsity is denoted by θ_r^h and θ_r^t, their corresponding sparse matrices differ. Consequently, the output transition matrices from the self-attention model are also different. In this case, the projected vectors for the nodes are expressed as:

$$h_p = Z_r^h(\theta_r^h)h, t_p = Z_r^t(\theta_r^t)t \tag{8}$$

3.4 Optimization

Let R be the set of all edge types (relations). Nodes are projected into their respective semantic space specific to edge types (relations) and the proximity is measured using Euclidean distance. For node h and t, when $(h,\, t)$ is a training sample and each sample has a label y_i indicating whether it is a positive sample ($y_i = 1$) or a negative one ($y_i = 0$), the distance between h and t is:

$$d_r(h,t) = ||h_p - t_p||, r \in R \tag{9}$$

For edge e_{ijr}, Eq. (9) is also a scoring function. The score is expected to be small when node v_i and v_j are connected by an edge of type r(i.e. a positive sample), and large for a negative sample. For each relation r, we define the following margin-based loss function as the training objective:

$$L_r = \sum_{(v_i,v_j)\in D_r} \sum_{(v_i,v_k)\notin D_r} [margin + d_r(v_i,v_j)^2 - d_r(v_i,v_k)^2]_+ \tag{10}$$

where D_r is the set of edges with relation r, $[x]_+ = max(0, x)$, and the margin value is the boundary that distinguishes positive samples from negative samples.

The overall loss is:

$$L = \sum_{r \in R} \sum_{(v_i, v_j) \in D_r} \sum_{(v_i, v_k) \notin D_r} [margin + d_r(v_i, v_j)^2 - d_r(v_i, v_k)^2]_+ \tag{11}$$

Therefore, TSAS-HetNE can be transformed into the following optimization problem:

$$\begin{aligned} &\underset{h_p, t_p, Z_r}{minimize} \quad L \\ &subject\ to \quad ||h_p|| \leq 1;\ ||t_p|| \leq 1;\ ||Z_r|| \leq 1 \end{aligned}$$

The whole method is summarized as Algorithm 1.

Algorithm 1: TSAS-HetNE

---

input : A heterogeneous network $G(V, E)$, structured and unstructured sparse pattern settings for the head nodes and tail nodes, minimum sparsity θ_{min}, learning rate l and margin.

output: Embeddings of network nodes

Randomly initialize the node embeddings, the parameter matrix S in the self-attention model and the values of the elements at the non-zero positions of the sparse matrices used;

for *every batch* **do**

  $\Gamma_{batch} \leftarrow sample(\Gamma, b)$; `// Sample a minibatch of size` b

  $T_{batch} \leftarrow \phi$

  for $(h, t) \in \Gamma_{batch}$ *and* $(h, t) \in D_r$ **do**

    `// Assign the negative samples`

    if *flag="h"* **then**

      $T_{batch} \leftarrow T_{batch} \cup \{(h, t), (k, t)\}$ and $(k, t) \notin D_r$;

    if *flag="t"* **then**

      $T_{batch} \leftarrow T_{batch} \cup \{(h, t), (h, k)\}$ and $(h, k) \notin D_r$;

    Update the node embeddings and parameter matrix S in the self-attention model w.r.t. equation(10) with the learning rate l;

    Plug the updated S into the self-attention model(equation(4)) to obtain the projection transition matrix for each node;

  if $||Z_r|| > 1$ **then**

    Normalize Z_r;

  for *every node* v_i **do**

    if $||v_i|| > 1$ **then**

      $v_i \leftarrow v_i / ||v_i||$;

  `// constrains:` $||h_p|| \leq 1, ||t_p|| \leq 1, ||Z_r|| \leq 1$

---

4 Experiments

In this section, we first introduce three real datasets adopted and other state-of-the-art methods used for comparison. Then we focus on the task of link prediction to evaluate the performance of TSAS-HetNE.

4.1 Datasets

The datasets used in the experiments include: the Amazon[1] dataset with products as nodes, Twitter[2] and Youtube[3] datasets with users as nodes. The data statistics for the three datasets are summarized in Table 1.

Amazon: We use metadata for electronic products in Amazon, including product attributes (price, sales rank, brand, category), co-viewing and co-purchasing connections between products.

Twitter: The Twitter dataset consists of tweets related to the discovery of the Higgs boson between July 1st and 7th, 2012. It comprises four types of relationships between Twitter users: re-tweet, reply, mention, and friend/colleague relationships.

Youtube: The Youtube dataset consists of five types of interactions among users, including contacts, shared friends, shared subscriptions, shared subscribers, and shared favorite videos.

Table 1. Dataset Statistics

Dataset	Number of Nodes	Edge Types	Number of Edges	#Train	#Valid	#Test
Amazon	10099	also_bought	88,342	126,535	14,756	29,492
		also_viewed	82,441			
Twitter	9900	re-tweet	377,845	282,115	32,926	65840
		reply	1,719			
		mention	163			
		follower	1,152			
Youtube	2000	contacts	10,445	1,114,025	131,023	262,014
		shared friend	406,474			
		shared subscription	574,338			
		shared subscriber	185,268			
		shared favorite video	330,538			

[1] http://jmcauley.ucsd.edu/data/amazon.
[2] https://snap.stanford.edu/data/higgs-twitter.html.
[3] http://socialcomputing.asu.edu/datasets/YouTube.

4.2 Performance Analysis

Link prediction is a common task for assessing the quality of network embeddings involving learning latent information from a small number of known links in a graph to predict hidden links, which finds wide applications in recommendation scenarios. In this paper, we create a validation/testing set, each containing 5%/10% positive edges, with an equal number of negative edges selected for each edge type. The validation set is used for hyperparameter tuning, while the testing set is used for evaluation. Only the hyperparameters tuned on the validation set are used for a single experiment on the testing set.

Some other heterogeneous graph embedding methods are used as baseline: Line [7], DeepWalk [8], Node2vec [9], Hin2vec [10], GATNE [11].

Quantitative Results. In our experiments, some commonly used evaluation metrics are utilized for prediction accuracy, namely ROC curve (ROC-AUC) [12], PR curve (PR-AUC) [13], and F1 score [14]. Experiments are conducted on the aforementioned three datasets, and the experimental results are listed in Table 2. Compared with the state-of-the-art heterogeneous graph representation methods, TSAS-HetNE achieves superior prediction results. We find that GATNE produces suboptimal solutions in most cases. But through t-test for TSAS-HetNE and GATNE(listed in Table 3), we determine that TSAS-HetNE's superiority is significant, based on the P-value of 0.025.

Table 2. Three evaluation metrics result of different methods

	Amazon			Twitter			Youtube		
	ROC	PR	F1	ROC	PR	F1	ROC	PR	F1
Line-1st	95.34	<u>96.30</u>	91.39	60.28	67.85	56.28	78.20	74.50	71.28
Line-2nd	86.76	86.25	78.65	79.00	79.99	76.87	84.10	79.72	76.87
DeepWalk	94.52	93.71	88.11	68.30	67.26	62.24	79.63	75.77	72.77
Node2vec	79.08	78.39	74.80	60.58	71.46	58.83	76.15	75.67	70.34
Hin2vec	85.32	86.90	82.85	77.19	68.83	62.42	<u>85.27</u>	<u>88.81</u>	<u>83.48</u>
GATNE	<u>96.66</u>	95.92	<u>91.97</u>	<u>81.20</u>	<u>82.11</u>	<u>78.08</u>	83.72	81.59	76.57
TSAS-HetNE	**97.03**	**96.44**	**94.23**	**81.91**	**82.96**	**80.04**	**89.18**	**88.93**	**87.05**

Table 3. Main results of t-test

			95% confidence interval	
degree of freedom	t	Sig.(2-tailed)	Lower	Upper
8	2.763	0.025	0.55053	6.10502

Parameter Sensitivity. The sensitivity of different hyperparameters in TSAS-HetNE is investigated, including the node embedding dimension n and the minimum sparsity threshold $\theta_{\min}$. Figure 3 illustrates the performance of TSAS-HetNE on datasets when changing the node embedding dimension n or the minimum sparsity threshold $\theta_{\min}$ from the default settings ($n = 100$, $\theta_{\min} = 80\%$).

From the Fig. 3, it can be concluded that regarding the setting of the node embedding dimension, performance tends to degrade when the dimension is either too low or too high. Low dimension cannot be able to contain rich information, while the high dimension produces redundancy. As for the setting of the minimum sparsity threshold, the changes in $\theta_{\min}$ have a minor impact on ROC-AUC results on Amazon, while they have a significant impact on Twitter and Youtube. In fact, this aligns with our expectations: Amazon exhibits minimal difference in the number of different types of edges, and that is to say, the issue of imbalance in Amazon is not significant, so the sparsity of different nodes is roughly the same, with a small difference from $\theta_{\min}$. Conversely, strong imbalance exists in Twitter and Youtube, where the proper setting of $\theta_{\min}$ is crucial, as evidenced by the sensitivity of the prediction results to this parameter. Additionally, $\theta_{\min}$ here should be set at a relatively high level to emphasize sparsity.

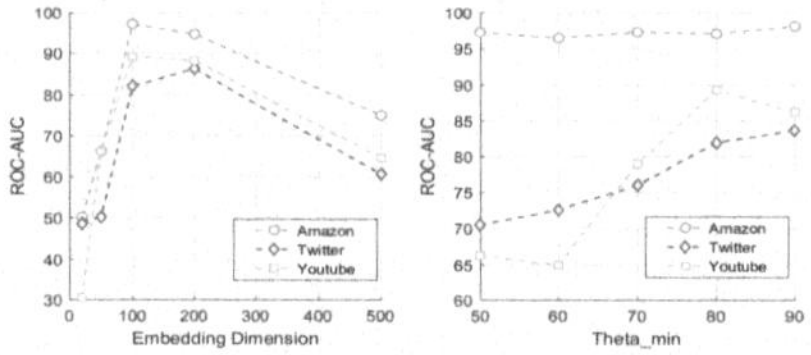

Fig. 3. The sensitivity of different hyperparameters in TSAS-HetNE

5 Conclusion

In this paper, we address the problem of heterogeneous networks embedding and propose a new method TSAS-HetNE. Based on the idea of projection, nodes of different types are mapped into different semantic spaces. For the transition matrix, we exploit the sparsity of matrices to characterize the imbalance of heterogeneous graphs and input sparse matrices of two sparse patterns into a self-attention model. In addition, the effectiveness of the fusion by the self-attention model is demonstrated mathematically. The method proposed in this paper achieves better performance than previous state-of-the-art methods on link prediction tasks in three real-world datasets and then the impacts of hyperparameters in TSAS-HetNE are explored. But the proposed algorithm still lacks extensive experiments, and in the future we plan to extend the proposed model for more tasks like node classification and node clustering.

References

1. Chen, H., Yin, H., Wang, W., Wang, H., Hung, N.Q.V., Li, X.: PME: projected metric embedding on heterogeneous networks for link prediction. In: Proceedings of the 24th ACM SIGKDD International Conference on Knowledge Discovery and Data Mining, pp. 1177–1186. Association for Computing Machinery, London (2018)
2. Saad, Y.: Iterative Methods for Sparse Linear Systems, 2nd edn. Society for Industrial and Applied Mathematics, Philadelphia (2000)
3. Ji, G., Liu, K., He, S., Zhao, J.: Knowledge graph completion with adaptive sparse transfer matrix. In: Proceedings of the 13th AAAI Conference on Artificial Intelligence, pp. 985–991. AAAI Press, Phoenix (2016)
4. Xu, L., Wei, X., Cao, J., Yu, P.S.: Embedding of Embedding (EOE): joint embedding for coupled heterogeneous networks. In: Proceedings of the Tenth ACM International Conference Web Search and Data Mining, pp. 741–749. Association for Computing Machinery, Cambridge (2017)
5. Zhang, C., Song, D., Huang, C., Swami, A., Chawla, N.V.: Heterogeneous graph neural network. In: Proceedings of the 25th ACM SIGKDD International Conference on Knowledge Discovery and Data Mining, pp. 793–803. Association for Computing Machinery, Anchorage (2019)
6. Wang, X., et al.: Heterogeneous graph attention network. In: Proceedings of the World Wide Web Conference, pp. 2022–2032. Association for Computing Machinery, San Francisco (2019)
7. Tang, J., Qu, M., Wang, M., Zhang, M., Yan, J., Mei, Q.: LINE: large-scale information network embedding. In: Proceedings of the 24th International Conference on World Wide Web, pp. 1067–1077. International World Wide Web Conferences Steering Committee, Switzerland (2015)
8. Bryan, P., Rami, A.-R., Steven, S.: DeepWalk online learning of social representations. In: Proceedings of the 20th ACM SIGKDD International Conference on Knowledge Discovery and Data Mining, pp. 701–710. Association for Computing Machinery, New York (2014)
9. Grover, A., Leskovec, J.: node2vec: scalable feature learning for networks. In: Proceedings of the 22nd ACM SIGKDD International Conference on Knowledge Discovery and Data Mining, pp. 855–864. Association for Computing Machinery, San Francisco (2016)
10. Fu, T., Lee, W., Lei, Z.: HIN2Vec: explore meta-paths in heterogeneous information networks for representation learning. In: Proceedings of the 2017 ACM on Conference on Information and Knowledge Management, pp. 1797–1806. Association for Computing Machinery, Singapore (2017)
11. Cen, Y., Zou, X., Zhang, J., Yang, H., Zhou, J., Tang, J.: Representation learning for attributed multiplex heterogeneous network. In: Proceedings of the 2017 ACM on Conference on Information and Knowledge Management, pp. 1358–1368. Association for Computing Machinery, Anchorage (2019)
12. Hanley, J.A., McNeil, B.J.: The meaning and use of the area under a receiver operating characteristic (ROC) curve. Radiology **143**(1), 29–36 (1982)
13. Davis, J., Goadrich, M.: The relationship between Precision-Recall and ROC curves. In: Proceedings of the 23rd International Conference on Machine Learning, pp. 233–240. Association for Computing Machinery, Pittsburgh (2006)
14. Chinchor, N.: MUC-4 evaluation metrics. In: Proceedings of the 4th Conference on Message Understanding, pp. 22–29. Association for Computational Linguistics, Virginia (1992)

Author Index

S. Ghosh and Z. Zhang (Eds.): AAIM 2024, LNCS 15179, pp. 287–288, 2024.
https://doi.org/10.1007/978-981-97-7798-3

The manufacturer's authorised representative in the EU is Springer Nature Customer Service Centre GmbH, Europaplatz 3, 69115 Heidelberg, Germany. If you have any concerns regarding our products, please contact ProductSafety@springernature.com

Printed and bound by CPI Group (UK) Ltd, Croydon, CR0 4YY
01/10/2026
02225908-0006